OCR
A LEVEL

D0547871

CHEMISTRY

John Older

Mike Smith

HODDER
EDUCATION
AN HACHETTE UK COMPANY

Orders: please contact Bookpoint Ltd, 130 Milton Park, Abingdon, Oxon OX14 4SB. Telephone: +44 (0)1235 827720. Fax: +44 (0)1235 400454. Lines are open 9.00a.m.–5.00p.m., Monday to Saturday, with a 24-hour message answering service. Visit our website at www.hoddereducation.co.uk

© John Older, Mike Smith 2015

First published in 2015 by

Hodder Education,

An Hachette UK Company

Carmelite House,

50 Victoria Embankment

London EC4Y 0DZ

Impression number   10 9 8 7 6 5 4 3 2 1

Year   2019 2018 2017 2016 2015

Cover photo © Nneirda – Fotolia

Typeset in 10.5/12 pt Bliss Light by Integra Software Services Pvt. Ltd., Pondicherry, India

Printed in Italy

A catalogue record for this title is available from the British Library

ISBN 9781471827181

Contents

Get the most from this book

Welcome to the **OCR A Level Chemistry 2 Student's Book**! This book covers Year 2 of the OCR A Level Chemistry specification.

The following features have been included to help you get the most from this book.

Prior knowledge

This is a short list of topics that you should be familiar with before starting a chapter. The questions will help to test your understanding.

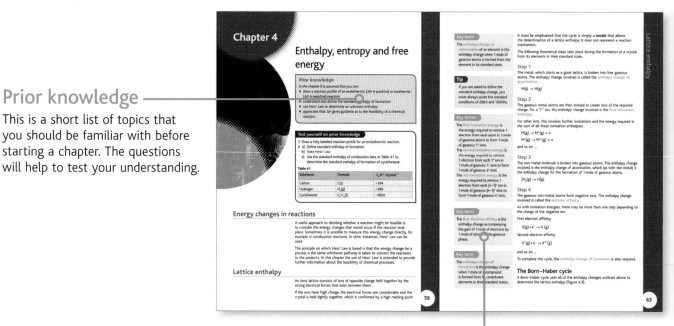

Key terms and formulae

These are highlighted in the text and definitions are given in the margin to help you pick out and learn these important concepts.

Tips

These highlight important facts, common misconceptions and signpost you towards other relevant chapters.

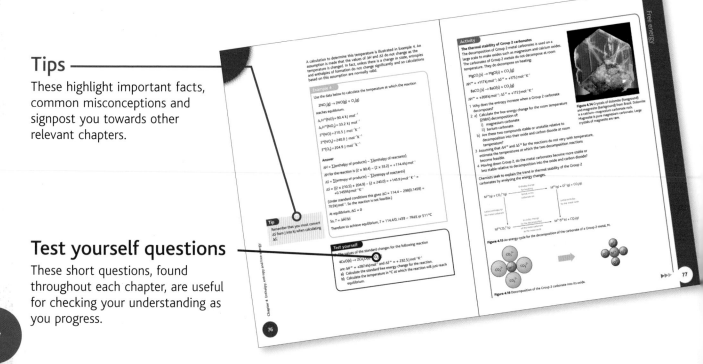

Test yourself questions

These short questions, found throughout each chapter, are useful for checking your understanding as you progress.

Activities

These practical-based activities will help consolidate your learning and test your practical skills.

Examples

Examples of questions or calculations are included to illustrate chapters and feature full workings and answers.

Practice questions

You will find Practice questions, including multiple-choice questions, at the end of every chapter. These follow the style of the different types of questions with short and longer answers that you might see in your examination, and they are colour coded to highlight the level of difficulty. Challenge questions are also provided.

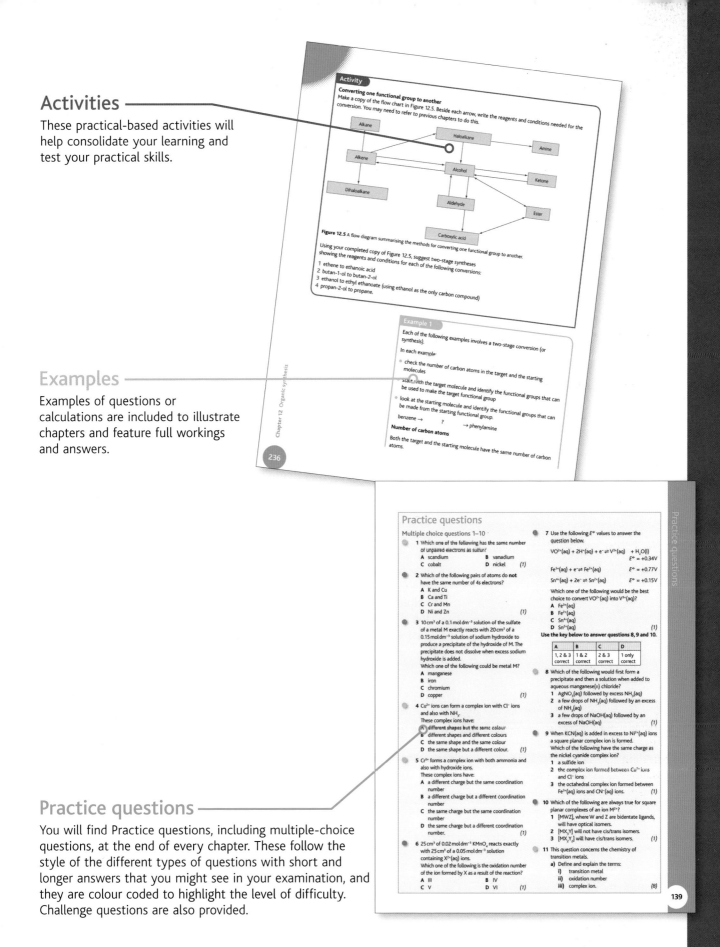

A dedicated chapter for developing your **Maths** can be found at the back of this book.

The questions in this book

The 'Test yourself' questions within the text of each chapter are designed to check that you have understood the short section of work that has just been covered. They are intended to be used during the course but can also be helpful while you are revising. The 'Activities' provide a chance to tackle a more extended question and also provide a check that you are able to understand and interpret experiments. The 'Practice' questions at the end of the chapters cover a broader section of work. They include multiple choice, structured and a smaller number of free response questions. They also sometimes focus on the evaluation of some experimental results. They are graded as follows:

- Basic questions that everyone must be able to answer without difficulty within the exam.

- Questions which cover work that are a regular feature of exams and that all competent candidates should be able to handle.

- More demanding questions that sometimes go beyond the normal requirements of the exam but which the best candidates should be able to do.

Challenge These are questions for the most able candidates to test their full understanding and sometimes their ability to use ideas in a novel situation.

It must be emphasised that although these questions cover the skills and knowledge required to be successful in the examination, they are not exam questions as such. Many only cover components of questions which on the exam paper may come from more than one section of the specification. When you are confident that your knowledge base and your understanding is sufficient you must start practising the past examination questions which are available from the OCR website. Mark schemes are also available and these are very helpful in making it clear what points examiners are looking for and the depth that they expect. The best practice for the exam is doing past exams.

Chapter 1

How fast? Rates of reaction

Prior knowledge

In this chapter it is assumed that you understand that:
- reactions occur because the reactants collide with sufficient energy to exceed the activation energy required for the conversion to the products
- the energies of particles are spread so that some have higher energy than the average and some have lower energy; the spread of energies is shown by the Boltzmann distribution
- the Boltzmann distribution of energies is temperature dependent and, as the temperature is raised, more particles will have an energy greater than the activation energy for a reaction
- catalysts provide a route for a reaction which has a lower activation energy
- tangents can be used to find the rate of reaction from a concentration/time graph.

As a reminder of the last point, the graphs in Figure 1.1 and Figure 1.2 illustrate the procedure.

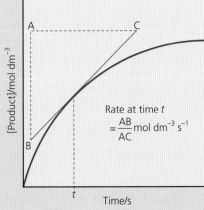

Rate at time t
$= \dfrac{AB}{AC}$ mol dm^{-3} s^{-1}

Figure 1.1

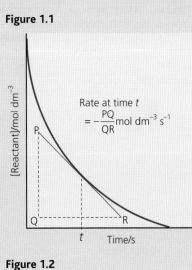

Rate at time t
$= -\dfrac{PQ}{QR}$ mol dm^{-3} s^{-1}

Figure 1.2

Reaction rates

The rate at which chemical reactions take place varies widely. Explosions are reactions that occur so rapidly that energy is released almost instantaneously. Other processes, such as the conversion of diamond into the more stable graphite, are so slow that they appear not to be happening at all. It is not unreasonable to say that diamonds are forever! The controlling factor is the activation energy of a reaction. This chapter considers reaction rates in more detail and focuses on the route by which the final products are created. It is possible to take a quantitative view of the role of the reactants and this can sometimes provide an insight into the way a reaction takes place.

Obtaining practical data

There is a wide variety of methods that can be used to determine the rate of a chemical reaction. Although you may be asked to suggest a possible procedure, there is no need to learn the details of any particular experiment. Wherever possible, a method is chosen that does not interfere with the reaction taking place as this could lead to confusing results. Possible procedures that might be selected are shown in Figures 1.3, 1.4 and 1.5.

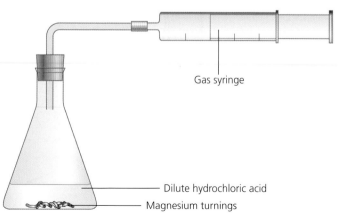

Gas syringe

Dilute hydrochloric acid
Magnesium turnings

Figure 1.3

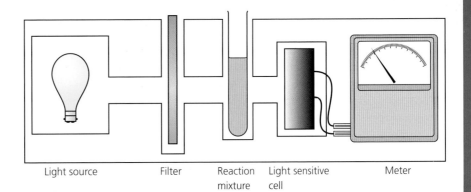

Light source   Filter   Reaction   Light sensitive   Meter
mixture   cell

Figure 1.4

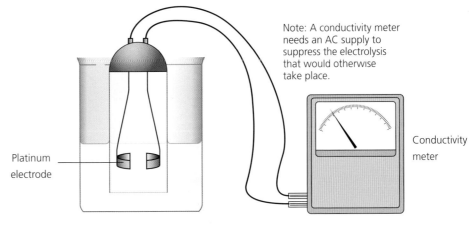

Note: A conductivity meter
needs an AC supply to
suppress the electrolysis
that would otherwise
take place.

Platinum
electrode

Conductivity
meter

Figure 1.5

Sometimes, samples of the reacting mixture are withdrawn at regular
intervals during the course of a reaction. The concentrations of the
reactants or products present are measured at these time intervals. These
concentrations can then be analysed to determine how far the reaction has
proceeded. Usually, it is only necessary to determine the concentration of
one of the reactants to deduce what has happened. A problem with this
method is that the reaction will continue to take place in the sample that
has been withdrawn. Therefore, some way has to be found to stop the
reaction in the sample. It might be possible to stop the reaction by cooling
the sample rapidly in an ice-bath. Or perhaps, if the reaction depended on
an acid being present, the reaction could be stopped quickly by neutralising
the acid present with an alkali prior to carrying out an analysis.

Test yourself

1 Suggest a suitable method for measuring the rate of each of these
reactions:
a) $Br_2(aq) + HCOOH(aq) \rightarrow 2HBr(aq) + CO_2(g)$
b) $CH_3COOCH_3(l) + H_2O(l) \rightarrow CH_3COOH(aq) + CH_3OH(aq)$
c) $C_4H_9Br(l) + H_2O(l) \rightarrow C_4H_9OH(l) + HBr(aq)$

Activity

Investigating the effect of concentration on the rate of a reaction

Bromine oxidises methanoic acid in aqueous solution to carbon dioxide. The reaction is catalysed by hydrogen ions:

$$Br_2(aq) + HCOOH(aq) \rightarrow 2Br^-(aq) + 2H^+(aq) + CO_2(g)$$

The reaction can be followed using a colorimeter (Figure 1.6). Table 1.1 shows some typical results. The concentration of methanoic acid was kept constant throughout the experiment by having it present in large excess.

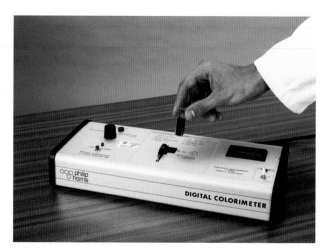

Figure 1.6 A colorimeter can be used to follow the changes in concentration of coloured chemicals during a reaction.

1. Explain why it is possible to follow the rate of this reaction using a colorimeter.
2. Suggest a suitable chemical to use as the catalyst for the reaction.
3. Explain the purpose of adding a large excess of methanoic acid.
4. Plot a graph of concentration against time using the results in Table 1.1.
5. Draw tangents to the graph and measure the gradient to obtain values for the rate of reaction at two points during the experiment. Use the values at 100 s and 500 s. (Remember when calculating the gradients that the bromine concentrations are 1000 times smaller than the numbers in the table.)
6. Table 1.2 shows values for the reaction rate obtained by drawing gradients at other times on the concentration–time graph. Plot a graph of rate against concentration using the two values you calculated in question 5 and the values in Table 1.2.
7. How does the bromine concentration change with time?
8. How does the rate of reaction change with time?
9. How does the rate of reaction depend on the bromine concentration?

Table 1.1 Results of an experiment to investigate the rate of reaction of bromine with methanoic acid. Note that the bromine concentrations are multiplied by 1000. The actual bromine concentration at 90 seconds, for example, was 0.0073 mol dm^{-3}.

Time/s	Concentration of bromine/10^{-3} mol dm^{-3}
0	10.0
10	9.0
30	8.1
90	7.3
120	6.6
180	5.3
240	4.4
360	2.8
480	2.0
600	1.3

Table 1.2 Rate values obtained by measuring gradients of tangents to the concentration–time graph. Note that the rate is multiplied by 100 000. The actual rate at 300 seconds, for example, was 1.2×10^{-5} mol dm^{-3} s^{-1}.

Time/s	Concentration of bromine/10^{-3} mol dm^{-3}	Rate of reaction from gradients to the concentration time graph/10^{-5} mol dm^{-3} s^{-1}
200	5.0	1.7
300	3.5	1.2
400	2.5	0.8

Orders of reaction

You will probably have assumed at GCSE that changing the concentration of a reagent will change the rate of reaction proportionately. Sometimes this is true but, in practice, the effect of a change in a concentration may not be as simple as that. In some reactions, doubling the concentration of one chemical has no effect on the rate, while doubling the concentration of another reagent can double, or even quadruple, the rate of reaction. The effect that each reagent has on the rate is summarised by attributing an **order** to each reagent.

Where increasing the concentration of a reagent increases the rate proportionately, the reaction is said to be **first order** with respect to that reagent. If increasing the concentration of a reagent does not affect the rate, the reaction is said to be **zero order** with respect to that reagent. A reaction can be **second order** with respect to a reagent; in which case, doubling the concentration of this reagent causes the rate to increase four fold. An example of a second-order reaction is the breakdown of nitrogen(IV) oxide into nitrogen(II) oxide and oxygen:

$$2NO_2 \rightarrow 2NO + O_2$$

The reaction is second order with respect to nitrogen(IV) oxide.

However, it is worth emphasising that it is not possible to judge the order of a reaction from the equation. The breakdown of nitrogen(V) oxide into nitrogen(IV) oxide and oxygen

$$2N_2O_5 \rightarrow 4NO_2 + O_2$$

is first order with respect to the nitrogen(V) oxide.

A summary to explain the effects of the different orders of reaction is shown in Table 1.3.

Table 1.3

Change in concentration of reagent	Change in rate due to change in concentration	Order of reaction with respect to the reagent
doubled	rate remains the same	0
doubled	rate doubles	1
doubled	rate quadruples	2

The rate equation and the rate constant

The **rate equation** for a reaction between three reagents, A, B and C takes the form:

$$\text{rate} = k[A]^a[B]^b[C]^c$$

where the square brackets represent the concentrations of the reagents (in molar units such as $mol\,dm^{-3}$), and a, b and c are the orders of the reaction with respect to the reagents.

The **rate constant**, k, is a constant of proportionality.

For the examples above, the rate equation for the breakdown of nitrogen(IV) oxide into nitrogen(II) oxide and oxygen is:

$$\text{rate} = k[NO_2]^2$$

The rate equation for the breakdown of nitrogen(v) oxide into nitrogen(iv) oxide and oxygen is:

$$\text{rate} = k[N_2O_5]^1$$

Propanone, CH_3COCH_3, and iodine, when catalysed by hydrogen ions, form iodopropanone, CH_3COCH_2I, and hydrogen iodide as products. The equation for the reaction is:

$$I_2 + CH_3COCH_3 \xrightarrow{H^+} CH_3COCH_2I + HI$$

Iodine has no effect on the rate of reaction and so is zero order. The rate equation for the reaction in the presence of hydrogen ions is:

$$\text{rate} = k[CH_3COCH_3]^1[H^+]^1[I_2]^0$$

which is the same as:

$$\text{rate} = k[CH_3COCH_3]^1[H^+]^1$$

The overall order of a reaction is sometimes required. This is the sum of the individual orders with respect to the reagents. In a general case, this is $a + b + c + \ldots$. The reaction of iodine, hydrogen ions and propanone has, therefore, an overall order of 2.

The orders of reaction are nearly always whole numbers (integers) and are usually 0, 1 or 2.

The value of the rate constant, k, varies from reaction to reaction and reflects the ease with which the reaction takes place. For a fast reaction, the value of k is large, and, for a slow reaction, k is small. In most circumstances, the rate of reaction is controlled more by the value of the rate constant than by the concentrations of the reagents. k is temperature dependent and, as the temperature rises,:

- the value of k increases and the reaction proceeds more quickly

- more collisions between the reagents involved in the rate-determining step have sufficient energy to overcome the activation energy barrier.

There is a rule of thumb which states that a temperature rise of 10 °C doubles the rate of reaction. If the reaction is extremely slow, this might not be apparent but, in most instances, the effect of heating is obvious.

The units of the rate constant

The rate of a reaction is usually followed by observing either a reduction in one of the reactants or the formation of one of the products.

The rate of a reaction is usually expressed as a change in concentration over a period of time. The units are usually $mol\,dm^{-3}\,s^{-1}$.

The units of the rate constant, k, depend on the orders of reaction of the reagents.

If a reaction is first order overall, the rate equation is:

$$\text{rate} = k[A]^1$$

If the rate is measured in $mol\,dm^{-3}\,s^{-1}$ and $[A]$ is in $mol\,dm^{-3}$, the unit of k is s^{-1}.

> **Tip**
>
> If the reaction is first order with respect to a particular reactant, it is not necessary to include the power '1'. The rate equation for the reaction of iodine and propanone in the presence of hydrogen ions could be written as:
> $$\text{rate} = k[CH_3COCH_3][H^+]$$

The important thing to remember is that the units of the rate constant, k, ensure that the units on the left-hand side of the equation are the same as those on the right-hand side.

In the example above, the units are:

$$\frac{mol\ dm^{-3}}{s} = \frac{1}{s} \times mol\ dm^{-3}$$
$$rate = k[A]$$

If the overall reaction is second order, as in rate $= k[A]^1[B]^1$, the units of k are different in order to keep the units consistent.

$$\frac{mol\ dm^{-3}}{s} = \frac{1}{mol\ dm^{-3}s} \times mol\ dm^{-3} \times mol\ dm^{-3}$$
$$rate = k[A][B]$$

In this example, k has the units $mol^{-1}\,dm^3\,s^{-1}$ or $dm^3\,mol^{-1}\,s^{-1}$

Test yourself

2 The rate of reaction of 1-bromopropane with hydroxide ions is first order with respect to the haloalkane, and first order with respect to hydroxide ions.
   a) Write the rate equation for the reaction.
   b) What is the overall order of reaction?
   c) What are the units of the rate constant?
3 For the reaction A + B + C → 2E + F, the rate equation is rate $= k[A]^2[C]^2$.
   a) Why is [B] not included in the rate equation?
   b) What is the overall order of reaction?
   c) What will be the units of the rate constant k?
   d) What would happen to the rate of reaction if:
      i) [A] was doubled
      ii) [C] was trebled
      iii) the concentrations of all three chemicals were doubled
      iv) [A] and [C] were both halved?

Determining orders of reaction

There are two distinct methods that can be used to determine the order with respect to a reactant:

1 In one method, the concentration of each reactant is altered in turn to see what effect the change has on the rate at the start of the reaction. The start of the reaction is chosen because this is the only point in the reaction at which the concentrations of the reactants are definitely known. This type of experiment is called the 'initial rates' method.

2 The second method involves monitoring the reaction throughout its course. In this procedure, the amount of a reactant or product is established at various time intervals throughout the course of the reaction.

Many chemical reactions occur rapidly. The distinguished chemist George Porter (Figure 1.7) was, for many years, Director of the Royal Institution in London, where both Davy and Faraday made important discoveries in the nineteenth century. Porter invented an ingenious and important method for identifying the short-lived radicals produced in gaseous photochemical reactions. By using pulses of light of shorter duration than the existence of the radicals, he was able to identify the spectra that the latter produced. For this work, he shared the Nobel Prize for Chemistry in 1967.

Figure 1.7 George Porter (1920–2002).

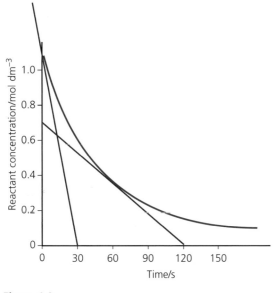

Figure 1.8

The initial rates method

A typical reaction can be represented graphically as shown in Figure 1.8.

The reaction proceeds quickly at the start and slows down towards the finish. The rate of the reaction is fastest where the curve is steepest and the slope then decreases until the curve becomes horizontal, which indicates that the reaction is complete. To provide a numerical value for the rate, the steepness of the curve is established by finding its gradient. This is achieved by drawing a tangent to the curve. In the figure, tangents have been drawn at two points on the graph. The numerical values of their gradients show how much the reaction has slowed down from the initial rate (the gradient at the start of the reaction) to a point about halfway through the reaction.

From the graph:

- the initial rate, at time $t = 0$ seconds, is about 1.07/30 = 0.036 $mol\,dm^{-3}\,s^{-1}$

- the rate after 60 seconds is 0.7/120 = 0.006 $mol\,dm^{-3}\,s^{-1}$.

These values show that the rate slows considerably as the reaction progresses.

If the concentrations of the reagents are known at the start of the reaction, a measurement of the initial gradient ($t = 0$) gives a numerical value of the rate at these concentrations.

If the experiment is repeated with the concentration of one of the reactants doubled, while the other concentrations are kept the same, a new initial rate can be established. Therefore, it is possible to see the effect of this reactant on the overall rate. If it is found that the gradient has doubled, then the reaction is first order with respect to that reactant. If, however, the rate increased four fold, it would be second order with respect to the reagent. Changing the concentrations of each reactant in turn allows the order of reaction with respect to each reagent to be determined.

Example 1

Nitrogen(II) oxide and bromine react together as shown by the following equation:

$$2NO(g) + Br_2(g) \rightarrow 2NOBr(g)$$

Some data for the initial rates of reaction are shown in Table 1.4.

Table 1.4

Experiment	[NO]/mol dm⁻³	[Br₂]/mol dm⁻³	Initial rate/mol dm⁻³ s⁻¹
1	0.01	0.01	0.011
2	0.01	0.02	0.022
3	0.02	0.01	0.044
4	0.03	0.03	0.297

a) Use the results of experiments 1–3 to determine the order of reaction with respect to:
  i) bromine
  ii) nitrogen(II) oxide.
b) Write the rate equation for the reaction.
c) Use the results from experiment 4 to confirm that your orders for bromine and nitrogen(II) oxide are correct.
d) Calculate the value of the rate constant and give its units.

Answer

a) i) It can be seen from experiments 1 and 2, that doubling the concentration of bromine while the concentration of NO is unchanged doubles the rate. Therefore, the order with respect to bromine is 1.
  ii) Experiments 1 and 3 show that doubling the concentration of nitrogen(II) oxide while the concentration of Br_2 remains the same quadruples the rate. Therefore, the order with respect to nitrogen(II) oxide is 2.
b) rate = $k[NO(g)]^2[Br_2(g)]$
c) If the equation is correct, the rate should increase three fold as a result of the change in the bromine concentration and nine fold as a result of the change in the nitrogen(II) oxide concentration. Therefore, the overall increase in rate for experiment 4 should be $3 \times 9 = 27$ fold. This is confirmed as the initial rate in experiment 4 ($0.297 \, mol \, dm^{-3} s^{-1}$) is 27 × the initial rate in experiment 1 ($0.011 \, mol \, dm^{-3} s^{-1}$).
d) Using the results of experiment 1:
  $0.011 = k[0.01]^2[0.01]$
  $0.011 = k(0.000001)$
  $k = 11000 = 1.1 \times 10^4$
  The units of k are obtained by comparing the units on either side of the rate equation.
  Therefore, the units of k are $\dfrac{1}{s \times (mol \, dm^{-3})^2} = dm^6 \, mol^{-2} \, s^{-1}$

Tip

Remember that the units of k can also be calculated from $[mol \, dm^{-3}]^{1-n} [time]^{-1}$ where n is the overall order of the reaction. In this reaction, $1 - n = -2$.

4 At 107°C, the rate of decomposition of di(benzenecarbonyl) peroxide:

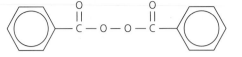

is first order with respect to the peroxide.
Calculate the rate constant for the reaction at 107°C if the rate of decomposition of the peroxide at this temperature is $7.4 \times 10^{-6}\,mol\,dm^{-3}\,s^{-1}$ when the concentration of peroxide is $0.02\,mol\,dm^{-3}$. Give the units of the rate constant.

5 Hydrogen gas reacts with nitrogen monoxide gas to form steam and nitrogen.
Doubling the concentration of hydrogen doubles the rate of reaction.
Tripling the concentration of nitrogen monoxide increases the rate by a factor of nine.
   a) Write the balanced equation for the reaction.
   b) Write the rate equation for the reaction.

6 The initial rate method was used to study the reaction:
$$BrO_3^-(aq) + 5Br^-(aq) + 6H^+(aq) \rightarrow 3Br_2(aq) + 3H_2O(l)$$

Table 1.5

Experiment	Initial concentration of BrO_3^-/mol dm^{-3}	Initial concentration of Br^-/mol dm^{-3}	Initial concentration of H^+/mol dm^{-3}	Initial rate of reaction/ mol dm^{-3} s^{-1}
1	0.1	0.10	0.10	1.2×10^{-3}
2	0.2	0.10	0.10	2.4×10^{-3}
3	0.1	0.30	0.10	3.6×10^{-3}
4	0.2	0.10	0.20	9.6×10^{-3}

Use the table of results, Table 1.5, to:
   a) determine the rate equation for the reaction
   b) calculate the value of the rate constant and give its units.

The experimentation involved in using the method described on page 8 to provide data for the initial rate of a reaction is time consuming. For this reason, an approximation to obtain an initial rate is often used. Given that most reactions lead to an integer order, the approximation is usually acceptable. The time taken to reach a specific point in the reaction (which should be soon after the reaction has started) is recorded. The experiment is then repeated with different concentrations to see how the time taken to reach this point changes. The rate to this point is taken to be proportional to 1/time taken for each reaction, as the following example should make clear.

Sodium thiosulfate reacts with dilute hydrochloric acid to form a precipitate of sulfur:

$$Na_2S_2O_3(aq) + 2HCl(aq) \rightarrow 2NaCl(aq) + SO_2(g) + S(s) + H_2O$$

It is possible to measure the time taken to a point at which a fixed small amount of sulfur has been formed. Provided that this fixed point can be identified each time the reaction is carried out, the experiment can be repeated with different initial concentrations of the reactants and the time taken to reach this point can be measured.

Suppose the '*distance*' through the reaction to this point is x and the time taken is t, then the rate can be expressed as:

$$rate = x/t$$

Tip

This type of reaction is known as a clock reaction.

Repeating the experiment with different concentrations of reactants allows the effect of each to be established, and the order of reaction with respect to each to be determined. To do this, we have to compare the initial rate for which we know the concentrations of all the reagents. This requires that, in each case, the reaction has not progressed very far – in this example, the amount of precipitate must be small.

It might be possible to determine the orders of reaction from just a few measurements. However, it is usual to record a range of results and plot a graph of 1/time (i.e. rate) against concentration. Doing this helps to take into account the approximations that have been made.

Determining the orders of reaction from graphs

Figure 1.9 Rate–concentration graph for a zero-order reaction.

Using rate–concentration graphs

The order of reaction determines the shape of a graph of concentration against rate.

If the reaction is **zero order** with respect to a particular reactant, it doesn't mean that the reactant is not required for the reaction! It means the concentration of the reactant has no effect on the rate and the graph appears as in Figure 1.9.

If the reaction is **first order** with respect to a particular reactant, the rate is proportional to the concentration and the graph appears as in Figure 1.10. The graph is a straight line that **always** goes through the origin (if there is no reactant, there can be no reaction).

If the reaction is **second order** with respect to a particular reactant then the graph is a curve (Figure 1.11). This is because the rate is proportional to the square of the concentration of the reactant.

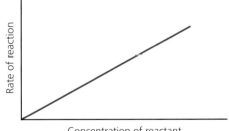

Figure 1.10 Rate–concentration graph for a first-order reaction.

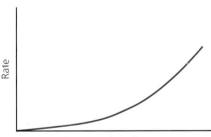

Figure 1.11 Rate–concentration graph for a second-order reaction.

Using concentration–time graphs

It is sometimes possible to obtain the order of reaction for a particular reagent by recording data throughout the course of the reaction. The reaction must either involve only one reagent or be set up so that the

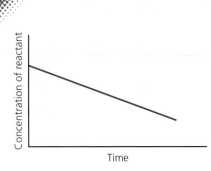

Figure 1.12 Concentration–time graph for a zero-order reaction.

reaction rate depends solely on that reagent. In the latter case, reaction mixtures are made up in which all the other reactants are present in large excess. The assumption is that throughout the reaction there will be such an excess of the other reactants that their concentration will be nearly constant and they will have little or no effect on the change in the reaction rate. Therefore, whatever is found will be due to the reagent at low concentration.

The interpretation of the results depends on analysing the shape of the graph of concentration of the reactant against time. The theory that explains the likely shapes of these graphs is mathematical and is not required for A Level chemistry.

For a **zero-order reaction**, the graph of concentration against time appears as in Figure 1.12. The reactant is consumed in the reaction and, therefore, its concentration decreases. However, the reactant has no effect on the rate and so the line is not curved.

For a **first-order reaction**, the graph of concentration against time is shown in Figure 1.13.

Key term

The **half-life** of a reaction is the time taken for the concentration of one of the reactants to fall by half.

Tip

Although you do not need to know how the rate-constant equation is derived for the exam, mathematicians might appreciate that it follows from the integration of the first-order rate equation of a reaction involving a substance with a concentration [A].

It is expressed as $-d[A]/dt = k[A]$ (the minus sign indicates that the concentration of A decreases as it reacts).

This integrates to the equation $-\ln[A] = kt + C$ (where C is the constant of integration).

Suppose at the start of the reaction where $t = 0$, the concentration is $[A]_0$ and $C = -\ln[A]_0$

The equation for a first-order reaction then becomes

$-\ln[A] = kt - \ln[A]_0$

or $kt = \ln[A]_0 - \ln[A]$

or $kt = \ln([A]_0/[A])$

The half-life, $t_{1/2}$, is the time taken for $[A]_0$ to become $[A]_0/2$

Therefore $kt_{1/2} = \ln 2$ or $k = \ln 2/t_{1/2}$

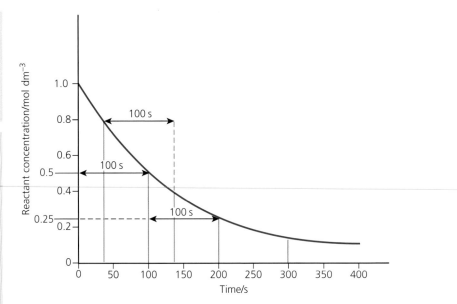

Figure 1.13 Following the course of a first-order reaction.

The graph is a curve, showing how the reactant influences the reaction. It is used up quickly at the start and then its effect diminishes steadily as the reaction proceeds. There is a more subtle feature of the curve that is not immediately apparent but which is a useful way of distinguishing it from other curves, such as that obtained for a second-order reaction. This is that the time taken for the concentration of the reactant to halve is the same, whatever the starting concentration. This time is known as the half-life of the reaction. You may be familiar with this term, since it is used to indicate the stability of radioactive isotopes; these decay by a first-order mechanism.

Referring to Figure 1.13, it can be seen that the time taken for the concentration to fall to half its original value is 100 s. For example, it takes 100 s for the initial concentration of $1.0 \, mol \, dm^{-3}$ to fall to $0.5 \, mol \, dm^{-3}$. It also takes 100 s for the concentration to fall from $0.5 \, mol \, dm^{-3}$ to $0.25 \, mol \, dm^{-3}$ and for $0.8 \, mol \, dm^{-3}$ to fall to $0.4 \, mol \, dm^{-3}$. Whichever starting value is taken, the time taken to halve this concentration is constant. This proves the reaction is first order.

Determining the rate constant from the half-life of a first-order reaction

A useful equation links the half-life of a first-order reaction to its rate constant. This is:

$k = \ln 2/t_{\frac{1}{2}}$ (ln is the natural logarithm which is to the base e (Chapter 14))

So that if, at a particular temperature, a reaction has a half-life of 72 s then the value of the rate constant is

$k = \ln 2/72 = 0.693/72 = 0.0096\,s^{-1}$

For a **second-order reaction** the graph of concentration against time is shown in Figure 1.14.

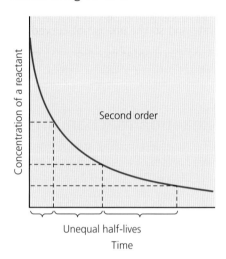

Figure 1.14

The half-life for a second-order reaction is not constant.

Test yourself

7 Table 1.6 gives the change in concentration of a reactant every 5 minutes during the course of a reaction.

Table 1.6

Concentration of reactant/mol dm^{-3}	4.0	3.4	2.9	2.5	2.1	1.8	1.5	1.3	1.1	0.93
Time /min	0	5	10	15	20	25	30	35	40	45

Plot a graph of concentration against time and use the graph to determine:

a) the initial rate of reaction

b) the rate after 30 minutes

c) the order of reaction with respect to the reactant. Justify your answer.

8 If a reaction is first order and has a rate constant of 0.056 s^{-1}, what is its half-life?

The rate-determining step

Although it might be supposed that a reaction could proceed via several slow steps, in practice, in almost all cases, it is found that just one step controls the overall rate. This step is called the rate-determining step.

Although it might be supposed that a reaction could proceed via several slow steps, in practice, in almost all cases, it is found that just one step controls the overall rate. This step is called the rate-determining step.

It is important to appreciate that:

- not all the reactants are necessarily involved in the rate-determining step

- the numbers of particles involved in the rate-determining step are not directly related to the balanced equation (quite often they are different).

There is a direct relationship between the orders of reaction and the rate-determining step, in that the order for a particular reactant relates the amount in moles of that reactant to the rate-determining step. So if reactant X has an order of 1, reactant Y has an order of 2 and reactant Z has an order of 0, it can be concluded that the rate-determining step involves 1 mole of X, 2 moles of Y and that Z is not involved in the rate-determining step at all.

> **Key term**
>
> The rate-determining step in a reaction that occurs in several steps is the slowest step in the reaction. It has the highest activation energy and, therefore, controls the overall rate of the reaction.

Reaction mechanisms

The rate-determining step is important because it may give a clue as to the mechanism of a reaction. For instance, the two isomers 1-chlorobutane, $CH_3CH_2CH_2CH_2Cl$, and 2-chloro-2-methylpropane, $(CH_3)_3CCl$, are both hydrolysed by aqueous sodium hydroxide to give the corresponding alcohol. However, the rate equations are different.

The rate equation for 1-chlorobutane is

$$r = k[CH_3CH_2CH_2CH_2Cl][OH^-]$$

The rate equation for 2-chloro-2-methylpropane is

$$r = k[(CH_3)_3CCl]$$

For 1-chlorobutane, the rate equation indicates that the slow rate-determining step involves both the chloroalkane and the sodium hydroxide, whereas for 2-chloro-2-methylpropane only the chloroalkane is involved in the slow step.

A few more examples are included in Table 1.7.

Table 1.7

Equation	Rate equation	Reagents involved in rate determining step
$CH_3CH_2CH_2CH_2Cl + OH^- \rightarrow$ $CH_3CH_2CH_2CH_2OH + Cl^-$	$r = k[CH_3CH_2CH_2CH_2Cl][OH^-]$	1 mol of $CH_3CH_2CH_2CH_2Cl$ 1 mol of OH^-
$(CH_3)_3CCl + OH^- \rightarrow (CH_3)_3COH + Cl^-$	$r = k[(CH_3)_3CCl]$	1 mol of $(CH_3)_3CCl$ and no OH^-
$NO_2 + CO \rightarrow NO + CO_2$	$r = k[NO_2]^2$	2 mols of NO_2 and no CO
$H_2 + 2ICl \rightarrow 2HCl + I_2$	$r = k[H_2][ICl]$	1 mol of H_2 and 1 mol of ICl
$2H_2 + 2NO \rightarrow 2H_2O + N_2$	$r = k[H_2][NO]^2$	1 mol of H_2 and 2 mols of NO

You may be asked in an examination to suggest a mechanism that is consistent with the orders of reaction that you have calculated. This involves providing a sequence that has:

- a rate-determining step consistent with those orders

- another step that, when the steps are combined, gives the overall equation.

For example, nitrogen(IV) oxide and carbon monoxide react to form nitrogen(II) oxide and carbon dioxide:

$$NO_2 + CO \rightarrow NO + CO_2$$

The rate equation for this reaction is:

$$\text{rate} = k[NO_2]^2$$

The rate equation indicates that the slow step involves only two molecules of NO_2.

This might suggest to you a mechanism with a rate-determining step such as:

$$2NO_2 \rightarrow 2NO + O_2$$

or

$$2NO_2 \rightarrow N_2O_4$$

Both of these are valid suggestions.

Whichever route you suggest, it is then essential that the sum of the steps in the mechanism add up to the overall balanced equation.

So if $2NO_2 \rightarrow 2NO + O_2$ is suggested as the slow step, a second step can be deduced from the overall equation. It is necessary, in this step, that one of the NO molecules and the O_2 react together and that CO is introduced into the reaction. This suggests:

$$NO + O_2 + CO \rightarrow CO_2 + NO_2$$

This can be checked by considering whether the slow step and fast step combined give the balanced equation.

$$\begin{array}{ll} 2NO_2 \rightarrow 2NO + O_2 & \text{slow} \\ NO + O_2 + CO \rightarrow CO_2 + NO_2 & \text{fast} \\ \hline NO_2 + CO \rightarrow NO + CO_2 & \end{array}$$

This mechanism is, in fact, incorrect. However, it is consistent with the information given and would be allowed in an examination. In fact, the following version is believed to be correct:

● rate-determining step: $2NO_2 \rightarrow NO + NO_3$

● fast step: $NO_3 + CO \rightarrow NO_2 + CO_2$

The important point is to make a suggestion that is supported by the balanced equation and the rate equation.

Test yourself

9 Hydrogen reacts with iodine(I) chloride to form iodine and hydrogen chloride.
The rate equation for the reaction is:
Rate = $k[H_2][ICl]$
a) Write a balanced equation for the reaction.
b) The reaction occurs by a two-step process. Suggest a mechanism for the reaction and indicate which is the slow step.

The effect of temperature on rate constants – the Arrhenius Equation

Tip

Make sure you know how to use your calculator!

As the temperature of a reaction is raised, more of the particles taking part have energy greater than the activation energy required for the reaction to take place. This can be expressed quantitatively in an important relationship known as the Arrhenius equation:

$$k = Ae^{-E_a/RT}$$

You will not be expected to remember this equation as it will be provided on the exam data sheet

In this equation:

A is a constant of proportionality known as the pre-exponential factor

e is a number which forms the basis of what is known as natural logarithms

E_a is the activation energy of the reaction

R is the gas constant

T is measured in K (**not** °C)

If you are not studying maths or physics, this equation may look rather intimidating but do not be too concerned as you will not be expected to use it extensively. You may be familiar with logarithms (log) – they are a means of expressing numbers as a power of 10 so that log 100 = 2 because 10^2 =100. You can also express numbers in what are known as natural logarithms expressed as ln(n) where n is a number whose natural logarithm is required (see Chapter 14). Both the functions 'e' and 'ln' can be found on scientific calculators.

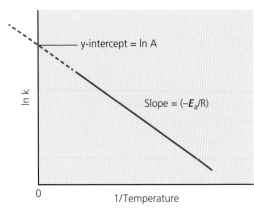

Figure 1.15

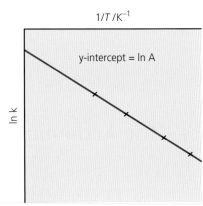

Figure 1.16

You could also plot a graph of log k against 1/T and the gradient would then be equal to $-E_a/2.3R$.

The constant A takes account of a number of factors that affect a reaction, such as the effectiveness of the collisions that take place, but you will not be expected to provide details. Every reaction has a different value for A and its size has a considerable effect on the rate of reaction. The units of the pre-exponential factor A are identical to those of the rate constant and vary depending on the order of the reaction.

R is another constant and arises as a constant of proportionality relating the pressure and volume to the temperature of gases. The value of R is provided on the data sheet in the exam.

Although you will not be able to understand how this equation is derived, you must be able to use it to derive the value of the activation energy for a reaction.

If natural logarithms (ln) of the Arrhenius equation are taken, the equation can be written as:

$$\ln k = \ln A - E_a/RT$$

You will recall that a straight line graph can be described by the equation $y = mx + c$ where m is the gradient and c is the intercept

$$\ln k = \ln A - E_a/RT$$

$$(y = c + mx)$$

This means that a graph of ln k against 1/T will be a straight line with a gradient of $-E_a/R$ and an intercept of ln A. It might look like Figure 1.15.

But if ln k is negative it would look like Figure 1.16.

The gradient, once measured, will be equal to $-E_a/R$.

Example 2

What is the value of the activation energy for a reaction at 410 °C which has a pre-exponential factor of 2.70×10^{11} and a rate constant of $0.00450 \, dm^3 \, mol^{-1} \, s^{-1}$?

Answer

$$k = Ae^{-E_a/RT}$$

and so

$$\ln k = \ln A - E_a/RT$$

The temperature must be in K, so $T = 273 + 410 = 683 K$

and $R = 8.314 \, J \, K^{-1} \, mol^{-1}$

Therefore,

$$\ln(0.00450) = \ln(2.70 \times 10^{11}) - E_a/(8.314 \times 683)$$

$$-5.40 = 26.32 - E_a/5678.5$$

$$E_a/5678.5 = 31.725$$

$$E_a = 31.725 \times 5678.5/1000 = 180 \, kJ \, mol^{-1} \text{ (to 3 significant figures)}$$

Example 3

At 550 K the rate constant for a reaction is $5.8 \times 10^{-4} \, s^{-1}$.

At 700 K the rate constant for the same reaction is $1.2 \times 10^{-2} \, s^{-1}$.

a) Determine the activation energy of the reaction.
b) Calculate the value of the pre-exponential factor, A.

Answer

a) $k = Ae^{-E_a/RT}$

and $\ln k = \ln A - E_a/RT$

$R = 8.314 \, J \, K^{-1} \, mol^{-1}$

Therefore at 550 K,
$\ln(5.8 \times 10^{-4}) = \ln A - E_a/(8.314 \times 550)$
and at 700K,
$\ln(1.2 \times 10^{-2}) = \ln A - E_a/(8.314 \times 700)$

Subtracting these two equations,
$\ln(5.8 \times 10^{-4}) - \ln(1.2 \times 10^{-2}) = -E_a/(8.314 \times 550) - (-E_a/(8.314 \times 700)$
$-7.4525 - (-4.4228) = -E_a/4572.7 + E_a/5819.8$
$-3.0297 = -0.000046862E_a$
$E_a = 64\,651.6 \, J \, mol^{-1}$ or $64.65 \, kJ \, mol^{-1}$

b) Substituting into the first equation,
$\ln(5.8 \times 10^{-4}) = \ln A - 64651.6/(8.314 \times 550)$
$-7.4525 = \ln A - 14.139$
$\ln A = 6.6861$ and $A = 801 \, s^{-1}$ (Remember A has the same units as k)

Example 4

The rate constant, k, is measured at five different temperatures for the hydrolysis of bromoethane with aqueous sodium hydroxide. The results are shown in Table 1.8.

Calculate the activation energy for the hydrolysis and determine a value for the pre-exponential factor.

Answer

We will use $k = Ae^{-E_a/RT}$ and $\ln k = \ln A - E_a/RT$. The data gives values for T and for k. R is given on the data sheet and A and E_a can be deduced by plotting a graph of $\ln k$ against $1/T$.

First use the data to calculate the temperature (T) in K and then determine $1/T$ and $\ln k$ (Table 1.9).

Table 1.8

Temperature/°C	k/s^{-1}
30	8.60×10^{-5}
40	3.35×10^{-4}
50	1.00×10^{-3}
60	3.35×10^{-3}
70	1.64×10^{-2}

Table 1.9

Temperature /°C	T/K	$1/T$/K^{-1}	k/s^{-1}	$\ln k$
30	303	0.00330	8.60×10^{-5}	−9.36
40	313	0.00319	3.35×10^{-4}	−8.00
50	323	0.00310	1.00×10^{-3}	−6.91
60	333	0.00300	3.35×10^{-3}	−5.70
70	343	0.00292	1.64×10^{-2}	−4.11

Then plot a graph of $\ln k$ against $1/T$ as shown in Figure 1.17.

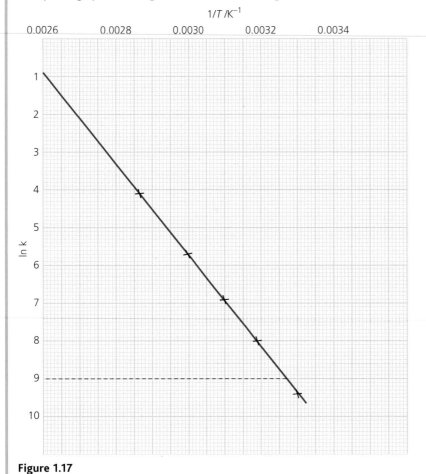

Figure 1.17

The gradient is obtained using the blue line on the graph and is equal to $-E_a/R$ (R is the gas constant = $8.314\,J\,K^{-1}\,mol^{-1}$)

y-value is $(-9.0) - (-0.90) = -8.1$

x-value is $0.00327 - 0.0026 = 0.00067$

So, the gradient = $-8.1/0.00067 = -12\,090$ and this equals $-E_a/R$

Therefore $E_a = 12\,090 \times 8.314 = 100\,512\,J\,mol^{-1} = 100.5\,kJ\,mol^{-1}$

The intercept on the y-axis is used to determine the value of the pre-exponential factor, A. However, this should be when $x = 0$ and the axes as used do not include this point. Therefore, in this case, the value of the intercept at $(0.0026, -0.09)$ must be substituted into the equation:

$$\ln k = \ln A - E_a/RT$$

$$\text{or } \ln A = \ln k + E_a/RT$$

Remembering that $E_a/R = 12\,089$

$\ln A = -0.9 + (12\,089 \times 0.0026) = -0.9 + 31.43 = 30.53$ (you could use any other values that lie on the graph)

and $A = 1.82 \times 10^{13}\,s^{-1}$

Test yourself

10 The rate constant, k, is measured at six different temperatures for the elimination of hydrogen iodide from iodoethane,

$$C_2H_5I \rightarrow H_2C=CH_2 + HI$$

Table 1.10

Temperature/K	k/s^{-1}
650	4.02×10^{-4}
675	1.68×10^{-3}
700	6.40×10^{-3}
725	2.19×10^{-2}
750	6.97×10^{-2}
800	5.66×10^{-1}

Use the results in Table 1.10 to determine the activation energy and the pre-exponential factor for the elimination of HI from C_2H_5I.

Practice questions

Multiple choice questions 1–10

1 The table below shows the results of three experiments to determine the orders of reaction of carbon monoxide and chlorine in a reaction to make carbonyl chloride ($COCl_2$).

Experiment number	Initial concentration of CO/mol dm^{-3}	Initial concentration of Cl$_2$/mol dm^{-3}	Initial rate of reaction/ mol dm^{-3} min^{-1}
1	0.80	0.40	6.4×10^{-24}
2	0.40	0.80	6.4×10^{-24}
3	0.80	0.80	1.28×10^{-23}

What will be the initial rate of reaction in mol dm^{-3} min^{-1} if the initial concentration of CO is 0.80 mol dm^{-3} and the initial concentration of Cl$_2$ is 0.20 mol dm^{-3}?

A 1.6×10^{-24} B 3.2×10^{-24}

C 6.4×10^{-24} D 1.28×10^{-23} *(1)*

2 A substance has a concentration of 2.4 mol dm^{-3}. Its decomposition is first order. If the half-life for this decomposition is 120 s, how long would it take it to decompose until its concentration was 0.15 mol dm^{-3}?

A 360 s   B 480 s

C 600 s   D 720 s *(1)*

3 Under some conditions the decomposition of $N_2O(g)$ into $N_2(g)$ and $O_2(g)$ in the presence of a hot platinum wire catalyst is zero order.
If a graph of rate (*y*-axis) against concentration (*x*-axis) is drawn, which of the options A–D represents the shape of the graph that will be obtained? *(1)*

A  B

C D

4 The decomposition of a gas, **X**, is second order. When its initial concentration is 1 mol dm^{-3} the initial rate of reaction is measured as 0.016 dm^3 mol^{-1} s^{-1}.
What will be the initial rate of reaction when the concentration of **X** is 0.25 mol dm^{-3}?

A 0.001 dm^3 mol^{-1} s^{-1}

B 0.004 dm^3 mol^{-1} s^{-1}

C 0.009 dm^3 mol^{-1} s^{-1}

D 0.012 dm^3 mol^{-1} s^{-1} *(1)*

Questions 5 and 6 refer to the following reaction.
Propanone and iodine react together in the presence of a hydrogen ion catalyst to form iodopropane.

$$CH_3COCH_3 + I_2 \xrightarrow{H^+} CH_3COCH_2I + HI$$

The rate equation for the reaction is:

$$\text{rate} = k[CH_3COCH_3][H^+]$$

5 What will be the units of the rate constant, *k*?

A mol dm^{-3} s^{-1}

B mol^{-1} dm^3 s^{-1}

C mol^2 dm^{-6} s^{-1}

D mol^{-2} dm^6 s^{-1} *(1)*

6 The reaction was carried out with the following concentrations:
$[CH_3COCH_3] = 0.1$ mol dm^3, $[I_2] = 0.1$ mol dm^3 and $[H^+] = 1$ mol dm^3.
The initial rate of reaction is measured as R mol dm^{-3} s^{-1}.
Which of the following mixtures would give an initial rate that was greater than R?

A $[CH_3COCH_3] = 0.05$ mol dm^{-3}, $[I_2] = 0.1$ mol dm^{-3} and $[H^+] = 2$ mol dm^{-3}

B $[CH_3COCH_3] = 0.05$ mol dm^{-3}, $[I_2] = 0.2$ mol dm^{-3} and $[H^+] = 2$ mol dm^{-3}

C $[CH_3COCH_3] = 0.1$ mol dm^{-3}, $[I_2] = 0.3$ mol dm^{-3} and $[H^+] = 1$ mol dm^{-3}

D $[CH_3COCH_3] = 0.1$ mol dm^{-3}, $[I_2] = 0.05$ mol dm^{-3} and $[H^+] = 2$ mol dm^{-3} *(1)*

7 The following reaction involves two gaseous molecules labelled **G** and **H** that react together to form products labelled **J**, **K** and **L**.

The equation for the reaction is $2G + H \rightarrow J + K + L$

The reaction has two steps. The slow step is $G + H \rightarrow I + L$.

Which of the following equations gives the equation for the second step?

A $I + L \rightarrow J + K$

B $I + L + G \rightarrow J + K$

C $G + I \rightarrow J + K$

D $G + H \rightarrow I + J + K$ *(1)*

Use the key below to answer questions 8, 9 and 10.

A	B	C	D
1, 2 & 3 correct	1 & 2 correct	2 & 3 correct	1 only correct

8 Tertiary haloalkanes react with hydroxide ions to form an alcohol. The rate of reaction is first order with respect to the tertiary haloalkane and zero order with respect to the hydroxide ion.

Which of the following apply to this reaction?

1 Doubling the concentration of the tertiary halide will double the rate of the reaction.

2 Doubling the concentration of the hydroxide ion will have no effect on the rate of the reaction.

3 Raising the temperature will lower the activation energy of the reaction. *(1)*

9 At room temperature, ozone reacts with nitrogen(II) oxide to form nitrogen(IV) oxide and oxygen.

In a determination of the order of reaction with respect to nitrogen(II) oxide, a small concentration of nitrogen(II) oxide is reacted with a large excess of ozone and the concentration of nitrogen(II) oxide is measured over a period of time.

Why is it necessary to have a **large** excess of ozone present?

1 So that the effect of ozone on the rate of reaction will be negligible.

2 So that the order of reaction with respect to ozone will be zero.

3 To make sure that there is ozone present throughout the reaction. *(1)*

10 The table below shows the results of three experiments to determine the orders of reaction of **X**(aq) and **Y**(aq) in the following reaction:

$$X(aq) + Y(aq) \rightarrow Z(aq)$$

Experiment number	Initial concentration of X/mmoldm^{-3}	Initial concentration of Y/mmoldm^{-3}	Initial rate of reaction/ mmoldm^{-3}s^{-1}
1	0.10	0.10	0.08
2	0.10	0.20	0.32
3	0.20	0.20	0.32

Which of the following can be deduced from this data?

1 The order with respect to X is zero.

2 The order with respect to Y is 2.

3 The rate constant is $8.0\,dm^3\,mmol^{-1}\,s^{-1}$ *(1)*

11 The gaseous oxide N_2O_5 decomposes to NO_2 gas and oxygen.

a) Write a balanced equation for the reaction.

b) If the rate of disappearance of N_2O_5 is $3.5 \times 10^{-4}\,mol\,dm^{-3}\,s^{-1}$, what is the

   i) rate of formation of NO_2

   ii) rate of formation of O_2? *(3)*

12 The hydrolysis of methyl ethanoate in alkali is first order with respect to the ester and first order with respect to hydroxide ions. The rate of reaction is $0.000\,69\,mol\,dm^{-3}\,s^{-1}$, at a given temperature, when the ester concentration is $0.050\,mol\,dm^{-3}$ and the hydroxide ion concentration is $0.10\,mol\,dm^{-3}$.

Write out the rate equation for the reaction and calculate the rate constant. *(3)*

13 If concentrations are measured in $mol\,dm^{-3}$ and rate as $mol\,dm^{-3}\,s^{-1}$, deduce the units of k for a reaction that has the rate equation:

a) rate = $k[A][B][C]$

b) rate = $k[A]^2[B]$. *(3)*

14 The table gives data for the reaction:

$$2NOCl(g) \rightarrow 2NO(g) + Cl_2(g)$$

Initial concentration NOCl/mol dm^{-3}	Initial rate of reaction/mol dm^{-3} s^{-1}
0.1	4×10^{-10}
0.2	1.6×10^{-9}
0.4	6.4×10^{-9}

a) Calculate the order with respect to NOCl.
b) Calculate the value of the rate constant and give its units.
c) What is the initial rate of reaction when the initial concentration of NOCl is 0.15 mol dm⁻³? [7]

15 Aqueous mercury(II) chloride can be reduced by aqueous ethanedioate ions. A precipitate of mercury(I) chloride is obtained.
The equation is:

$$2HgCl_2(aq) + (COO)_2^{2-}(aq) \rightarrow Hg_2Cl_2(s) + CO_2(g) + 2Cl^-(aq)$$

The initial rates of reaction at various concentrations of aqueous mercury(II) chloride and aqueous ethanedioate ions were obtained by measuring the amount of mercury(I) chloride precipitated. The initial rate can be expressed in units of mol dm⁻³ min⁻¹ of mercury(II) chloride that has reacted.
The results are shown in the table.

Initial concentration of $HgCl_2(aq)/$ mol dm⁻³	Initial concentration of $(COO)_2^{2-}(aq)/$ mol dm⁻³	Initial rate/ mol dm⁻³ min⁻¹
0.08	0.02	0.48 × 10⁻⁴
0.08	0.04	1.92 × 10⁻⁴
0.04	0.04	0.96 × 10⁻⁴

a) Determine the orders of reaction with respect to:
   i) $(COO)_2^{2-}(aq)$
   ii) $HgCl_2(aq)$.
b)
   i) Determine the rate equation for this reaction.
   ii) Calculate the value of the rate constant. Quote the units. (7)

16 The table gives data for the reaction:

$$BrO_3^-(aq) + 5Br^-(aq) + 6H^+(aq) \rightarrow 3Br_2(aq) + 3H_2O(l)$$

$[BrO_3^-(aq)]/$ mol dm⁻³	$[Br^-(aq)]/$ mol dm⁻³	$[H^+(aq)]/$ mol dm⁻³	Initial rate/ mol dm⁻³ s⁻¹
0.1	0.2	0.1	1.64 × 10⁻³
0.2	0.1	0.1	1.64 × 10⁻³
0.2	0.2	0.1	3.28 × 10⁻³
0.2	0.1	0.2	6.56 × 10⁻³
0.25	0.25	0.25	x

a) Determine the rate equation for this reaction.
b) Calculate the value of k, including its units.
c) What is the initial rate of reaction, x, for the initial concentrations shown in the last row of the table? (9)

17 a) The data in the table below gives some results of the measurement of the rate of reaction of 1-bromobutane with hydroxide ions.

Experiment number	Concentration of 1-bromobutane/ mol dm⁻³	Concentration of hydroxide/ mol dm⁻³	Initial rate of reaction/ mol dm⁻³ s⁻¹
1	0.0100	0.0100	0.300
2	0.0300	0.0200	1.80
3	0.0100	0.0025	0.075

   i) Give the equation for the reaction.
   ii) Deduce the rate equation for the reaction.

b) The data in the table below gives some results of the measurement of the rate of reaction of 2-bromo-2-methylpropane with hydroxide ions.

Experiment number	Concentration of 2-bromo-2-methylpropane/ mol dm⁻³	Concentration of hydroxide/ mol dm⁻³	Initial rate of reaction/ mol dm⁻³ s⁻¹
1	0.0100	0.0100	7.500
2	0.0300	0.0200	22.50
3	0.0100	0.0025	7.500

   i) Give the equation for the reaction.
   ii) Deduce the rate equation for the reaction.
c)
   i) What type of reaction is taking place in these two experiments?
   ii) How do the mechanisms of the two reactions differ?
   iii) By how much do the values of the rate constant differ in the two reactions? (18)

18 Explain how the results given in the table indicate that the reaction is first order with respect to the reactant, B. (4)

Concentration of B/mol dm⁻³	0.800	0.640	0.512	0.410	0.328	0.262
Time/s	0	20	40	60	80	100

19 Two reagents **X** and **Y** react together according to the equation:

$$X + 2Y \rightarrow products$$

a) The concentration of **X** is measured in the presence of such a large excess of **Y** that it can be assumed that **Y** has no effect on the reaction rate.
The results obtained are shown in the table.

Concentration of X /mol dm^{-3}	0.50	0.43	0.36	0.29	0.22	0.15
Time/s	0	30	60	90	120	150

Plot a graph to deduce the order of reaction with respect to **X**.

b) The experiment is repeated, but this time **X** is present in sufficient excess for the rate of reaction to depend only on **Y**. The results obtained are shown in the table.

Concentration of Y/mol dm^{-3}	1.00	0.70	0.49	0.34	0.24	0.17
Time/s	0	30	60	90	120	150

Deduce the order of reaction with respect to **Y**.

c) Using your answers to parts **a** and **b**:
   i) Write the overall rate equation for the reaction.
   ii) Deduce a value for the initial rate of the reaction.
   iii) Deduce a value for the rate constant. *(11)*

20 a) Hydrogen peroxide decomposes according to the equation:

$$2H_2O_2(aq) \rightarrow 2H_2O(l) + O_2(g)$$

The following results were obtained in an experiment to measure the decomposition of hydrogen peroxide in the presence of an acid catalyst.

Concentration of H$_2$O$_2$/ mol dm^{-3}	0.10	0.09	0.08	0.07	0.06	0.05	0.04	0.03
Time/min	0	792	1678	2682	3841	5210	6888	9051

Show, by plotting a graph, that the acid-catalysed decomposition of hydrogen peroxide is first order.

b) Use an expression relating the rate constant and the half-life to determine a value for the rate constant. *(4)*

21 Sulfuryl chloride decomposes in an organic solvent as follows:

$$SO_2Cl_2 \rightarrow SO_2 + Cl_2$$

The decomposition of sulfuryl chloride is followed by measuring its concentration every 30 min. The results obtained are shown in the table.

Concentration of SO$_2$Cl$_2$/ mol dm^{-3}	1.00	0.750	0.563	0.422	0.316	0.237
Time/min	0	30	60	90	120	150

a) Determine the order of reaction.
b) Calculate the rate constant.
c) Confirm your answer to part **b** using an equation which relates the rate constant and the half-life to determine a value for the rate constant. *(10)*

22 Ammonia gas decomposes to nitrogen and hydrogen in the presence of a hot platinum wire. Experiments show that the reaction continues at a constant rate until all the ammonia has disappeared.
a) Sketch a concentration–time graph for the reaction and deduce the rate.
b) Write the balanced chemical equation and the rate equation for this reaction.
c) Give the units of the rate constant. *(6)*

23 The rate equation for the reaction
$2NO_2(g) + F_2(g) \rightarrow 2NO_2F(g)$ is
rate $= k[NO_2(g)][F_2(g)]$.
Suggest a two-step mechanism for the reaction which is consistent with the rate equation. Indicate which is the rate-determining step. *(3)*

24 The decomposition of hydrogen peroxide into oxygen and water can be catalysed by iodide ions. When this occurs, the rate equation for the reaction is rate $= k[H_2O_2(aq)][I^-(aq)]$.
Suggest a two-step mechanism for the reaction which is consistent with the rate equation. Indicate which is the rate-determining step. *(3)*

25 Hydrogen reacts with nitrogen(II) oxide to form water and a gas with relative molecular mass $= 28$. The rate equation is $r = k[H_2][NO]^2$.
a) Identify the gas with relative molecular mass of 28.
b) Write a balanced equation for the reaction.
c) The reaction occurs by a two-step process. Suggest a mechanism for the reaction and indicate which is the slow step. *(5)*

26 Three possible mechanisms for the reaction of nitrogen(II) oxide and oxygen to make nitrogen(IV) oxide ($2NO + O_2 \rightarrow 2NO_2$) have been suggested:

Mechanism X:

$2NO + O_2 \rightarrow 2NO_2$ Slow

Mechanism Y:

$NO + O_2 \rightarrow NO_3$ Slow

$NO_3 + O \rightarrow 2NO_2$ Fast

Mechanism Z:

$O_2 \rightarrow 2O$ Slow

$2NO + 2O \rightarrow 2NO_2$ Fast

a) Write the rate equation for each mechanism.

b) When this reaction is investigated using the initial rates procedure, the initial partial pressures of the gases are used as measures of their concentrations.

The results obtained are shown in the table.

Initial pressure of NO/kPa	Initial pressure of O_2/kPa	Initial rate/ kPa hr^{-1}
50	50	0.100
50	20	0.040
100	60	0.480

i) Calculate the orders of reaction with respect to NO and O_2.

ii) Which of the above mechanisms **X**, **Y** and **Z** is compatible with your answers to part **i)**? (9)

27 Calculate the activation energy (in kJ mol^{-1}) for a reaction which has a pre-exponential factor of 3.05×10^5 mol dm^{-3} s^{-1} and a rate constant of 0.0450 mol dm^{-3} s^{-1} at 650 K. (3)

28 Calculate the value of the pre-exponential factor if a reaction at 600 °C has rate constant of 0.0082 min^{-1} and an activation energy of 96 kJ mol^{-1}. (5)

29 At 400 K the rate constant for a reaction is 3.2×10^{-4} s^{-1}.

At 500 K the rate constant for the same reaction is 8.4×10^{-3} s^{-1}.

a) Determine the activation energy of the reaction?

b) Calculate the value of the pre-exponential factor, A. (10)

30 Iodine(I) chloride reacts with hydrogen to form iodine and hydrogen chloride.

$2ICl + H_2 \rightarrow I_2 + 2HCl$

The rate constant for this reaction is measured at various temperatures and the results are shown in the table.

Temperature/K	k/dm^6 mol^{-2} s^{-1}
500	0.128
510	0.277
515	0.367
520	0.579
525	0.829

Calculate the activation energy and the pre-exponential factor for this reaction. (8)

31 The rate constant, k, for the dissociation of hydrogen iodide varies with temperature.

$2HI(g) \rightleftharpoons H_2(g) + I_2(g)$ $\Delta H = 25.9$ kJ mol^{-1}

Temperature T/°C	$k \times 10^5$/dm^3 mol^{-1} s^{-1}
400	4.130
450	42.05
500	186.5
550	1498
600	5213

a) Plot ln k against $1/T$ and calculate the activation energy for the forward reaction.

b) Calculate the activation energy for the reverse reaction. (9)

Challenge

32 Alcohol is absorbed through the stomach lining into the bloodstream. It is then slowly removed by oxidation in an enzyme-catalysed reaction.

a) A person consumes a pint (568 cm^3) of beer containing 4.1 % by volume of ethanol. The density of ethanol is 0.789 g cm^{-3}. Show that the amount of ethanol that has been consumed is 0.40 mol.

b) As soon as the beer has been drunk, ethanol is transferred to the blood from the stomach. The amount of ethanol (in moles) remaining in the stomach over a period of time is shown in the table.

Amount of ethanol/ mol	0.40	0.34	0.29	0.24	0.17	0.076	0.014	0.0063
Time/ min	0	1	2	3	5	10	20	30

    i)   Show that the transfer of ethanol from the stomach to the blood is first order with respect to ethanol.

    ii)   What is the half-life of the transfer?

    iii) Calculate the value of the rate constant.

c) Assume that no oxidation takes place during the transfer of the ethanol. Copy and complete the second table to show the concentration of ethanol in the blood. Assume that there is 40 dm³ of fluid, including blood, in the body and that the amount of ethanol is distributed uniformly throughout.

Time/min	0	1	2	3	5	10	20	30
Concentration of ethanol/mol dm⁻³								

d) However oxidation does begin to take place as soon as ethanol enters the fluid. The oxidation is zero order with respect to ethanol and it has a rate constant equal to 0.2 g dm⁻³ h⁻¹.

    i)   Show that the value of the rate constant is equivalent to 7.25×10^{-5} mol dm⁻³ min⁻¹.

ii) Copy and complete the table below to adjust the values you calculated in c) to take into account the ethanol removed by oxidation.

Time/min	0	1	2	3	5	10	20	30
Adjusted concentration of ethanol / mol dm⁻³								

e) Plot a graph of concentration of ethanol against time. Determine the time when the concentration of ethanol in the body fluid, including the blood, is at its highest value.

f) In parts of the UK, the legal maximum concentration of ethanol allowed in the blood for a person to drive a vehicle is 80 mg per 100 cm³ of blood. Taking your answer from part e), with this level of ethanol, would the person be fit to drive a car according to the law?   (16)

It should be added that this calculation is based on a number of assumptions that differ from person to person and that any amount of alcohol interferes with the capacity to drive safely.

Chapter 2

How far? Equilibrium

Prior knowledge

In this chapter it is assumed that you understand that:
- a reaction in which the reactants and products are in a closed system (i.e. one in which none of the component particles can escape) may reach dynamic equilibrium; the reaction then proceeds in both directions
- le Chatelier's principle can be applied to determine how a change in conditions affects the balance of reactants and products
- the equilibrium constant, K_c, controls the position of equilibrium between the reactants and products.

Test yourself on prior knowledge

1 N_2O_4 forms an equilibrium with NO_2 as shown below.

$$N_2O_4(g) \rightleftharpoons 2NO_2(g) \quad \Delta H = 58.0\,kJ$$

colourless   brown

  a) Write an expression for the equilibrium constant for the equilibrium mixture.

  b) Explain what would be the effect on the equilibrium mixture of:

    i) increasing the pressure

    ii) increasing the temperature

    iii) adding a catalyst?

    In each case state what you would see?

2 What can you conclude about the position of the equilibrium in each of these examples?

  a) $Zn(s) + Cu^{2+}(aq) \rightleftharpoons Zn^{2+}(aq) + Cu(s)$ $K_c = 1 \times 10^{37}$ at 298 K

  b) $2HBr(g) \rightleftharpoons H_2(g) + Br_2(g)$ $K_c = 1 \times 10^{-10}$ at 298 K

Chemical equilibria

The principles of chemical equilibria are essential to an understanding of the nature of chemical reactions. Some of the ideas in this chapter have been considered before, but these are now extended to include the quantitative relationships between the reactants and products in equilibrium mixtures.

Units of the equilibrium constant, K_c

You will probably have met K_c earlier in the course but may not, at that time, have considered its units. In fact, the units of K_c vary according to the equilibrium being considered. Each of the examples in question 2 above has no units, because the numbers of moles on each side of the equilibrium are the same. But for the equilibrium:

$$N_2O_4(g) \rightleftharpoons 2NO_2(g) \quad K_c = \frac{[NO_2(g)]^2}{[N_2O_4(g)]}$$

and this will have units of $\dfrac{(mol\ dm^{-3})^2}{(mol\ dm^{-3})}$ which is $mol\ dm^{-3}$.

Each equilibrium must be considered individually when deciding on the correct units for K_c.

Test yourself

1 Write the K_c expressions for the following equations and state the units of the equilibrium constant for each example:
   a) $CO_2(g) + H_2(g) \rightleftharpoons CO(g) + H_2O(g)$
   b) $N_2(g) + 3H_2(g) \rightleftharpoons 2NI_3(g)$
   c) $2O_3(g) \rightleftharpoons 3O_2(g)$
   d) $4PF_5(g) \rightleftharpoons P_4(g) + 10F_2(g)$

Calculations involving K_c

The relationship between the value of the equilibrium constant, K_c, and the concentrations of the components of an equilibrium mixture makes it possible to obtain quantitative information about these components. Calculations to illustrate this are shown below.

Example 1

At equilibrium, the concentrations of sulfur dioxide and oxygen in the mixture

$$2SO_2(g) + O_2(g) \rightleftharpoons 2SO_3(g)$$

are found to be $[SO_2] = 0.020\ mol\ dm^{-3}$ and $[O_2] = 0.010\ mol\ dm^{-3}$.

If the value of K_c is $1.28 \times 10^4\ dm^3\ mol^{-1}$, what is the equilibrium concentration of sulfur trioxide?

Answer

$$K_c = \frac{[SO_3(g)]^2}{[SO_2(g)]^2[O_2(g)]}$$

So, $[SO_3(g)]^2 = K_c[SO_2(g)]^2[O_2(g)]$

$[SO_3(g)]^2 = 1.28 \times 10^4 \times (0.020)^2 \times 0.010 = 0.0512\ mol^2\ dm^{-6}$

Therefore, the equilibrium concentration of sulfur trioxide $= \sqrt{0.0512} = 0.226\ mol\ dm^{-3}$

Example 2

At 473 K, the value of K_c for the decomposition of PCl_5 is $8 \times 10^{-3}\,mol\,dm^{-3}$.

$$PCl_5(g) \rightleftharpoons PCl_3(g) + Cl_2(g)$$

A sample of pure PCl_5 is heated to 473 K in a vessel containing no other chemicals. At equilibrium the concentration of PCl_5 is $5 \times 10^{-2}\,mol\,dm^{-3}$.

What are the equilibrium concentrations of PCl_3 and Cl_2?

Answer

The key to this is realising that, every time a molecule of PCl_3 is made, a molecule of Cl_2 is also formed.

Therefore $[PCl_3(g)] = [Cl_2(g)] = x$

At equilibrium:

$$K_c = 8 \times 10^{-3} = \frac{\left[PCl_3(g)\right]\left[Cl_2(g)\right]}{\left[PCl_5(g)\right]} = \frac{x^2}{5 \times 10^{-2}}$$

Therefore, $x^2 = (8 \times 10^{-3})(5 \times 10^{-2}) = 40 \times 10^{-5}$

And $x = 2 \times 10^{-2}\,mol\,dm^3$

So the concentrations of PCl_3 and Cl_2 are both $0.02\,mol\,dm^{-3}$

> **Tip**
>
> You should always check your answers if possible and, in this case, it is quick to do. The equilibrium constant should be $(2 \times 10^{-2})^2/5 \times 10^{-2} = 4/5 \times 10^{-2} = 8 \times 10^{-3}$, confirming your calculation is correct.

Example 3

0.50 mol of carbon dioxide and 0.80 mol of hydrogen were mixed together in a 10 dm³ flask and allowed to reach equilibrium as shown below.

$$CO_2(g) + H_2(g) \rightleftharpoons CO(g) + H_2O(g)$$

When analysed, the equilibrium mixture was found to contain 0.04 mol of carbon monoxide.

Calculate the value of K_c for this reaction.

Answer

You may meet a variety of problems of this type, so it is important to have a strategy for approaching them. Table 2.1 summarises what happens.

Table 2.1

	CO_2	H_2	CO	H_2O
Initial amount (mol)	a	b	c	d
Equilibrium amount (mol)	$(a - x)$	$(b - x)$	$(c + x)$	$(d + x)$
Initial amount	0.50	0.80	0	0
Equilibrium amount			0.04	

The initial amount of CO = 0 mol and the equilibrium amount of CO = 0.04 mol. Therefore, $x = 0.04$ mol.

The equilibrium amount of each reactant and product can now be calculated.

In making the 0.04 mol of CO, both the CO_2 and H_2 will have lost 0.04 mol of their original amount so,

> **Tip**
>
> Calculations of this type are frequently set in exams and you should make sure you understand the procedure.

the amount (in moles) of CO_2 at equilibrium = 0.50 − 0.04 = 0.46 mol

the amount (in moles) of H_2 = 0.80 − 0.04 = 0.76 mol.

And, as the 0.04 mol of CO is made, so is 0.04 mol of H_2O.

The equilibrium constant, K_c, requires concentrations of gases.

The concentration of the gases = amount in moles/volume.

Since the volume of the container was 10 dm³, the amounts in moles are divided by 10.

Therefore,

$$K_c = \frac{0.004 \times 0.004}{0.046 \times 0.076} = 4.6 \times 10^{-3}$$

> **Tip**
>
> You may notice that, in this case, you would get the correct answer if you didn't divide by 10. This will only be the case if the number of molecules on either side of the equation is the same.

Example 4

1.68 mol of $PCl_5(g)$ and 0.36 mol of $PCl_3(g)$ are mixed in a 2.0 dm³ container and allowed to come to equilibrium at 570 K. The mixture is then analysed and found to contain 1.44 mol of $PCl_5(g)$.

Calculate K_c for the reaction $PCl_5(g) \rightleftharpoons PCl_3(g) + Cl_2(g)$.

Answer

Table 2.2

	PCl_5	PCl_3	Cl_2
Initial amount (mol)	a	b	c
Equilibrium amount (mol)	$(a - x)$	$(b + x)$	$(c + x)$
Initial amount	1.68	0.36	0
Equilibrium amount	1.44		

The initial amount of PCl_5 = 1.68 mol and the equilibrium amount is 1.44 mol. Therefore, x = 0.24 mol

The equilibrium amount of each reactant and product can now be calculated.

PCl_3 = 0.36 + 0.24 = 0.60 mol

Cl_2 = 0.24 mol

The equilibrium constant, K_c, requires concentrations of gases.

The concentration of the gases = amount in moles/volume.

Since the volume of the container was 2 dm³, the amounts in moles are divided by 2.

Therefore:

$$K_c = \frac{0.30 \times 0.12}{0.72} = 0.050$$

and the units of K_c are mol dm⁻³

2 Nitrogen(II) oxide and oxygen form an equilibrium mixture with nitrogen(IV) oxide.

$$2NO(g) + O_2(g) \rightleftharpoons 2NO_2(g)$$

An equilibrium mixture is analysed and found to contain the following concentrations:

$[NO] = 0.200\,mol\,dm^{-3}$, $[O_2] = 0.050\,mol\,dm^{-3}$ and $[NO_2] = 0.180\,mol\,dm^{-3}$
Calculate K_c for this equilibrium and give its units, if any.

3 Ethanoic acid and ethanol react to form an equilibrium mixture with ethyl ethanoate and water:

$$CH_3COOH(l) + C_2H_5OH(l) \rightleftharpoons CH_3COOC_2H_5(l) + H_2O(l)$$

When 0.50 mol of $CH_3COOH(l)$ is dissolved in 0.5 dm³ of an organic solvent with 0.09 mol of $C_2H_5OH(l)$ and allowed to come to equilibrium at 293 K, the amount of the ester, $CH_3COOC_2H_5(l)$, formed at equilibrium is 0.086 mol.

Calculate the value of K_c for the reaction and give its units, if any.

Experimental methods for determining equilibrium constants

A variety of experiments can be carried out to determine equilibrium constants, although if the equilibrium constant is very large or very small these are difficult to do without specialist equipment. A feature of any method is that the equilibrium must not be disturbed by any measurement taken. This usually means that the position of equilibrium must be determined by measuring a physical property of one of the components. For example, the acidity of a component might be measured before and after equilibrium has been established. However, this would need to be done using a pH meter which would not affect the position of equilibrium. A titration would not be appropriate because the removal of the hydrogen ions during the titration would cause the equilibrium to rebalance. The only circumstances in which a titration could be used are those in which the equilibrium is so slow to form that the titration would be completed before much movement in the equilibrium mixture had taken place.

Other properties that might be used are a change in colour or electrical conductivity. Each equilibrium has to be considered individually to establish what might be possible.

Activity

Testing the equilibrium law

The reversible reaction involving hydrogen, iodine and hydrogen iodide has been used to test the equilibrium law experimentally. In a series of six experiments, samples of the chemicals were sealed in reaction tubes and then heated at 731 K until the mixtures reached equilibrium. Four of the tubes started with different mixtures of hydrogen and iodine. Two of the tubes started with just hydrogen iodide. Once the tubes had reached equilibrium, they were rapidly cooled to stop the reactions. Then the contents of the tubes were analysed to find the compositions of the equilibrium mixture. The results for six of the tubes are shown in Table 2.3.

Table 2.3

Tube	Initial concentrations/mol dm⁻³			Equilibrium concentrations/mol dm⁻³		
	$[H_2(g)]$	$[I_2(g)]$	$[HI(g)]$	$[H_2(g)]$	$[I_2(g)]$	$[HI(g)]$
1	2.40×10^{-2}	1.38×10^{-2}	0	1.14×10^{-2}	0.12×10^{-2}	2.52×10^{-2}
2	2.40×10^{-2}	1.68×10^{-2}	0	0.92×10^{-2}	0.20×10^{-2}	2.96×10^{-2}
3	2.44×10^{-2}	1.98×10^{-2}	0	0.77×10^{-2}	0.31×10^{-2}	3.34×10^{-2}
4	2.46×10^{-2}	1.76×10^{-2}	0	0.92×10^{-2}	0.22×10^{-2}	3.08×10^{-2}
5	0	0	3.04×10^{-2}	0.345×10^{-2}	0.345×10^{-2}	2.35×10^{-2}
6	0	0	7.58×10^{-2}	0.86×10^{-2}	0.86×10^{-2}	5.86×10^{-2}

1 Write the equation for the reversible reaction to form hydrogen iodide from hydrogen and iodine.

2 Show that the equilibrium concentration of:
   a) hydrogen in tube 1 is as expected given the value of $[I_2(g)]_{eqm}$
   b) hydrogen iodide in tube 2 is as expected given the value of $[I_2(g)]_{eqm}$.

3 Explain why $[H_2(g)]_{eqm} - [I_2(g)]_{eqm}$ in tubes 5 and 6.

4 For each of the tubes work out the value of:

   a) $\dfrac{[HI(g)]_{eqm}}{[H_2(g)]_{eqm}[I_2(g)]_{eqm}}$

   b) $\dfrac{[HI(g)]^2_{eqm}}{[H_2(g)]_{eqm}[I_2(g)]_{eqm}}$

   c) Enter your values in a table and comment on the results.

5 What is the value of K_c for the reaction of hydrogen with iodine at 731 K?

The effect of a change in conditions on the value of K_c

The equilibrium constant is dependent on temperature. It is **not** changed by altering any other condition.

K_c is **unaffected** by:

● a change in concentration

● the presence of a catalyst

● a change in pressure.

The fact that K_c is unaffected by a change in pressure may seem rather surprising and, at first glance, seems to contradict le Chatelier's principle. It is worth explaining in more detail.

Consider the reaction:

$$N_2(g) + 3H_2(g) \rightleftharpoons 2NH_3(g)$$

$$K_c = \frac{[NH_3(g)]^2}{[N_2(g)][H_2(g)]^3}$$

The square brackets indicate concentrations. Therefore, if the reaction takes place in a reacting volume, $V\,dm^3$, the equilibrium expression can be rewritten in terms of the amount n (in moles) of the three gases present. In each case, the concentration is the amount (in moles) divided by the volume and, since they are in the same container, the volume, V, will be the same for each gas.

$$K_c = \frac{(n_{NH_3(g)}/V)^2}{(n_{N_2(g)}/V)(n_{H_2(g)}/V)^3}$$

This simplifies to

$$K_c = \frac{(n_{NH_3(g)})^2 \times V^2}{(n_{N_2(g)})(n_{H_2(g)})^3}$$

Tip

The simplification of the expression worries some students, but any mathematician will confirm it is correct!

Suppose the pressure is increased. The volume, V, will then decrease. The value of K_c, however, stays constant because the amount of moles of ammonia $(n_{NH_3(g)})$ formed increases and the amount of moles of nitrogen and hydrogen decrease. Therefore, le Chatelier is correct and more ammonia is obtained and K_c does remain constant.

It is important, therefore, to realise that K_c controls the position of an equilibrium mixture as a result of a change in the pressure or concentration. However, it needs to be stressed that K_c does not give any information on how quickly equilibrium will be achieved. An increase in pressure or the presence of a catalyst both cause the equilibrium to be established faster, as both increase the rate of reaction.

To determine the effect of a change in temperature also requires careful consideration. If the temperature is raised, equilibrium is established faster but, unlike the other changes in conditions, the *value* of K_c also changes. Whether it increases or decreases depends on whether the forward reaction is endothermic or exothermic.

The manufacture of ammonia from nitrogen and hydrogen is exothermic in the forward direction:

$$N_2(g) + 3H_2(g) \rightleftharpoons 2NH_3(g) \quad \Delta H = -93\,kJ\,mol^{-1}$$

The equilibrium constant for this reaction is:

$$K_c = \frac{[NH_3(g)]^2}{[N_2(g)][H_2(g)]^3}$$

Le Chatelier's principle predicts that an increase in temperature favours the breakdown of ammonia and the formation of nitrogen and hydrogen. This means that the concentration of ammonia decreases and the concentrations of nitrogen and hydrogen both increase. This does affect the value of the equilibrium constant, which will decrease.

Test yourself

4 At 473 K, the value of K_c for the decomposition:

$$PCl_5(g) \rightleftharpoons PCl_3(g) + Cl_2(g)$$

is $8 \times 10^{-3} \, mol \, dm^{-3}$ whereas at 570 K, $K_c = 5 \times 10^{-2} \, mol \, dm^{-3}$. Explain whether this indicates that the decomposition of $PCl_5(g)$ into $PCl_3(g) + Cl_2(g)$ is exothermic or endothermic.

Example 5

$$H_2(g) + I_2(g) \rightleftharpoons 2HI(g)$$

At 220 °C, the value of the equilibrium constant, K_c, is 160; at 450 °C it is 49.

Explain whether the breakdown of hydrogen iodide into hydrogen and iodine is exothermic or endothermic.

Answer

The value of K_c decreases as the temperature rises from 220 °C to 450 °C. This means that as the temperature increases less HI is present.

The equilibrium is, therefore, being moved to the left-hand side of the equation as the temperature increases.

From le Chatelier's principle, we know that this occurs when the forward reaction is exothermic.

Therefore, the breakdown of hydrogen iodide is endothermic.

Equilibrium constant from partial pressures, K_p

If an equilibrium is based on solutions or liquids it would be natural to measure the equilibrium constant based on the concentrations of the components present. However, for gases the more normal quantities to measure are temperature, volume and pressure.

The pressure of a gas depends on the number of particles present in a given volume. For a mixture of gases, the pressure of an individual gas within the mixture is directly related to the total number of particles of that gas present. The total pressure is then the sum of the individual pressures contributed by each gas. Since the mole is a measure of a number of particles, the overall pressure of a mixture depends on the sum of the amount in moles of each gas present.

So for gases, an equilibrium constant, K_p, is defined based on the pressure that each individual component gas is contributing to the overall pressure. This pressure of each component gas is known as its partial pressure.

For a reaction

$$aW(g) + bX(g) \rightleftharpoons cY(g) + dZ(g)$$

$$K_p = \frac{p_Y^c \times p_Z^d}{p_W^a \times p_X^b}$$

where 'p' is the partial pressure of the gas.

So for the equilibrium

$$2SO_2(g) + O_2(g) \rightleftharpoons 2SO_3(g)$$

$$K_p = \frac{p_{SO_3}^2}{p_{SO_2}^2 \times p_{O_2}}$$

Key term

The partial pressure is the pressure that would be exerted by a gas in a mixture of gases if it occupied the same volume on its own at the same temperature.

<table>
</table>

> **Tip**
>
> Don't put square brackets round the partial pressures as this implies concentration.

Equilibria based on partial pressures are widely used industrially to monitor gaseous reactions such as the Haber process for the production of ammonia from nitrogen and hydrogen.

Remember that partial pressures only apply to gases. If an equilibrium contains a solid, liquid or solution, the partial pressure of these is taken as 1. So K_p for an equilibrium such as

$$CaCO_3(s) \rightleftharpoons CaO(s) + CO_2(g)$$

is simply $K_p = p_{CO_2}$

Mole fractions

> **Tip**
>
> The mole fractions of all the gases in an equilibrium mixture must always add up to 1.

> **Key term**
>
> The **mole fraction** is the amount in moles of a component in a mixture divided by the total amount of all components in moles in the mixture.

In order to determine partial pressures, we need to know the contribution to the total pressure that is being made by an individual gas in the equilibrium mixture. If the equilibrium mixture contains a total amount in moles equal to N and there are n moles of an individual gas in the mixture then its contribution to the total pressure, P, will be $\frac{n}{N}P$.

The ratio n/N is known as the mole fraction of the gas and the

partial pressure of a gas = mole fraction of the gas × the total pressure of the equilibrium mixture.

It is inevitably the case that:

the sum of the partial pressures of the components of an equilibrium is equal to the total pressure.

Units of the equilibrium constant, K_p

The units for K_p are worked out in the same way as those for K_c and, like K_c, are dependent on the amounts in moles on both sides of the equation. Pressures can be measured in several different units but kPa is the correct SI unit. For the equilibrium:

$$N_2(g) + 3H_2(g) \rightleftharpoons 2NH_3(g)$$

$K_p = \dfrac{(pNH_3(g))^2}{(pN_2(g))(pH_2(g))^3}$ and this will have units of $\dfrac{(kPa)^2}{(kPa)(kPa)^3}$ which is kPa^{-2}.

Calculations involving K_p

> **Example 6**
>
> Write the expression for K_p for the following equilibrium:
>
> $$5CO(g) + I_2O_5(s) \rightleftharpoons I_2(g) + 5CO_2(g)$$
>
> **Answer**
>
> $$K_p = \frac{p_{I_2} \times p_{CO_2}^5}{p_{CO}^5}$$
>
> the point to remember is that as I_2O_5 is a solid it is excluded from the expression for K_p.

Example 7

If steam is passed over heated carbon at 900 K the following equilibrium is formed.

$$H_2O(g) + C(s) \rightleftharpoons H_2(g) + CO(g) \quad K_p = 58\,kPa$$

When 2.0 mol of steam is heated in a container to 900K it is found that 1.2 mol of hydrogen is present in the equilibrium mixture formed.

Calculate the pressure that is exerted in the container by the equilibrium mixture.

Answer

For this type of calculation you start by summarising what is happening, just as you would for a K_c calculation.

If 1.2 mol of H_2 is present at equilibrium then 1.2 mol of CO will also be formed and the amount of H_2O remaining will be $(2.0 - 1.2) = 0.8$ mol.

The total number of moles present at equilibrium is, therefore, $1.2 + 1.2 + 0.8 = 3.2$ mol.

The partial pressures are the (mole fraction × pressure). Therefore,

$$p_{H_2} = (1.2/3.2) \times P = 0.375P$$

(where P is the pressure exerted).

$$p_{CO} = (1.2/3.2) \times P = 0.375P$$

$$p_{H_2O} = (0.8/3.2) \times P = 0.25P$$

(Always check that mole fractions add up to 1:
$0.375 + 0.375 + 0.25 = 1.0$)

$$K_p = \frac{p_{H_2} \times p_{CO}}{p_{H_2O}} = \frac{(0.375P)\,(0.375P)}{0.25P} = 0.5625P$$

As $K_p = 58$,

$$0.5625P = 58$$

and $P = 103\,kPa$

Table 2.4

	H$_2$O	H$_2$	CO
Initial amount (mol)	a		
Equilibrium amount (mol)	(a – x)	x	x
Initial amount (mol)	2.0	0	0
Equilibrium amount (mol)	0.8	1.2	1.2

Example 8

When 5.00 mol of $SO_2(g)$ is mixed with 2.00 mol of $O_2(g)$ and allowed to come to equilibrium at a high temperature and it is found that 3.00 mol of sulfur trioxide has formed and the pressure in the container is 200 kPa.

$$2SO_2(g) + O_2(g) \rightleftharpoons 2SO_3(g)$$

Calculate K_p for this reaction.

Answer

$2SO_2(g) + O_2(g) \rightleftharpoons 2SO_3(g)$ can be written as
$SO_2(g) + \frac{1}{2}O_2(g) \rightleftharpoons SO_3(g)$.

Table 2.5

	SO$_2$	O$_2$	SO$_3$
Initial amount (mol)	a	b	c
Equilibrium amount (mol)	(a – x)	(b – ½ x)	x
Initial amount (mol)	5.00	2.00	0
Equilibrium amount (mol)	(5.00 – 3.00)	(2.00 – 1.50)	3.00
=	2.00	0.50	3.00

So 1 mol of SO_2 reacts with ½ mol of O_2 to form 1 mol of SO_3.

As 3.00 mol of SO_3 is formed, $x = 3.00$, so the mol of SO_2 at equilibrium will be (5.00 − 3.00) mol and the mol of O_2 will be 0.50 mol.

The total number of moles at equilibrium is, therefore,
2.00 + 0.50 + 3.00 = 5.50 mol.

The partial pressures are the (mole fraction × pressure). Therefore,

$$p_{SO_3} = (3.00/5.50) \times 200 = 109.1 \, kPa$$

$$p_{SO_2} = (2.00/5.50) \times 200 = 72.7 \, kPa$$

$$p_{O_2} = (0.50/5.50) \times 200 = 18.2 \, kPa$$

(Always double check that the partial pressures add up to the total pressure 109.1 + 72.7 + 18.2 = 200.)

$$K_p = \frac{P_{SO_3}^2}{P_{SO_2}^2 \times P_{O_2}} = \frac{(109.1)^2}{(72.7)^2 \times (18.2)} = 0.124 \, kPa^{-1}$$

The pressure exerted by each gas is dependent on the number of particles of that gas present and, in a fixed volume, this is related directly to its concentration. Therefore K_c and K_p are also directly related and, in fact, if the total number of moles of the reactants equals the total number of moles of the products in an equilibrium then $K_c = K_p$.

Test yourself

5 Write an expression for K_p for each of the following equations and give its units.
   a) $H_2(g) + I_2(g) \rightleftharpoons 2HI(g)$
   b) $2O_3(g) \rightleftharpoons 3O_2(g)$
   c) $H_2O(g) + C(s) \rightleftharpoons H_2(g) + CO(g)$
6 K_p for the equilibrium $H_2(g) + CO_2(g) \rightleftharpoons CO(g) + H_2O(g)$ at 700 K is 12.3 kPa. If the partial pressures of the CO(g) and $H_2O(g)$ are both 35 kPa and the partial pressure of $H_2(g)$ is 18 kPa what is the partial pressure of $CO_2(g)$ and what is the total pressure of the mixture of gases?
7 The following equilibrium can be formed at 470 K

   $PCl_5(g) \rightleftharpoons PCl_3(g) + Cl_2(g)$

   When 0.15 mol of PCl_5 is heated, the equilibrium is formed and when it is analysed it is found to contain 0.11 mol of PCl_3 and the pressure of the mixture is 110 kPa.
   Calculate K_p.

Practice questions

Multiple choice questions 1–10

1 Which one of the following might affect the value of an equilibrium constant?
- **A** The presence of a catalyst
- **B** An increase in pressure
- **C** A decrease in the concentration of the reactants
- **D** A decrease in the temperature. *(1)*

2 If an equilibrium $A(g) \rightleftharpoons 2B(g)$ has a value for $K_c = 1\,mol\,dm^{-3}$, this will mean that at equilibrium:
- **A** $[A] = [B]$
- **C** $[B] = [A]^2$
- **B** $[A] = 2[B]$
- **D** $[B] = \sqrt{[A]}$ *(1)*

3 The equilibrium
$H_2(g) + CO_2(g) \rightleftharpoons H_2O(g) + CO(g)$ is established by allowing 1.0 mol each of $H_2(g)$ and $CO_2(g)$ to react.
When the equilibrium mixture is analysed it is found to contain 0.60 mol of $CO(g)$. What is the total number of moles of gases present in the equilibrium mixture?
- **A** 1.2 mol
- **C** 3.2 mol
- **B** 2.0 mol
- **D** 3.6 mol *(1)*

4 If pressures are measured in kPa, the units for K_p for the equilibrium $2Cl_2O_5(g) \rightleftharpoons 2Cl_2(g) + 5O_2(g)$ are:
- **A** kPa^{-5}
- **C** kPa^2
- **B** kPa^{-2}
- **D** kPa^5 *(1)*

5 When analysed, the equilibrium
$CO_2(g) + C(s) \rightleftharpoons 2CO(g)$ is found to contain $CO(g)$ with a mole fraction of 0.50.
If K_p for the equilibrium is 600 kPa which one of the following is the pressure of the equilibrium mixture?
- **A** 24 000 kPa
- **C** 6000 kPa
- **B** 12 000 kPa
- **D** 3000 kPa *(1)*

6 The equilibrium mixture
$CO(g) + 3H_2(g) \rightleftharpoons CH_4(g) + H_2O(g)$ is set up by reacting together 0.02 mol of $CO(g)$ and 0.025 mol of $H_2(g)$ in a 5 dm³ container.
When the equilibrium is formed, it is found that it contains 0.012 mol of $CO(g)$. Which one of the following describes the concentrations of the gases present?
- **A** $[CH_4] = [H_2O] > [CO] > [H_2]$
- **B** $[CO] > [H_2] > [CH_4] = [H_2O]$
- **C** $[CO] > [CH_4] = [H_2O] > [H_2]$
- **D** $[H_2] > [CO] > [CH_4] = [H_2O]$ *(1)*

7 Which one of the following could be the units of K_c for the equilibrium
$2N_2O_5(g) \rightleftharpoons 4NO_2(g) + O_2(g)$?
- **A** $mol^3\,dm^{-9}$
- **B** $mol^4\,dm^{-12}$
- **C** $dm^6\,mol^{-3}$
- **D** $dm^9\,mol^{-3}$ *(1)*

Use the key below to answer questions 8, 9 and 10.

A	B	C	D
1, 2 & 3 correct	1 & 2 correct	2 & 3 correct	1 only correct

8 Which of the following affect the rate at which a reaction proceeds towards equilibrium?
1. The presence of a catalyst
2. An increase in pressure
3. A decrease in temperature

9 For the equilibrium $N_2O_4(g) \rightleftharpoons 2NO_2(g)$,
$\Delta H = 58.0\,kJ\,mol^{-1}$.
$N_2O_4(g)$ is heated to a temperature $T_1\,°C$ and an equilibrium is established with $NO_2(g)$.
The experiment is then repeated at a higher temperature, $T_2\,°C$.
Compared with the equilibrium at $T_1\,°C$, the equilibrium at $T_2\,°C$ will:
1. have an equilibrium constant which has a larger value
2. have a greater total amount (in mol) of gases
3. be formed faster. *(1)*

10 The equilibrium $2SO_2(g) + O_2(g) \rightleftharpoons 2SO_3(g)$ is established by allowing 1.0 mol each of $SO_2(g)$ and $O_2(g)$ to react at 900 K.
When the equilibrium mixture is analysed, it is found to contain 0.70 mol of $SO_3(g)$.
The equilibrium mixture is heated and at 1000 K it is found to contain 0.48 mol of $SO_3(g)$.
Which of the following can be deduced from the change that occurs as a result of heating the equilibrium mixture?
1. The reaction $2SO_2(g) + O_2(g) \rightarrow 2SO_3(g)$ is exothermic.
2. The total amount (in mol) of the gases in the equilibrium mixture increases as a result of heating.
3. $SO_3(g)$ has weaker sulfur–oxygen bonds than $SO_2(g)$. *(1)*

11 Give the expression for the equilibrium constant, K_c, for each of the following reactions.

  a) $N_2O_4(g) \rightleftharpoons 2NO_2(g)$
  b) $2N_2O(g) \rightleftharpoons 2N_2(g) + O_2(g)$
  c) $2C(s) + O_2(g) \rightleftharpoons 2CO(g)$
  d) $2CO(g) + O_2(g) \rightleftharpoons 2CO_2(g)$ *(4)*

12 An equilibrium is established between ethanol in the blood and ethanol vapour:

$$C_2H_5OH(blood) \rightleftharpoons C_2H_2OH(g)$$

$$K_c = 4.5 \times 10^{-4}$$

Calculate the concentration of ethanol vapour in equilibrium with ethanol in the blood in a person who has a concentration of 80 mg of ethanol in 100 cm^3 of blood. *(4)*

13 Explain what is wrong with each of the following statements. To what extent, if any, are these statements true?

  a) Once a reaction mixture reaches equilibrium there is no further reaction.
  b) Adding more of one of the reactants to an equilibrium mixture increases the yield of products because the value of K_c increases.
  c) Adding a catalyst to make a reaction go faster can increase the amount of product at equilibrium.
  d) Raising the temperature to make a reaction go faster can increase the amount of product at equilibrium. *(11)*

14 Nitrogen(I) oxide decomposes to nitrogen and oxygen according to the equation:

$$2N_2O(g) \rightleftharpoons 2N_2(g) + O_2(g)$$

In an experiment, 1 mol of nitrogen(I) oxide was heated in a 1 dm^3 container until equilibrium was established. The mixture was then analysed and found to contain 0.1 mol of nitrogen(I) oxide.

  a) Calculate the concentrations of nitrogen and oxygen present in the equilibrium mixture.
  b) Calculate the equilibrium constant, K_c.
  c) If the experiment were repeated using 1 mol of nitrogen(I) oxide in a 2 dm^3 container, how would the value of K_c change? *(8)*

15 The dissociation of chlorine into its atoms (radicals) can be accomplished in the laboratory by heating chlorine gas to 1200 °C.

$$Cl_2(g) \rightleftharpoons 2Cl(g)$$

The equilibrium constant for this reaction increases as the temperature is raised.

  a) State whether the dissociation is exothermic or endothermic. Explain your answer.
  b) The dissociation occurs more readily high in the upper atmosphere in the region called the stratosphere. When produced at this height, the presence of chlorine radicals is damaging.
  i) Explain why the breakdown of chlorine occurs more readily in the stratosphere.
  ii) Explain why the production of chlorine radicals is damaging. *(5)*

16 A mixture of 0.200 mol NO, 0.100 mol H$_2$ and 0.200 mol of H$_2$O was placed in a 2.00 dm^3 container and allowed to come to equilibrium. Initially no N$_2$ was present.
Once equilibrium had been established, the amount of NO present in the mixture was measured and found to be 0.124 mol.
Determine the value of the equilibrium constant, K_c, for the reaction $2NO(g) + 2H_2(g) \rightleftharpoons N_2(g) + 2H_2O(g)$. *(8)*

17 a) A flask contains an equilibrium mixture of hydrogen gas (0.01 mol dm^{-3}), iodine gas (0.01 mol dm^{-3}) and hydrogen iodide gas (0.07 mol dm^{-3}) at a constant temperature. Calculate K_c for the reaction of hydrogen with iodine to form hydrogen iodide at this temperature.

  b) Enough hydrogen is added to the mixture in part **a)** to suddenly double the hydrogen concentration in the flask to 0.02 mol dm^{-3}. After a while, the mixture settles down with a new iodine concentration of 0.007 mol dm^{-3} at the same temperature as before. What are the new concentrations of hydrogen and hydrogen iodide? *(8)*

18 Give the expression for the equilibrium constant, K_p, for each of the following reactions and state its units.

  a) $N_2O_4(g) \rightleftharpoons 2NO_2(g)$
  b) $2N_2O(g) \rightleftharpoons 2N_2(g) + O_2(g)$
  c) $2C(s) + O_2(g) \rightleftharpoons 2CO(g)$
  d) $2CO(g) + O_2(g) \rightleftharpoons 2CO_2(g)$ *(4)*

19 Potassium chlorate(V) forms an equilibrium mixture when heated in a closed container.

$$2KClO_3(s) \rightleftharpoons 2KCl(s) + 3O_2(g)$$

The pressure of the mixture is measured as 36 kPa. What is K_p for the equilibrium mixture? *(4)*

20 K_p for the equilibrium
$2N_2O_5(g) \rightleftharpoons 4NO_2(g) + O_2(g)$ is 60 kPa.
In an equilibrium mixture, the partial pressure of NO_2 is measured as 10 kPa and the partial pressure of O_2 is 4 kPa.
What is the partial pressure of N_2O_5? (3)

21 When carbon dioxide is heated with carbon at high temperatures, an equilibrium is formed with the carbon monoxide produced.

$CO_2(g) + C(s) \rightleftharpoons 2CO(g)$

If 1.0 mol of carbon dioxide is allowed to react, the equilibrium mixture has a pressure of 25 kPa and 1.4 mol of carbon monoxide is formed.
Calculate K_p for the equilibrium at that temperature. (7)

22 K_p for the equilibrium $2SO_2(g) + O_2(g) \rightleftharpoons 2SO_3(g)$ at 700 K is 3.0×10^6 kPa.
In an experiment, 1.0 mol of $SO_2(g)$ is reacted with 0.50 mol of $O_2(g)$ and allowed to come to equilibrium. Analysis shows that the equilibrium mixture obtained contains 0.98 mol of SO_3.
Calculate the pressure of the equilibrium mixture. (10)

23 When 1 mol of $P_4(g)$ is heated to 1100 K, the following equilibrium is formed.

$P_4(g) \rightleftharpoons 2P_2(g)$

It is found that the equilibrium mixture contains 0.9 mol of P_4 and the pressure in the container is 40 kPa.
Calculate the value of K_p at 1100 K. (8)

Challenge

24 The equilibrium constant for the acid-catalysed reaction between ethanol and ethanoic acid to form ethyl ethanoate and water can be determined in the laboratory.

$C_2H_5OH(l) + CH_3COOH(l) \rightleftharpoons CH_3COOC_2H_5(l) + H_2O(l)$

In an experiment, 20.0 cm³ of ethanoic acid, 20.0 cm³ ethanol and 30.0 cm³ of ethyl ethanoate are mixed together.
10.0 cm³ of 0.500 mol dm⁻³ sulfuric acid is then added. This provides the necessary acid catalyst and 10.0 cm³ of water.
The mixture is left for 1 week to come to equilibrium.

1.0 cm³ of the mixture is then pipetted into a small conical flask. 15.0 cm³ of water is added to increase the volume and the mixture is then titrated against a 0.250 mol dm⁻³ solution of sodium hydroxide. It is found that 14.50 cm³ of sodium hydroxide solution is required to neutralise the acid in the mixture.

a) Given the following densities, calculate the amount (in moles) of ethanol, ethanoic acid, ethyl ethanoate and water that were present at the start of the experiment:
$C_2H_5OH(l) = 0.79$ g cm⁻³
$CH_3COOH(l) = 1.05$ g cm⁻³
$CH_3COOC_2H_5(l) = 0.92$ g cm⁻³
$H_2O(l) = 1.00$ g cm⁻³

b) Use the titration result to determine the amount (in moles) of ethanoic acid that is present in the 1.00 cm³ sample taken after equilibrium had been established.

c) Use your answer from part b) to determine the amount, in moles, of ethanoic acid present in the full 80 cm³ of the equilibrium mixture.

d) From your answer to part c) calculate the amount, in moles, of ethanol, ethyl ethanoate and water present in the equilibrium mixture.

e) Calculate the value of the equilibrium constant for this reaction.

f) Explain why the calculation can be carried out using the amounts in moles, rather than the concentrations in mol dm⁻³.

g) Calculate the enthalpy change for the forward reaction, given the following values for the enthalpies of formation.
$\Delta_f H(C_2H_5OH) = -277.7$ kJ mol⁻¹
$\Delta_f H(CH_3COOH) = -484.5$ kJ mol⁻¹
$\Delta_f H(CH_3COOC_2H_5) = -485.8$ kJ mol⁻¹
$\Delta_f H(H_2O) = -285.9$ kJ mol⁻¹

h) Use your answer to part g) to comment on the value of the equilibrium constant.

i) The answer obtained in part e) is smaller than the equilibrium constant at 25 °C that is usually quoted in data books. It is suggested that a reason for the difference was that the experiment described above was carried out at 20 °C. Could this account, at least in part, for the difference in the equilibrium constant obtained? Explain your answer. (20)

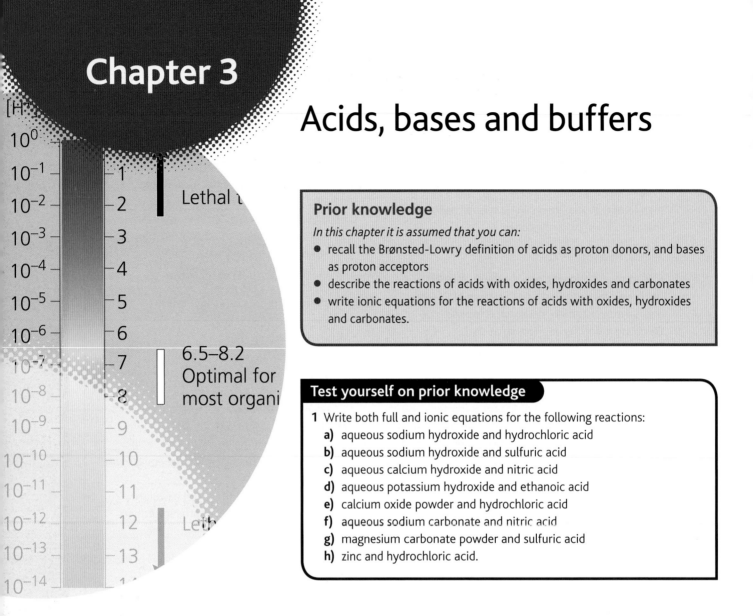

Chapter 3

Acids, bases and buffers

Prior knowledge

In this chapter it is assumed that you can:

- recall the Brønsted-Lowry definition of acids as proton donors, and bases as proton acceptors
- describe the reactions of acids with oxides, hydroxides and carbonates
- write ionic equations for the reactions of acids with oxides, hydroxides and carbonates.

Test yourself on prior knowledge

1 Write both full and ionic equations for the following reactions:
   a) aqueous sodium hydroxide and hydrochloric acid
   b) aqueous sodium hydroxide and sulfuric acid
   c) aqueous calcium hydroxide and nitric acid
   d) aqueous potassium hydroxide and ethanoic acid
   e) calcium oxide powder and hydrochloric acid
   f) aqueous sodium carbonate and nitric acid
   g) magnesium carbonate powder and sulfuric acid
   h) zinc and hydrochloric acid.

Conjugate acids and bases

An important class of reactions is that of acids and bases. Control of acidity is important in biochemical systems, soil science, the cosmetics industry and many other areas of life. You will be familiar with some reactions of acids and bases, but this chapter considers the behaviour of acids and bases in more detail.

As explained in Chapter 2, many reactions are more accurately described as an equilibrium between the reactants and products. An example is the dissolving of ammonia in water. Ammonium ions and hydroxide ions are produced:

$$NH_3(g) + H_2O(l) \rightleftharpoons NH_4^+(aq) + OH^-(aq)$$

In the forward reaction, the water donates a proton to the ammonia. Therefore, in this context, water is considered to be an acid and, since ammonia accepts a proton, it is a base.

For the reverse reaction, the ammonium ion is the acid and the hydroxide ion is the base.

In summary:

$$NH_3(g) + H_2O(l) \rightleftharpoons NH_4^+(aq) + OH^-(aq)$$

base 1   acid 2   acid 1   base 2

If acid–base reactions are considered as equilibria, then each side of the equation contains an acid and a base. The related species are called **conjugate acid–base pairs**.

In the example above, the NH_4^+ and NH_3 are a conjugate acid–base pair; NH_4^+ is the acid and NH_3 is its conjugate base. The H_2O is an acid and OH^- is its conjugate base. The species with the extra proton is *always* the conjugate acid and the species without the proton is *always* the conjugate base.

In some circumstances, NH_3 is able to lose a proton to form the ion NH_2^-. In this case, the NH_3 and NH_2^- are a conjugate acid–base pair, with NH_3 as the conjugate acid and NH_2^- as its conjugate base.

Another example is the formation of the nitration mixture for a reaction with benzene (see page 148). A mixture of concentrated sulfuric and nitric acids is used, which forms the equilibrium:

$$H_2SO_4 + HNO_3 \rightleftharpoons HSO_4^- + H_2NO_3^+$$

acid 1   base 2   base 1   acid 2

In this reaction:

- H_2SO_4 is an acid; its conjugate base is HSO_4^-.
- HNO_3 is a base; its conjugate acid is $H_2NO_3^+$.

It may seem strange to refer to nitric acid as a base, but in this circumstance that is what it is.

The Brønsted–Lowry definition extends the idea of acids into a more comprehensive theory of reactions involving proton transfer. This is analogous to the way that the definitions of oxidation and reduction are extended beyond the addition or removal of oxygen to the theory of electron transfer.

Mono-, di- and tribasic acids

In the Year 1 book it was mentioned that acids can form more than one type of salt if they can donate more than one hydrogen ion. This means they can also form more than one conjugate base. Phosphoric acid, H_3PO_4 for example, has three possible conjugate bases: $H_2PO_4^-$, HPO_4^{2-} and PO_4^{3-}. The one that forms depends on the reaction conditions and the nature of the base to which the acid is added.

Strong and weak acids

Acids vary considerably in the ease with which they are able to release hydrogen ions. When the mineral acids (sulfuric, nitric and hydrochloric) are dissolved in water, they break up almost completely into their constituent ions. Although equilibria do exist, the acids are usually considered to be 100% ionised. They are referred to as **strong acids**. By contrast, when dissolved, organic acids (for example ethanoic acid) are only partially ionised and a significant concentration of molecular acid still exists. These acids are referred to as **weak acids**.

A similar distinction is made for bases, although you will encounter very few weak bases. Amines and ammonia are weak bases because they are protonated to only a small extent; all metal hydroxides are strong bases. For example, limewater (aqueous calcium hydroxide) is a strong base but, as it has low solubility, a solution of limewater is always dilute.

> **Tip**
>
> A given species may be an acid in one circumstance and a base in another.

> **Test yourself**
>
> 1 Give the formula and name the conjugate bases of these acids: HNO_3, CH_3COOH, H_2SO_4 and HCO_3^-.
> 2 Give the formula and name the conjugate acids of these bases: O^{2-}, OH^-, NH_3, CO_3^{2-}, HCO_3^- and SO_4^{2-}.

Do not muddle the use of the word 'strong' with 'concentrated', or 'weak' with 'dilute'.

- A **strong** acid is one which is highly ionised in aqueous solution.
- A **concentrated** acid is made by dissolving large amounts of the acid in a small volume of water.
- A **weak** acid is one which is only partially ionised in aqueous solution.
- A **dilute** acid is made by dissolving small amounts of the acid in a large volume of water.

The acid dissociation constant

The usual way of indicating the strength of an acid is to use the equilibrium constant for its ionisation in water. For example, ethanoic acid ionises as:

$$CH_3COOH(aq) \rightleftharpoons CH_3COO^-(aq) + H^+(aq)$$

As with other equilibria (Chapter 2), an equilibrium constant can be defined for this reaction. It is:

$$K_a = \frac{[CH_3COO^-(aq)][H^+(aq)]}{[CH_3COOH(aq)]}$$

In some books you will see H⁺(aq) written as H_3O^+. This indicates correctly that the 'aq' attachment represents one H_2O. This course does not require you to represent aqueous hydrogen ions as H_3O^+; it makes no difference to your understanding of the concepts under discussion.

To indicate that the reaction involves an acid, it is conventional to provide the equilibrium constant K with the subscript 'a'. K_a is called the **acid dissociation constant**.

For ethanoic acid, K_a has the value $1.7 \times 10^{-5}\,mol\,dm^{-3}$. This makes it clear that a solution of ethanoic acid consists largely of ethanoic acid molecules with relatively few ethanoate ions and hydrogen ions.

The K_a value of methanoic acid is $1.6 \times 10^{-4}\,mol\,dm^{-3}$, which is almost ten times larger than that for ethanoic acid. This tells us that methanoic acid, although it is a weak acid, is stronger than ethanoic acid.

The mineral acids have much larger K_a values. For example, the K_a for nitric acid is approximately $40\,mol\,dm^{-3}$ and that for sulfuric acid is often listed simply as 'very large'.

Calculating hydrogen ion concentration

Strong acids are assumed to be ionised completely when they are dissolved in water. This means that the hydrogen ion concentration is related directly to the amount of acid dissolved.

$$HCl(aq) \rightarrow H^+(aq) + Cl^-(aq)$$

For example, $0.0500\,mol\,dm^{-3}$ HCl will produce $0.0500\,mol\,dm^{-3}$ of H⁺(aq).

To determine the hydrogen ion concentration of a weak acid is more complicated because there is a significant amount of the undissociated (unionised) acid present in the solution. This means that it is necessary to refer to the value of K_a for the acid. Using ethanoic acid as an example, the hydrogen ion concentration present in a $0.050\,mol\,dm^{-3}$ solution is calculated as follows:

$$K_a = 1.7 \times 10^{-5} = \frac{[CH_3COO^-(aq)][H^+(aq)]}{[CH_3COOH(aq)]}$$

Tip

This does ignore a very small concentration of H⁺ ions ($10^{-7}\,mol\,dm^{-3}$) that come from the water (see page 45). This would only become significant if the acid was very dilute.

Every time a molecule of the acid dissociates, an H$^+$(aq) ion is formed together with a CH$_3$COO$^-$(aq) ion. This means that the concentration of CH$_3$COO$^-$(aq) is always the same as that of H$^+$(aq).

So, if [H$^+$(aq)] equals x mol dm^{-3}, then [CH$_3$COO$^-$(aq)] also equals x mol dm^{-3}, and the equation can be written as:

$$K_a = 1.7 \times 10^{-5} = \frac{x^2}{[CH_3COOH(aq)]}$$

The concentration of ethanoic acid in this equation is the equilibrium concentration. However, if the value of K_a is small (as it is in this case) then the concentration of undissociated acid at equilibrium will not be much less than the initial concentration. It is, therefore, reasonable to take the concentration at equilibrium as being the same as the initial concentration, unless the value of K_a is large (as it is for 'stronger' weak acids) when the amount the acid is dissociated might be significant.

Therefore,

$$x^2 = 0.050 \times 1.7 \times 10^{-5}$$

$$x = 9.2(2) \times 10^{-4} \text{ mol dm}^{-3}$$

Only for very accurate work would this difference be significant.

You can see that the hydrogen ion concentration in 0.05 mol dm^{-3} ethanoic acid is appreciably less than that present in 0.05 mol dm^{-3} hydrochloric acid.

For a weak acid, the calculation can be summarised thus: $[H^+] = \sqrt{K_a \times c}$ where K_a is the acid dissociation constant and c is the concentration of the solution.

It should be noted that this calculation has ignored the contribution of any hydrogen ions from the water. This will usually be a good approximation unless the acid is very weak or dilute.

Tip

If you doubt that this is a fair approximation, then it is possible to solve the precise equation:

$$K_a = 1.7 \times 10^{-5} = \frac{x^2}{0.050 - x}$$

If you carry out this calculation, you will find that the value using quadratic equations is $9.1(3) \times 10^{-4}$.

Test yourself

3 Calculate the hydrogen ion concentration of a 0.01 mol dm^{-3} solution of propanoic acid. K_a for the acid is 1.3×10^{-5} mol dm^{-3}.

pH and pK$_a$

The acidity of substances (see Figure 3.1) covers a range of values from concentrations as high as 10 mol dm^{-3} (and occasionally higher) to concentrations as low as 10^{-14} mol dm^{-3}. Although these can be expressed using [H$^+$] values, the pH scale is more convenient.

$$pH = -\log[H^+]$$

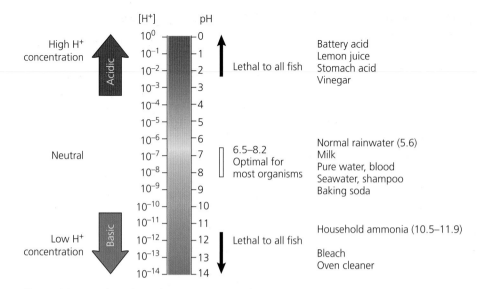

Figure 3.1 pH scale and pH of some common substances.

As it is assumed that all strong acids are totally ionised, the [H^+] concentration is obtained from the concentration of the acid.

The pH of 0.1 mol dm^{-3} HCl is therefore −log(0.1) = 1.

The pH of 0.05 mol dm^{-3} HCl is −log(0.05) = 1.3.

The pH of a weak acid is obtained from [H^+] which is calculated from the value of K_a.

The pH of 0.05 mol dm^{-3} ethanoic acid is $-\log \sqrt{K_a c}$

$$=-\log\sqrt{(1.7\times10^{-5})\times0.05}$$

$$= -\log (9.2 \times 10^{-4}) = 3.04$$

The pH scale is useful because it allows a wide range of hydrogen ion concentrations to be expressed as simple positive values.

The conversion of pH back to hydrogen ion concentration is obtained from

$$[H^+] = 10^{-pH}$$

For example, if a solution has a pH of 2.8 then [H^+] = $10^{-2.8}$ = 0.0016 or 1.6×10^{-3} mol dm^{-3}.

You will recall that, if the equilibrium constant K_c is small, the equilibrium lies to the left. The same is true for K_a. If K_a is small, the acid is weak. A similar scale can be used to give the values of the acid dissociation constant.

$$pK_a = -\log K_a$$

$$K_a = 10^{-pK_a}$$

Table 3.1 shows the K_a and pH values for some carboxylic acids.

Table 3.1 K_a and pH values for some carboxylic acids.

Acid	K_a	pK_a	
methanoic acid	1.6×10^{-4}	3.76	Like pH, pK_a gives an indication of strength of the acid; the lower the pK_a value, the more acidic the acid.
ethanoic acid	1.7×10^{-5}	4.78	
propanoic acid	1.3×10^{-5}	4.88	

Test yourself

4 Calculate the pH of a $0.08\,mol\,dm^{-3}$ solution of nitric acid.

5 What is the concentration of hydrogen ions in these solutions?
   a) orange juice with a pH of 3.3
   b) coffee with a pH of 5.4
   c) saliva with a pH of 6.7
   d) a suspension of an antacid in water with a pH of 10.5.

6 Calculate the pH of a $0.01\,mol\,dm^{-3}$ solution of hydrogen cyanide (HCN). K_a for the acid is $4.9 \times 10^{-10}\,mol\,dm^{-3}$.

The role of water

So far, the role of water in the determination of pH has not been considered. Yet it has a crucial role in controlling the pH of aqueous solutions. First, it is required to allow acids (which are covalent in their anhydrous state) to ionise. Second, it makes a contribution in its own right. Water is substantially covalent but does ionise to a small extent:

$$H_2O \rightleftharpoons H^+ + OH^-$$

Although it dissociates to produce hydrogen ions, it is neutral because the hydrogen ions are balanced by an equal number of hydroxide ions. It is this dissociation that defines what is meant by neutral on the pH scale. At room temperature, the concentration of hydrogen ions is $10^{-7}\,mol\,dm^{-3}$, which means that the pH is 7.

The equilibrium constant for this reaction is:

$$K = \frac{[H^+][OH^-]}{[H_2O]}$$

However, the water is present in such large excess that its concentration is regarded as being a constant and the expression for the equilibrium constant is simplified to:

$$K_w = [H^+][OH^-]$$

At 298 K, the equilibrium constant has the value 10^{-14} because $[H^+] = [OH^-] = 10^{-7}\,mol\,dm^{-3}$.

K_w is called the **ionic product of water**; the subscript 'w' indicates that the equilibrium constant refers to water.

Test yourself

7 The value of K_w varies with temperature. At 273 K its value is $1.1 \times 10^{-15}\,mol^2\,dm^{-6}$, while at 303 K it is $1.5 \times 10^{-14}\,mol^2\,dm^{-6}$.
   a) Is the ionisation of water an exothermic process or an endothermic process?
   b) What happens to the hydrogen ion concentration in pure water as the temperature increases? What happens to the pH?
   c) Does pure water stop being neutral if its temperature is above or below 298 K?

The pH of strong bases

It is important to realise that, in aqueous solutions, the ionic product of water may be the controlling influence on pH.

When an acid is dissolved in water, it releases so many hydrogen ions that the small contribution of hydrogen ions from the water is completely insignificant. This is particularly so because the addition of hydrogen ions would drive the equilibrium $H_2O \rightleftharpoons H^+ + OH^-$ to the left. On the other hand, water is the reason why even the most alkaline solutions contain some H^+ ions.

A $0.1\,mol\,dm^{-3}$ solution of sodium hydroxide contains $0.1\,mol\,dm^{-3}$ of OH^- ions. (It is a strong base and, therefore, dissociates fully.)

$$[H^+][OH^-] = K_w = 1 \times 10^{-14}\,mol^2\,dm^{-6}$$

So, $[H^+] \times 0.1 = 1 \times 10^{-14}\,mol\,dm^{-3} = 10^{-13}$

Therefore, this solution has a pH of 13.

Test yourself

8 At 25°C what is the pH of:
a) $0.02\,mol\,dm^{-3}$ NaOH
b) $0.001\,mol\,dm^{-3}$ Ba(OH)$_2$?

Buffer solutions

Key term

A **buffer solution** is a solution that resists a change in pH when a small quantity of acid or alkali is added.

A change in pH has a significant effect on a range of reactions. Examples of processes that depend on the maintenance of a particular pH include:

- the effectiveness of blood in transporting oxygen
- the stability of many molecules in living systems
- the effective use of chromatography to separate some components
- the safety of many products (Figure 3.2), for example shampoos.

There are natural processes that regulate pH in plants and animals, such as the hydrogencarbonate ion system discussed on page 49. It is also possible to construct mixtures in the laboratory that have the effect of resisting changes in pH. A solution that is able to do this is called a buffer solution.

Weak acid and salt of weak acid buffer mixture

There are many solutions that are capable of buffering. A simple example is the mixture of a weak acid and the soluble salt of the weak acid; for example a mixture of ethanoic acid and sodium ethanoate.

There are two points to note about this mixture:

- Sodium ethanoate is ionic and will be fully ionised, thereby contributing a large number of ethanoate ions to the mixture.
- The already weak ionisation of the ethanoic acid is reduced considerably when the ethanoate ions from the salt are added, because the equilibrium

$$CH_3COOH(aq) \rightleftharpoons CH_3COO^-(aq) + H^+(aq)$$

is forced to the left (le Chatelier's principle).

Therefore, the mixture will contain a large number of undissociated ethanoic acid molecules, $CH_3COOH(aq)$, and a large number of ethanoate ions, $CH_3COO^-(aq)$.

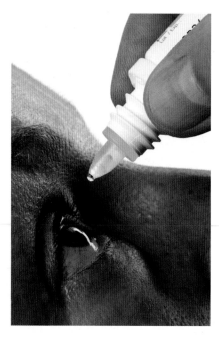

Figure 3.2 Eye drops contain a buffer solution to make sure that they do not irritate the sensitive surface of the eye.

The second bullet point above also means that, when sodium ethanoate is added, the pH of the ethanoic acid will be raised, because the concentration of hydrogen ions will be reduced.

The mixture then acts as a buffer solution:

● When hydrogen ions are added, ethanoate ions in the mixture combine with the hydrogen ions to form more undissociated ethanoic acid. By le Chatelier's principle, the equilibrium position shifts to the left resulting in only a small change to the overall pH.

● When hydroxide ions are added, they react with the existing hydrogen ions to form water molecules. However, because there is a reserve of ethanoic acid molecules in the mixture, these ionise to replace the lost hydrogen ions (by le Chatelier's principle, the equilibrium position shifts to the right). The pH is again changed very little.

The key points for a buffer solution of this type are summarised in Table 3.2.

Table 3.2

Large excess of undissociated acid molecules CH_3COOH		Large excess of acid salt anions CH_3COO^-	
These are a reservoir of H^+ ions that can be used when needed.		These are base ions with the capacity to accept H^+ ions.	
$CH_3COOH(aq)$ large excess	$\rightleftharpoons$		$CH_3COO^-(aq) + H^+(aq)$ large excess
Any OH^- added reacts with the H^+ on the right-hand side of the equilibrium to produce H_2O, but the excess CH_3COOH molecules on the left-hand side dissociate to replace the H^+ ions.		Any H^+ added reacts with the excess CH_3COO^- ions on the right-hand side, such that the equilibrium moves to the left-hand side.	

Excess weak acid and strong alkali buffer mixture

A buffer mixture can also be generated by mixing together a weak acid with a strong base, if the weak acid is in excess. For example if $50\,cm^3$ of $0.5\,mol\,dm^{-3}$ CH_3COOH (the weak acid) is mixed with $25\,cm^3$ of $0.5\,mol\,dm^{-3}$ NaOH (the strong base) (Table 3.3). The acid is in excess and all of the base will react to produce the salt of the weak acid. (See also the section on acid/base titrations on page 51.)

Table 3.3

	Before mixing	After mixing
Moles of acid, CH_3COOH	0.0250	0.0125
Moles of base, NaOH	0.0125	0
Moles of salt, $CH_3COO^-\,Na^+$	0	0.0125
Total volume of the mixture after mixing is $75\,cm^3$, so the concentration of the acid and the salt can be calculated		

There is a limit to the range in which the buffer solution will be effective, because its action is controlled by the size of the reserve of ethanoate ions and ethanoic acid molecules. However, buffer solutions are effective in providing control of the small changes that occur through the limited addition of hydrogen ions or hydroxide ions.

The effect of acidity on organisms in a lake

Lakes are sensitive to acidity. They need to be buffered if aquatic organisms are to be preserved. Figure 3.3 shows the ability of various organisms to survive a reduction of pH. Snails are sensitive to pH; frogs can exist in quite acidic water.

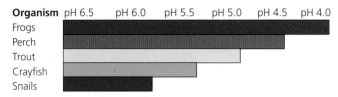

Organism	pH 6.5	pH 6.0	pH 5.5	pH 5.0	pH 4.5	pH 4.0
Frogs						
Perch						
Trout						
Crayfish						
Snails						

Figure 3.3 Organisms have differing abilities to survive pH changes.

The calculation of the pH of a buffer solution

The calculation of the pH of a buffer solution made from a weak acid and the salt of the weak acid requires an understanding that the pH depends on the behaviour of the equilibrium between the acid and its anion.

The weak acid in the buffer mixture is present substantially as undissociated molecules and the anions come almost wholly from the added salt. To calculate the pH of a buffer solution, the assumption is made that the concentration of the acid can be taken to be the original concentration of acid and concentration of the anion to be the same as the original concentration of the salt.

This is, in most cases, a valid approximation that greatly simplifies the calculation of the pH of buffer solutions.

$$K_a = \frac{[\text{salt}][\text{H}^+]}{[\text{acid}]}$$

Therefore,

$$\frac{K_a[\text{acid}]}{[\text{salt}]} = [\text{H}^+]$$

$$pH = -\log[\text{H}^+]$$

hence, $pH = -\log\left(\frac{K_a[\text{acid}]}{[\text{salt}]}\right)$

Example 2

Calculate the pH of a buffer solution containing $0.50\,\text{mol}\,\text{dm}^{-3}$ ethanoic acid and $0.10\,\text{mol}\,\text{dm}^{-3}$ sodium ethanoate. ($K_a = 1.7 \times 10^{-5}\,\text{mol}\,\text{dm}^{-3}$)

Answer

Substituting into the expression for K_a:

$$K_a = 1.7 \times 10^{-5} = \frac{[0.10][\text{H}^+]}{[0.50]}$$

Therefore,

$$[\text{H}^+] = 5.0 \times 1.7 \times 10^{-5}$$

$$= 8.5 \times 10^{-5}\,\text{mol}\,\text{dm}^{-3}$$

$$pH = -\log(8.5 \times 10^{-5})$$

$$= 4.07$$

A possible problem, when working out the pH of buffer solutions, is not remembering that when two solutions are mixed there is a **dilution**. For example, if 25 cm³ of 0.10 mol dm⁻³ sodium ethanoate were mixed with 75 cm³ of 0.5 mol dm⁻³ ethanoic acid, the total volume would be 100 cm³ and the concentrations of the components would be:

$$[CH_3COO^-] = (25/100) \times 0.10 = 0.025 \, \text{mol dm}^{-3}$$

and

$$[CH_3COOH] = (75/100) \times 0.50 = 0.375 \, \text{mol dm}^{-3}$$

Test yourself

9 a) What is the pH of a buffer solution containing 0.40 mol dm⁻³ methanoic acid and 1.0 mol dm⁻³ sodium methanoate? (K_a for methanoic acid 1.6×10^{-4} mol dm⁻³)

b) What is the pH of a buffer solution if 25 cm³ of 0.10 mol dm⁻³ sodium propanoate is mixed with 75 cm³ of 0.50 mol dm⁻³ propanoic acid? (K_a for propanoic acid = 1.3×10^{-5} mol dm⁻³)

c) What is the pH of a buffer solution if 25 cm³ of 0.50 mol dm⁻³ sodium hydroxide is mixed with 75 cm³ of 0.50 mol dm⁻³ propanoic acid? (K_a for propanoic acid = 1.3×10^{-5} mol dm⁻³)

The hydrogencarbonate ion buffer

An example of a natural buffering ion is the hydrogencarbonate ion, HCO_3^-. It is responsible for the maintenance of blood pH and for the pH in lakes and streams. The ion works as a buffer because, in water, it forms the following equilibrium system:

$$CO_2(g) + H_2O(l) \rightleftharpoons H_2CO_3(aq) \rightleftharpoons H^+(aq) + HCO_3^-(aq)$$

Addition of acid pushes the positions of both equilibria to the left. Carbonic acid (a weak acid) is formed, which can decompose to form carbon dioxide. If more acidity is required, the reverse process takes place and more HCO_3^- and H^+ ions are formed.

In blood plasma, the pH is held at a value of about 7.4 by these equilibria.

Activity

Blood buffers

In a healthy person, the pH of blood lies within a narrow range (7.35–7.45). Chemical reactions in cells tend to upset the normal pH. Respiration, for example, produces carbon dioxide all the time. The carbon dioxide diffuses into the blood, where it is mainly in the form of carbonic acid, H_2CO_3. However, the blood pH stays constant because it is stabilised by buffer solutions – in particular by the buffer system based on the equilibrium between carbon dioxide, water and hydrogencarbonate ions. This is the carbonic acid–hydrogencarbonate buffer.

$$CO_2(g) + H_2O(l) \rightleftharpoons H^+(aq) + HCO_3^-(aq)$$

Proteins in blood, including haemoglobin, can also contribute to the buffering of blood pH. This is because the molecules contain both acidic and basic functional groups. Two major organs help to control the total amounts of carbonic acid and hydrogencarbonate ions in the blood. The lungs (Figure 3.4)

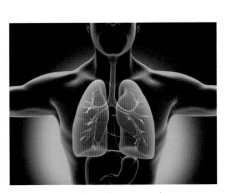

Figure 3.4 The lungs have a vital part to play in maintaining the pH of the blood.

remove excess carbon dioxide from the blood and the kidneys remove excess hydrogencarbonate ions. The brain responds to the level of carbon dioxide in the blood. During exercise, for example, the brain speeds up the rate of breathing. The consequences can be fatal if the blood pH moves outside the normal range. Patients who have been badly burned or suffered other serious injuries are treated quickly with a fluid drip into a vein (Figure 3.5). One of the purposes of an intravenous drip is to help to maintain the pH of the blood close to its normal value.

Figure 3.5 Paramedics transferring a patient into a helicopter ambulance. One of the paramedics is holding an intravenous drip bag.

1 a) Write an equation to show aqueous carbon dioxide reacting with water to form hydrogen ions and hydrogencarbonate ions.

b) Why do people breathe faster and more deeply when running?

2 a) Write the equilibrium expression for the equilibrium between carbon dioxide, water, hydrogen ions and hydrogencarbonate ions.

b) In a sample of blood, the concentration of hydrogencarbonate ions is $2.5 \times 10^{-2} \, mol \, dm^{-3}$. The concentration of aqueous carbon dioxide is $1.25 \times 10^{-3} \, mol \, dm^{-3}$. The value of $K = 4.5 \times 10^{-7} \, mol \, dm^{-3}$. Use this information to calculate:

   i) the hydrogen ion concentration in the blood

   ii) the pH of the blood.

   iii) What can you conclude about the person who gave the blood sample?

3 Explain why a mixture of carbon dioxide, water and hydrogencarbonate ions can act as a buffer solution.

4 Give one example each of an acidic functional group and a basic functional group that means that a protein can help to buffer blood pH.

5 Suggest reasons why people may need treatment to adjust their blood pH if they have been rescued after breathing thick smoke during a fire.

pH titration curves

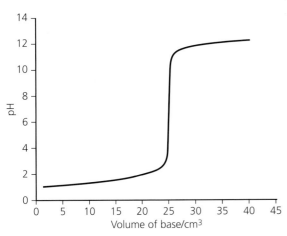

Figure 3.6 Titration of a strong acid with a strong base.

During an acid–base titration, the pH of the solution being titrated changes. The way in which it changes depends on the strength of the acid and alkali being used. The most straightforward case is that of a strong acid being titrated against a strong base.

Suppose that 25.0 cm³ of 0.100 mol dm⁻³ hydrochloric acid is titrated with 0.100 mol dm⁻³ sodium hydroxide solution. A plot of pH against the volume of sodium hydroxide solution added is shown in Figure 3.6.

The initial pH of 0.100 mol dm⁻³ HCl is 1 (–log 0.100). This begins to rise once the sodium hydroxide is added. What is perhaps surprising is that the increase in pH is quite gradual until about 22 cm³ of the alkali has been added. There is then a slightly more pronounced increase until, at 25.0 cm³, the line of the graph becomes almost vertical as the pH rises from about 3 to 11, indicating that the end point of the titration has been reached. The rise in pH then begins to tail off and reaches an almost steady figure of between 12 and 13.

The reaction is complete when the 25.0 cm³ of hydrochloric acid is neutralised by 25.0 cm³ of sodium hydroxide solution, so it would be expected that a pH of 7 would be recorded. However, an important feature of the graph is that the addition of a small amount of alkali after the end point results in an immediate increase in the pH.

> ## Tip
>
> Two terms are used for titrations:
>
> - The end-point is the point at which the pH changes rapidly.
> - The equivalence point is the point at which enough acid (or base) has been added to react completely with all of the base (or acid).

Activity

Titration of a strong acid with a strong base

Strong acids and strong bases are fully ionised in solution. Figure 3.7 shows the shape of the pH curve for a titration of a strong acid, such as hydrochloric acid, with a strong base, such as sodium hydroxide.

1 Show that pH = 1 for a solution of 0.1 mol dm⁻³ HCl(aq).
2 Why does the pH equal 7 at the equivalence point of a titration of a strong acid with a strong base?
3 Calculate the pH of 25 cm³ of a solution of sodium chloride after adding:
   a) 0.05 cm³ (1 drop) of 0.1 mol dm⁻³ of HCl(aq)
   b) 0.05 cm³ (1 drop) of 0.1 mol dm⁻³ of NaOH(aq).
   (In both instances assume that the volume change on adding the single drop is insignificant.)
4 Calculate the pH of the solution produced by adding 5 cm³ of 0.1 mol dm⁻³ NaOH(aq) to 25 cm³ of a solution of sodium chloride.
5 Show that your answers to questions 1, 2 and 3 are consistent with Figure 3.7.
6 What features of the curve plotted in Figure 3.7 are important to the practical accuracy of acid–base titrations of this kind?

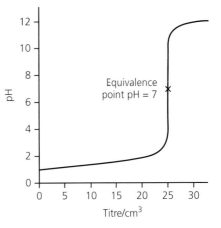

Figure 3.7 The pH change on gradually adding 0.1 mol dm⁻³ NaOH(aq) from a burette to 25 cm³ of 0.1 mol dm⁻³ HCl(aq).

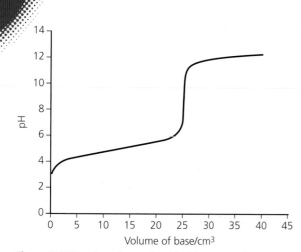

Figure 3.8 Titration of a weak acid with a strong base.

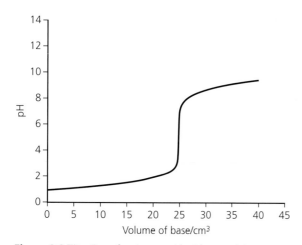

Figure 3.9 Titration of a strong acid with a weak base.

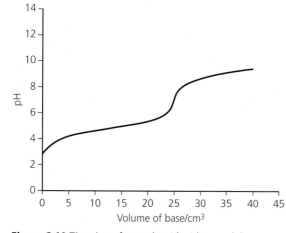

Figure 3.10 Titration of a weak acid with a weak base.

A graph illustrating the pH change during the titration of a weak acid with a strong base has a differently shaped curve.

Suppose that $25.0\,cm^3$ of $0.100\,mol\,dm^{-3}$ ethanoic acid are titrated with $0.100\,mol\,dm^{-3}$ sodium hydroxide, A plot of pH against volume of sodium hydroxide solution added is shown in Figure 3.8.

There is a sharp change of pH at the end point, but its range (from about 6.5 to 11) is not as great as when a strong acid is used. The shape after the acid is neutralised is identical to that in the strong acid–strong base example (Figure 3.6). However, before neutralisation, the pH rises quite steadily. It is worth looking at this in a little detail.

As soon as some sodium hydroxide is added, neutralisation begins and sodium ethanoate is formed. Therefore, a mixture of ethanoic acid and sodium ethanoate is present. The ethanoate ions from the sodium ethanoate cause the pH to rise as the ethanoate–acid equilibrium shifts to produce more ethanoic acid molecules. The fairly sharp rise in pH as the first drops of sodium hydroxide are added reflects this. Subsequently, the mixture acts as a buffer solution. Further rises in pH are contained until, close to the end point, the absence of remaining ethanoic acid means that the buffering effect is limited.

The titration curves of a strong acid with a weak base and of a weak acid with a weak base can be inferred from Figures 3.6 and 3.8.

A graph for the titration of a strong acid with a weak base is shown in Figure 3.9.

The acid section has the shape associated with a strong acid but, after the end point of the titration has been reached, the weak alkali section curves more gently and reaches a maximum of around 10. (The actual value depends on the weakness of the alkali.) The vertical section at the end point runs from about pH 3 to pH 7.5.

When both the acid and the base are weak, the change in pH is as shown in Figure 3.10. There is an indistinct change in pH at the end point.

Indicators

If a pH meter is not used, an acid–base titration requires an indicator to determine the end point.

Most indicators are weak acids. An indicator has the particular property that the molecular form is a different colour to that of the anion into which it dissociates,

for example:

- The molecular form of methyl orange is red; its anion is yellow.

- The molecular form of phenolphthalein is colourless; its anion is pink.

The equilibrium for methyl orange, represented as HIn, is:

$$HIn \rightleftharpoons H^+ + In^-$$

red   yellow

The end point is orange.

HIn and In⁻ are a conjugate acid–base pair.

If methyl orange is added to an acidic solution, the equilibrium shifts to favour the molecular form, HIn. Therefore, the solution appears red. In alkali, the H⁺ ions are removed from the equilibrium to form water. HIn then dissociates further and the indicator shows the colour of In⁻, which is yellow. At a halfway stage between red and yellow, it shows a combination of those colours and appears orange. The exact point at which the halfway stage is reached depends on the value of the equilibrium constant for the indicator, which is usually written as K_{In}.

For methyl orange, K_{In} is approximately $1 \times 10^{-4}\,mol\,dm^{-3}$ so

$$K_{In} = \frac{[H^+][In^-]}{[HIn]} = 1 \times 10^{-4}$$

The halfway point occurs when [HIn] = [In⁻]. Substituting into the expression for K_{In}:

$$K_{In} = \frac{[H^+][\cancel{In^-}]}{[\cancel{HIn}]} = 1 \times 10^{-4}$$

$$\text{therefore}\,[H^+] = 1 \times 10^{-4}$$

shows that this occurs when $[H^+] = 10^{-4}\,mol\,dm^{-3}$, i.e. when the pH is 4.

Therefore, during a titration, methyl orange changes colour at around pH 4. This may seem unsatisfactory, but if you return to the titration curves you will see that methyl orange will correctly show the end point if the colour change occurs on the vertical section of the graph where the pH is changing rapidly. Methyl orange is a suitable indicator for strong acid–strong base titrations and strong acid–weak base titrations. It is not suitable for weak acid–strong base titrations. The colour changes are shown in Figure 3.11.

For phenolphthalein, K_{In} is approximately $10^{-9}\,mol\,dm^{-3}$ so it changes colour at approximately pH 9. It is a suitable indicator for strong acid–strong base titrations and weak acid–strong base titrations (Figure 3.12).

If you look in data books you will find that there are many indicators, each of which changes colour at a particular pH. This enables a suitable indicator to be chosen for a particular titration. Usually, there are a number of indicators that could be used and the selection may depend simply on a preferred colour change. An indicator is given a pH range over which the colour change occurs because the eye is not capable of determining the moment when the two colours are exactly balanced.

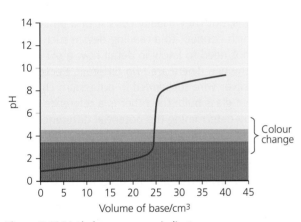

Colour change

Figure 3.11 Methyl orange as an indicator.

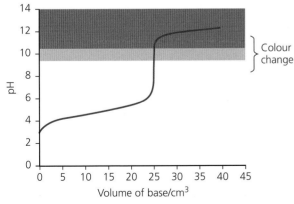

Colour change

Figure 3.12 Phenolphthalein as an indicator.

It can be seen from Figure 3.10 that there is no really decisive change of pH at the end point of a weak acid–weak base titration and no indicator can effectively be used.

Experiments to determine acidity

A pH meter is a useful device which can be used to measure acidity. There are various different designs, but they usually consist of a probe that contains two electrodes which connect to a reading device such as a digital scale. Although you will not need to know in detail how a pH meter works a brief description is worthwhile. There are two different electrodes in the probe. One consists of a silver wire suspended in potassium chloride solution and surrounded by a thin glass bulb; the other is a reference electrode. The pH is measured by determining the potential difference between the solution being tested and the glass-bulb electrode. This is then interpreted in terms of the concentration of hydrogen ions present. Before using a pH meter, it must be calibrated. This is done by inserting the probe into a buffer solution whose pH is known and adjusting the meter to that reading. Often comparison is made with two or three buffer solutions to increase the reliability of measurements. Many meters also have a temperature setting because pH is temperature dependent.

Once set up, a pH meter can be used to follow an acid–base titration through its course. Measuring the pH of a known concentration of an acid or a base would indicate whether it was strong or weak.

Practice questions

Multiple choice questions 1–10

1 The pH of a $0.01\,mol\,dm^{-3}$ sodium hydroxide is

  A 2

  B 10

  C 11

  D 12 *(1)*

2 What volume of water should be added to $100\,cm^3$ of hydrochloric acid with a pH of 2 to form a solution of pH whose pH is 3?

  A $100\,cm^3$

  B $200\,cm^3$

  C $900\,cm^3$

  D $1000\,cm^3$ *(1)*

3 Which one of the following indicators would be suitable to indicate the end point of a titration between $0.100\,mol\,dm^{-3}$ ethanoic acid and $0.100\,mol\,dm^{-3}$ sodium hydroxide?

  A bromophenol blue (pH range 2.8–4.6)

  B methyl red (pH range 4.2–6.3)

  C phenol red (pH range 6.8–8.4)

  D thymolphthalein (pH range 9.3–10.6) *(1)*

4 The indicator methyl yellow has a pH range of 2.9 to 4.0 and a colour change of red to yellow. The indicator bromocresol purple has a pH range of 5.2 to 6.8 and a colour change of yellow to purple. An aqueous solution is made and divided into two samples. Methyl yellow is added to one sample and bromocresol purple is added to the other. Which one of the following would produce the same colour in the indicator in both samples?

  A $0.01\,mol\,dm^{-3}$ HCl

  B $0.001\,mol\,dm^{-3}$ HCl

  C $0.0001\,mol\,dm^{-3}$ HCl

  D water. *(1)*

5 The pK_a for chlorous acid ($HClO_2$) is 2.0. The hydrogen ion concentration of a $1\,mol\,dm^{-3}$ solution of chloric acid will be approximately

  A 0.5   **B** 1.0

  C $\sqrt{2}$ **D** 2 *(1)*

6 The pK_a of a monobasic acid is 3. When $10\,cm^3$ of $1.0\,mol\,dm^{-3}$ sodium hydroxide is added to $20\,cm^3$ of a $1.0\,mol\,dm^{-3}$ solution of the acid the pH of the solution obtained is

  A 1/3 of the pH of a $1.0\,mol\,dm^{-3}$ solution of the acid

  B 1/2 of the pH of a $1.0\,mol\,dm^{-3}$ solution of the acid

  C 2/3 of the pH of a $1.0\,mol\,dm^{-3}$ solution of the acid

  D the same as the pK_a of a $1.0\,mol\,dm^{-3}$ solution of the acid. *(1)*

7 Like water other liquids can have an ionic product. The ionic product for liquid ammonia is:

$$[NH_4^+][NH_2^-] = 10^{-33}$$

If a solution of $0.001\,mol\,dm^{-3}$ sodium amide, $NaNH_2$, is dissolved in liquid ammonia what will be the concentration of NH_4^+ ions in the solution?

  A $10^{-36}\,mol\,dm^{-3}$

  B $10^{-30}\,mol\,dm^{-3}$

  C $10^{-15}\,mol\,dm^{-3}$

  D $10^{-11}\,mol\,dm^{-3}$ *(1)*

Use the key below to answer questions 8, 9 and 10.

A	B	C	D
1, 2 & 3 correct	1 & 2 correct	2 & 3 correct	1 only correct

8 The pK_a of propanoic acid is 4.9.
The pK_a of methanoic acid is 3.8.
Which of the following ions might be present in a mixture of liquid propanoic acid with liquid methanoic acid?

  1 $CH_3CH_2COOH_2^+$

  2 $HCOO^-$

  3 $HCOOH_2^+$ *(1)*

9 In which of the following reactions does nitric acid act as a base?

  1 $HNO_3 + HF \rightarrow H_2NO_3^+ + F^-$

  2 $HNO_3 + H_2SO_4 \rightarrow NO_2^+ + 2HSO_4^- + H_3O^+$

  3 $HNO_3 + HCOOH \rightarrow HCOOH_2^+ + NO_3^-$ *(1)*

10 Which of the following applies to a buffer solution made by mixing sodium ethanoate and ethanoic acid?

  1 The sodium ethanoate is highly ionised.

  2 The ethanoic acid is largely in the form of CH_3COOH molecules.

  3 Its pH is higher than the pH of the ethanoic acid from which it was made. *(1)*

11 For each of the following equilibria, identify the conjugate acid–base pairs:

  a) $HCO_3^- + H_2O \rightleftharpoons H_2CO_3 + OH^-$

  b) $HCO_3^- + OH^- \rightleftharpoons H_2O + CO_3^{2-}$

  c) $HCO_3^- + HCOOH \rightleftharpoons HCOO^- + H_2O + CO_2$ *(3)*

12 Calculate the pH of each of the following aqueous solutions:

a) $0.15\,mol\,dm^{-3}$ HNO_3

b) $0.15\,mol\,dm^{-3}$ HCN ($K_a = 4.8 \times 10^{-10}\,mol\,dm^{-3}$)

c) $0.15\,mol\,dm^{-3}$ NaOH

d) $0.15\,mol\,dm^{-3}$ Na_2SO_4. *(8)*

13 Calculate the pH of a mixture of $20.0\,cm^3$ of $1.00\,mol\,dm^{-3}$ HCl and $10.0\,cm^3$ of $1.00\,mol\,dm^{-3}$ NaOH. *(4)*

14 Calculate the hydrogen ion concentration of each of the following:

a) $0.5\,mol\,dm^{-3}$ KOH

b) an aqueous solution of pH 4.0

c) an aqueous solution of pH 2.7

d) an aqueous solution of pH 11.2. *(5)*

15 A sample of milk has a pH of 6.3. Calculate the concentration of hydrogen ions in the milk. *(1)*

16 A saturated solution of magnesium hydroxide has a pH of 10.5. Calculate the concentration of magnesium ions in the solution. *(4)*

17 A sample of lemon juice has a pH of 2.45.

a) What is the hydrogen ion concentration in $mol\,dm^{-3}$ in the lemon juice?

b) The acid in lemon juice is citric acid, which is tribasic (i.e. its formula is H_3X). Assuming the equilibrium:

$$H_3X \rightleftharpoons 3H^+ + X^{3-}$$

what is the concentration of citrate ion in the juice? *(3)*

18 Apple juice contains a monobasic acid.

a) When $25.0\,cm^3$ of apple juice is titrated with a $0.120\,mol\,dm^{-3}$ solution of sodium hydroxide, $22.90\,cm^3$ is required to reach the end point. Calculate the concentration of the acid in the apple juice.

b) The pH of apple juice is 3.5. Calculate the hydrogen ion concentration of the apple juice.

c) Use your results from parts a) and b) to calculate the value of K_a for the acid in apple juice. *(6)*

19 a) In aqueous solution, potassium ethanoate is slightly alkaline. Remembering that ethanoic acid is a weak acid, suggest the reason why aqueous potassium ethanoate has a pH greater than 7.

b) A $0.1\,mol\,dm^{-3}$ solution of potassium ethanoate has a lower pH than a $0.1\,mol\,dm^{-3}$ solution of potassium cyanide. Explain why this indicates that hydrocyanic acid is a weaker acid than ethanoic acid.

c) Suggest a possible pH for an aqueous solution of ammonium chloride. Explain your answer. *(10)*

20 a) Calculate the pH of a buffer solution made by mixing $50\,cm^3$ of $0.10\,mol\,dm^{-3}$ potassium propanoate with $50\,cm^3$ of $0.10\,mol\,dm^{-3}$ propanoic acid. (K_a for propanoic acid = $1.3 \times 10^{-5}\,mol\,dm^{-3}$)

b) What would happen to the pH of the buffer solution if more potassium propanoate were dissolved into it. *(7)*

21 a) The pH of a solution of ethanoic acid is 2.70. K_a for the acid is $1.7 \times 10^{-5}\,mol\,dm^{-3}$. Calculate the concentration of the ethanoic acid solution.

b) Calculate the mass of sodium ethanoate that must be added to the acid to create a buffer solution with a pH of 4.0. (Assume that the sodium ethanoate does not cause an increase in volume as it dissolves.) *(8)*

22 What is the pH of a buffer made by mixing $50.0\,cm^3$ of $0.500\,mol\,dm^{-3}$ CH_3COOH with $25.0\,cm^3$ of $0.500\,mol\,dm^{-3}$ NaOH. (K_a ethanoic acid = $1.7 \times 10^{-5}\,mol\,dm^{-3}$.) *(4)*

23 The pK_{In} of the indicator bromophenol blue is 4.1.

a) At what pH value will bromophenol blue show its 'neutral' colour?

b) Name two types of titration for which bromophenol blue would be a suitable indicator. *(4)*

24 The K_{In} value of chlorophenol red is 6.31×10^{-7}. Chlorophenol red is yellow in acid solution and red in alkaline solution.

a) Determine the pH which is the mid-point for its colour change.

b) Describe how an indicator works using chlorophenol red as your example.

c) What will be the colour of chlorophenol red when it is added to the following? Explain your answers.

i) $0.0001\,mol\,dm^{-3}$ hydrochloric acid

ii) pure water *(7)*

25 $25.0\,cm^3$ of a $0.020\,mol\,dm^{-3}$ solution of propanoic acid is titrated with $0.025\,mol\,dm^{-3}$ sodium hydroxide solution. The reaction is followed by measuring the pH as the sodium hydroxide is added.

a) Calculate the volume of sodium hydroxide that will be needed to reach the end point.

b) Calculate the pH of $0.020\,mol\,dm^{-3}$ propanoic acid. ($K_a = 1.3 \times 10^{-5}\,mol\,dm^{-3}$)

c) Calculate the pH of $0.025\,mol\,dm^{-3}$ sodium hydroxide.

d) Sketch the appearance of the titration curve that would be obtained by plotting the pH against the volume of sodium hydroxide added.

e) Suggest a suitable indicator for use in this titration. *(15)*

26 When $50.0\,cm^3$ of $1.00\,mol\,dm^{-3}$ sulfuric acid is added to $50.0\,cm^3$ of $1.00\,mol\,dm^{-3}$ sodium hydroxide at $19.0\,°C$ the temperature rises to $23.6\,°C$.

a) Write an equation for the reaction and determine which one of sulfuric acid or sodium hydroxide is present in excess.

b) Assume that the density of both solutions is $1\,g\,cm^{-3}$ and that the specific heat capacity is $4.18\,J\,g^{-1}\,K^{-1}$.
Calculate the enthalpy of neutralisation for this reaction. *(7)*

27 Here are three pairs of acids and bases that can react to form salts: $HBr/NaOH$, HCl/NH_3, CH_3COOH/NH_3.
Here are three values for standard enthalpy changes of neutralisation: $-50.4\,KJ\,mol^{-1}$, $-53.4\,KJ\,mol^{-1}$, $-57.6\,KJ\,mol^{-1}$.
Write the equations for the three neutralisation reactions, and match each with its corresponding value of ΔH neutralisation. Explain your answers. *(6)*

28 During the processing of apples (Figure 3.13), the skins may be loosened using aqueous sodium hydroxide at pH 12. The pH of the sodium hydroxide eventually drops to 11.5 and it becomes too dilute to be effective. This is still a very alkaline pH and so, before discarding the solution, it is reacted with hydrochloric acid to reduce the pH to the safer value of 10.8. To ensure this has been achieved, an indicator called benzaldehyde 3-nitrophenylhydrazone (NPB) is used. NPB is purple for solutions with a pH greater than 12 and yellow in those with a pH less than 11.

Figure 3.13 There's more than one way to peel an apple.

a) Calculate the change in the hydroxide ion concentration (in $mol\,dm^{-3}$) that occurs as the pH of the solution falls from pH 12 to pH 11.5.

b) Estimate a value for the equilibrium constant, K_{In}, for the indicator NPBH.

c) Estimate the ratio of the concentration of the unionised form of the indicator NPB to the concentration of the anion NPB^- at pH 10.8. *(10)*

57

29 Aspirin is an effective painkiller although its use has, to some extent, been discouraged because in some circumstances it can cause stomach bleeding. This appears to be triggered by the molecular form of aspirin dissolving in the covalent lipids of the stomach lining.

Aspirin contains a carboxylic acid group and an ester group. It is hydrolysed readily (Figure 3.14).

$$C_6H_4(OCOCH_3)CO_2H + H_2O \rightleftharpoons C_6H_4(OH)CO_2H + CH_3COOH$$

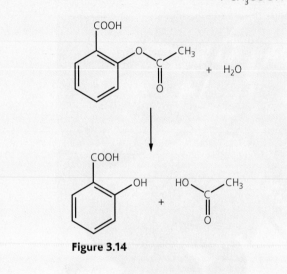

Figure 3.14

Because of the ease of hydrolysis, aspirin has a limited shelf-life.

a) Assuming that the pH of stomach acid is approximately 1, explain why stomach bleeding might be a problem. (K_a for aspirin is $3 \times 10^{-4}\,\text{mol dm}^{-3}$.)

b) The blood is buffered at pH 7.4. Calculate whether aspirin in the blood exists largely in its unionised molecular form or as an anion.

c) Aspirin is usually administered as a calcium salt since this is more soluble. However, as aspirin is hydrolysed rapidly above pH 8.5, care has to be taken in its preparation. If a solution of calcium hydroxide containing $0.741\,\text{g dm}^{-3}$ is used to create the calcium salt by a reaction with aspirin, is this likely to cause hydrolysis?

d) A 0.900 g sample of aspirin becomes damp and absorbs 0.100 g of water. An equilibrium is established and analysis shows that 0.117 g of ethanoic acid is present in the equilibrium mixture.

i) Calculate the value of the equilibrium constant for the hydrolysis of aspirin.

ii) What percentage of aspirin has been hydrolysed?
(The aspirin would, in fact, be unsafe to use.) *(18)*

Chapter 4

Enthalpy, entropy and free energy

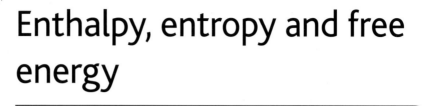

Prior knowledge

In this chapter it is assumed that you can:
- draw a reaction profile of an endothermic (ΔH is positive) or exothermic (ΔH is negative) reaction
- understand and define the standard enthalpy of formation
- use Hess' Law to determine an unknown enthalpy
- appreciate that ΔH gives guidance as to the feasibility of a chemical reaction.

Test yourself on prior knowledge

1 Draw a fully labelled reaction profile for an endothermic reaction.
2 a) Define standard enthalpy of formation.
   b) State Hess' Law.
   c) Use the standard enthalpy of combustion data in Table 4.1 to determine the standard enthalpy of formation of cyclohexane.

Table 4.1

Substance	Formula	$\Delta_c H^{\ominus}/\mathrm{kJ\,mol^{-1}}$
Carbon	C(s)	−394
Hydrogen	$H_2(g)$	−286
Cyclohexane	$C_6H_{12}(l)$	−3920

Energy changes in reactions

A useful approach to deciding whether a reaction might be feasible is to consider the energy changes that would occur if the reaction took place. Sometimes it is possible to measure this energy change directly, for example in combustion reactions. In other instances, Hess' Law can be used.

The principle on which Hess' Law is based is that the energy change for a process is the same whichever pathway is taken to convert the reactants to the products. In this chapter the use of Hess' Law is extended to provide further information about the feasibility of chemical processes.

Lattice enthalpy

An ionic lattice consists of ions of opposite charge held together by the strong electrical forces that exist between them.

If the ions have high charge, the electrical forces are considerable and the crystal is held tightly together, which is confirmed by a high melting point

The **lattice enthalpy** of an ionic substance is the energy released when one mole of an ionic compound is formed from its constituent ions in their gaseous state. Standard conditions of 298 K and 10 kPa are applied.

Tip

The definition that OCR uses means that lattice enthalpies always have a negative value. Clearly energy must be released as the ions combine together to form the solid.

This is the definition that you should use in your exams. However, you should be aware that some sources of data define lattice enthalpy as the energy change for the reverse reaction.

(for example, aluminium oxide). Although this fact can be appreciated in qualitative terms, it would be useful if the information could be obtained quantitatively. In order to do this, a definition of what is meant by the 'strength' of an ionic crystal must be given. This is known as the **lattice enthalpy** (Figure 4.1).

This can be summarised for an ionic crystal, A_xB_y, by the equation:

$$xA^{y+}(g) + yB^{x-}(g) \rightarrow A_xB_y(s)$$

This may look complicated, but for sodium chloride this is simply the enthalpy change that occurs for:

$$Na^+(g) + Cl^-(g) \rightarrow NaCl(s)$$

For sodium oxide, it is the enthalpy change that occurs for:

$$2Na^+(g) + O^{2-}(g) \rightarrow Na_2O(s)$$

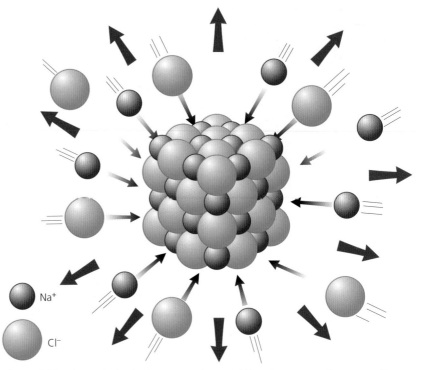

Figure 4.1 Lattice enthalpy is the energy that would be given out to the surroundings (red arrows) if one mole of a compound could be formed directly from free gaseous ions coming together (black arrows) and arranging themselves into a crystal lattice.

Factors affecting the size of lattice enthalpies

The lattice enthalpies of some ionic lattices are given in Table 4.2. The units of all the numbers in the table are $kJ\,mol^{-1}$.

The exact values of lattice enthalpies are often in some doubt because of the difficulty of obtaining precise values for the enthalpy changes used to calculate them.

Tip

When referring to lattice enthalpies, you must choose words carefully. For example, when comparing the lattice enthalpies of aluminium oxide ($-15\,900\,kJ\,mol^{-1}$) and magnesium oxide ($-3971\,kJ\,mol^{-1}$), it is very tempting to say that the lattice enthalpy of $Al_2O_3(s)$ is '**greater than**' that for MgO(s), but this would be incorrect as the numbers are negative. It is best to avoid words like 'greater than' or 'bigger than', instead stating that the lattice enthalpy of Al_2O_3 (s) is '**more negative**' or '**more exothermic**' than the lattice enthalpy of MgO(s).

Table 4.2 Table of approximate lattice enthalpies.

	O^{2-}	Cl^-	Br^-	I^-
Na^+	−2478	−780	−742	−705
Mg^{2+}	−3791	−2526	−2440	−2327
Al^{3+}	−15 900			
Ca^{2+}	−3401	−2258	−2176	−2074
Sr^{2+}	−3223	−2156	−2075	−1963

Table 4.2 illustrates two clear trends in the size of the lattice energies of ionic compounds relating to:

● ionic charge

● ionic radius.

Ionic charge

As the charge on an ion increases, the lattice enthalpy becomes more exothermic. (That is more energy is released when the lattice is formed and, hence, more energy is required to break the lattice.)

If you compare the value of the lattice enthalpy for sodium chloride with that of magnesium oxide, you will see that the magnesium oxide lattice is stronger (more energy is released when it is formed from its gaseous ions) than that of sodium chloride. This is a result of the stronger forces that exist within the crystal, because the charge on the oxide ion (O^{2-}) is higher than the charge on the chloride ion (Cl^-).

Likewise, if you compare the lattice enthalpies of the series Na_2O ($-2478\,kJ\,mol^{-1}$), MgO ($-3791\,kJ\,mol^{-1}$) and Al_2O_3 ($-15\,900\,kJ\,mol^{-1}$), the lattices become progressively stronger as the charge on the cation increases.

Ionic radii

If ions have the same charge, then smaller ions will form a stronger lattice. That is, the numerical value of the lattice enthalpy increases. If you compare the values of the lattice enthalpies of CaO ($-3401\,kJ\,mol^{-1}$), SrO ($-3223\,kJ\,mol^{-1}$) and BaO ($-3054\,kJ\,mol^{-1}$), you will see that the lattices become progressively weaker as the size of the cation increases. This is because the centres of the cation and anion are not as close, so the force of attraction between them is less. Ionic radius does not have such a pronounced effect as that caused by a difference in the charge of the ions, but it does make *some* difference.

You can see the same pattern when comparing the lattice strengths of the group(II) chlorides or the other halides of a group(II) metal.

Taken together, these factors have their most obvious effect on the melting points of compounds. In general, the more negative the lattice enthalpy, the higher the melting point.

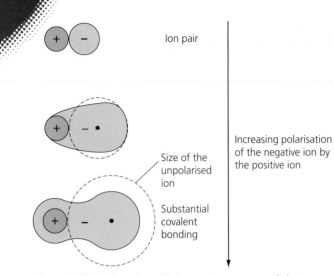

Ion pair

Increasing polarisation of the negative ion by the positive ion

Size of the unpolarised ion

Substantial covalent bonding

Figure 4.2 Ionic bonding with increasing amounts of electron sharing as a positive cation polarises neighbouring negative anions. (Dotted circles show the size of unpolarised ions.)

Polarisation of ions

There is one further factor that must be considered in relation to the strengths of ionic lattices. This is an effect called polarisation. In ionic compounds, positive metal ions attract the outermost electrons of negative ions, pulling these electrons into the spaces between the ions. This distortion of the electron clouds around anions by positively charged cations is an example of polarisation. As a result of polarisation, there is a significant degree of covalent bonding in some ionic compounds. This makes the size of the lattice enthalpy greater numerically than that expected from the purely ionic model, because the calculation does not allow for the extra attraction resulting from the distortion of the anions. Figure 4.2 shows three examples of ionic bonding with increasing degrees of electron sharing as a positive cation polarises the neighbouring negative ion. In general, results show that:

- the polarising power of a cation increases as its charge increases and as its radius decreases
- the polarisability of an anion increases as its radius increases.

As the negative anion gets larger, its outermost electrons are further from the attraction of its positive nucleus. Consequently, these outermost electrons are more readily attracted to a neighbouring positive ion and the anion is, therefore, more polarisable. This means that iodide ions are more polarisable than bromide ions, bromide ions are more polarisable than chloride ions and fluoride ions are very difficult to polarise. In fact, fluorine, with its small, singly charged fluoride ion, forms compounds which are more ionic than any other non-metal.

> **Key term**
>
> Polarisation is the distortion of the electron cloud in a molecule or ion by a nearby positive charge.

> **Test yourself**
>
> 1 Explain why the melting point of magnesium oxide (2852 °C) is so much higher than that of magnesium fluoride (1261 °C)?
> 2 The following are three values for lattice enthalpy in kJ mol^{-1}: −3791, −3299 and −2725.
>   The four ionic compounds to which these values relate are MgO, BaS and MgS.
>   Match the formulae with the values and justify your choices.
> 3 Place the following in order of decreasing lattice strength and explain how you decided on your answer:
>   NaCl, NaBr, NaI, Na$_2$O.

Calculating lattice enthalpy

It is not possible to measure lattice enthalpies directly by experiment because gaseous ions cannot be obtained separately from the ions of opposite charge. However, it is possible to obtain a numerical value indirectly from an energy cycle. The cycle has a number of steps.

Lattice enthalpy

Key term

The **enthalpy change of atomisation** of an element is the enthalpy change when 1 mole of gaseous atoms is formed from the element in its standard state.

Tip

If you are asked to define the *standard* enthalpy change, you must always quote the *standard* conditions of 298 K and 100 kPa.

Key terms

The **first ionisation energy** is the energy required to remove 1 electron from each atom in 1 mole of gaseous atoms to form 1 mole of gaseous 1^+ ions.

The **second ionisation energy** is the energy required to remove 1 electron from each 1^+ ion in 1 mole of gaseous 1^+ ions to form 1 mole of gaseous 2^+ ions.

The **nth ionisation energy** is the energy required to remove 1 electron from each $(n-1)^+$ ion in 1 mole of gaseous $(n-1)^+$ ions to form 1 mole of gaseous n^+ ions.

Key term

The **first electron affinity** is the enthalpy change accompanying the gain of 1 mole of electrons by 1 mole of atoms in the gaseous phase.

Key term

The **enthalpy change of formation** is the enthalpy change when 1 mole of a compound is formed from its constituent elements in their standard states.

It must be emphasised that the cycle is simply a **model** that allows the determination of a lattice enthalpy. It does not represent a reaction mechanism.

The following theoretical steps take place during the formation of a crystal from its elements in their standard state.

Step 1

The metal, which starts as a giant lattice, is broken into free gaseous atoms. The enthalpy change involved is called the enthalpy change of atomisation.

$$M(s) \rightarrow M(g)$$

Step 2

The gaseous metal atoms are then ionised to create ions of the required charge. For a '1^+' ion, the enthalpy change involved is the first ionisation enthalpy.

For other ions, this involves further ionisations and the energy required is the sum of all these ionisation enthalpies:

$$M(g) \rightarrow M^+(g) + e^-$$
$$M^+(g) \rightarrow M^{2+}(g) + e^-$$

and so on ...

Step 3

The non-metal molecule is broken into gaseous atoms. The enthalpy change involved is the enthalpy change of atomisation, which (as with the metal) is the enthalpy change for the formation of 1 mole of gaseous atoms.

$$\tfrac{1}{2}H_2(g) \rightarrow H(g)$$

Step 4

The gaseous non-metal atoms form negative ions. The enthalpy change involved is called the electron affinity.

As with ionisation energies, there may be more than one step depending on the charge of the negative ion.

First electron affinity:

$$X(g) + e^- \rightarrow X^-(g)$$

Second electron affinity:

$$X^-(g) + e^- \rightarrow X^{2-}(g)$$

and so on ...

To complete the cycle, the enthalpy change of formation is also required.

The Born–Haber cycle

A Born–Haber cycle uses all of the enthalpy changes outlined above to determine the lattice enthalpy (Figure 4.3).

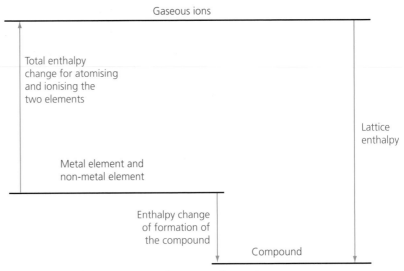

Figure 4.3 The overall structure of a Born–Haber cycle.

An example to determine the lattice enthalpy of magnesium chloride is shown in Figure 4.4.

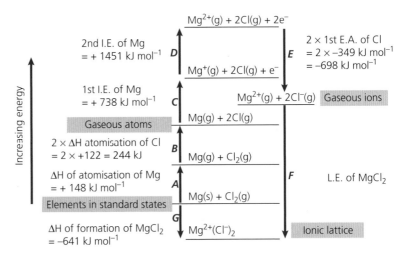

Figure 4.4 Born–Haber cycle for magnesium chloride.

Starting from the elements in their standard states, you will see that to create the ionic lattice:

● magnesium is atomised and then ionised to Mg^{2+} (both the first and the second ionisations are needed)

● chlorine is atomised – the atomisation enthalpy for 1 mol of atoms is +122 kJ mol^{-1} but, since 2 mol of atoms is required, this number must be doubled to +244 kJ mol^{-1}

● 2 mol of Cl^- are required, so the electron affinity of Cl must be doubled, giving a figure of −698 kJ mol^{-1}.

Notice that the electron affinity of chlorine is exothermic.

● The unknown lattice enthalpy can then be calculated by completing the cycle using the enthalpy of formation of magnesium chloride.

From the cycle in Figure 4.4:

$A + B + C + D + E + F = G$

$148 + 244 + 738 + 1451 + (−698) + F = −641$

Tip

Once you are familiar with them, Born–Haber cycles are easy to follow. However, they do require you to look closely at the steps. In exams, most errors are caused by carelessness with the signs and mistakes in the subsequent arithmetic. Practice at constructing the cycles is essential.

So, the lattice enthalpy, $\Delta_{Lattice}H$, of magnesium chloride is $-2524\,kJ\,mol^{-1}$.

You are not required to know the experimental details of how the energy changes in a Born–Haber cycle are determined. Most have been established quite accurately. It is worth mentioning, however, that a Born–Haber cycle is sometimes used, not to determine lattice enthalpies, but to provide values for electron affinities, which cannot be obtained readily.

Activity

The stability of ionic compounds

Almost all the compounds of metals with non-metals are regarded as ionic, and these compounds have standard enthalpy changes of formation which are exothermic. This means that the compounds are at a lower energy level and, therefore, are more stable than their elements.

Using a Born–Haber cycle with a theoretically calculated value for the lattice enthalpy, it is possible to estimate the standard enthalpy change of formation for compounds which do not normally exist. For example, consider the Born–Haber cycle for the hypothetical compound MgCl in Figure 4.5.

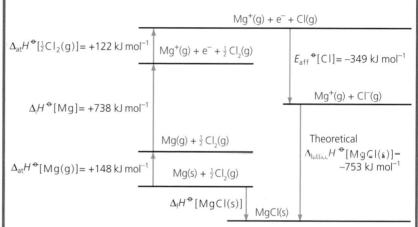

Figure 4.5 A Born–Haber cycle for the hypothetical compound MgCl.

1 Use Figure 4.5 to calculate a value for the standard enthalpy change of formation of MgCl(s).
2 What does your answer to question **1** suggest about the stability of MgCl(s)?
3 Using the Hess cycle in Figure 4.6, calculate the standard enthalpy change for the reaction

$$2MgCl(s) \rightarrow MgCl_2(s) + Mg(s)$$

knowing that $\Delta_f H\,[MgCl_2(s)] = -641\,kJ\,mol^{-1}$.

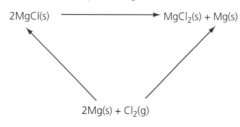

Figure 4.6 A Hess cycle for the reaction $2MgCl(s) \rightarrow MgCl_2(s) + Mg(s)$.

4 What does your result for question **3** tell you about the stability of MgCl(s)?

5 The Born–Haber cycle for the hypothetical compound $MgCl_3$ suggests that $\Delta_f H$ [$MgCl_3$(s)] = +3950 kJ mol^{-1}.
  a) What does the value of $\Delta_f H$ [$MgCl_3$(s)] tell you about the stability of $MgCl_3$(s)?
  b) Suggest why the value of $\Delta_f H$ [$MgCl_3$(s)] is so endothermic.
6 The estimated lattice enthalpy of $MgCl_3$(s) is −5440 kJ mol^{-1}.
  a) Write an equation to summarise the lattice enthalpy of $MgCl_3$.
  b) Why is the lattice enthalpy of $MgCl_3$ more exothermic than that of $MgCl_2$(s)?

Test yourself

4 Write equations for the following processes:
  a) the enthalpy change of formation of calcium oxide
  b) the lattice energy of calcium oxide
  c) the second ionisation energy of calcium
  d) the enthalpy change of atomisation of calcium
  e) the second electron affinity of oxygen.
5 Why does the lattice enthalpy of lithium fluoride indicate that the ionic bonding in lithium fluoride is stronger than that in sodium chloride?
  ($\Delta_{lattice} H^{\ominus}$ [LiF(s)] = −1031 kJ mol^{-1}; $\Delta_{lattice} H^{\ominus}$ [NaCl(s)] = −780 kJ mol^{-1})
6 Use Table 4.3 to construct a Born–Haber cycle to determine the lattice enthalpy of calcium fluoride.

Table 4.3

Enthalpy of formation of calcium fluoride	−1219.6 kJ mol^{-1}
Enthalpy of atomisation of calcium	+ 178.2 kJ mol^{-1}
Enthalpy of atomisation of fluorine	+ 79.0 kJ mol^{-1}
1st ionisation enthalpy of calcium	+ 590.0 kJ mol^{-1}
2nd ionisation enthalpy of calcium	+ 1145.0 kJ mol^{-1}
Electron affinity of fluorine	−328.0 kJ mol^{-1}

Enthalpy changes of hydration and solubility

Key term

The **enthalpy change of hydration** ($\Delta_{hyd} H$) of an ion is the enthalpy change that occurs when 1 mole of a gaseous ion is completely hydrated by water. It is, therefore, the enthalpy change for the process.

$$X^{n+}(g) \rightarrow X^{n+}(aq)$$

The standard enthalpy of hydration applies in the conditions of 25 °C and 100 kPa.

The concept of a Born–Haber cycle can be extended to provide a partial explanation of the solubility of substances in water. To understand this, another enthalpy term must be introduced. This is the enthalpy change of hydration of an ion.

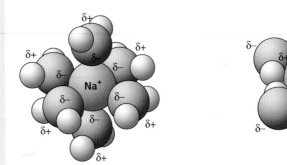
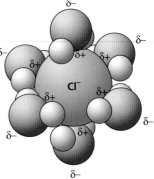

Figure 4.7 Sodium and chloride ions are hydrated when sodium chloride dissolves in water. Polar water molecules are attracted to both cations and anions.

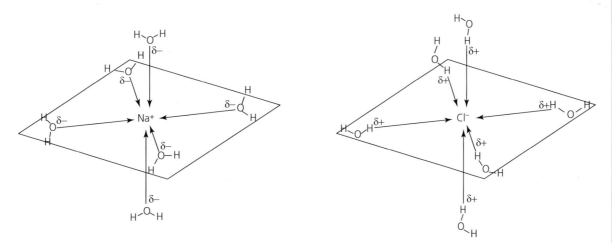

Figure 4.8 Sodium and chloride ions are hydrated when sodium chloride dissolves in water.

It has already been established that a small highly charged ion exerts a strong force on neighbouring ions of opposite charge. In the case of hydration, the attraction is either between a cation (Na^+) and the oxygen atom of a water molecule, or between an anion (Cl^-) and the hydrogen atom of the water molecule as shown in Figures 4.7 and 4.8. This occurs because of the charge separation across the $O^{\delta-}-H^{\delta+}$ bond.

As with lattice enthalpies, it is possible to appreciate trends in the size of enthalpy changes of hydration.

Some enthalpy changes of hydration are shown in Table 4.4.

Table 4.4 Some enthalpy changes of hydration.

Ion	Li^+	Na^+	K^+	Mg^{2+}	Ca^{2+}	Sr^{2+}	Al^{3+}	Cl^-	Br^-	I^-	O^{2-}
Enthalpy change of hydration/ $kJ\,mol^{-1}$	−545	−418	−351	−1923	−1653	−1482	−2537	−338	−304	−261	−937

Comparing values for the enthalpy change of hydration of the Group 1 ions, it can be seen that the larger the ionic radius, the smaller the numerical value of the hydration enthalpy. If the cation has a larger charge, this significantly increases the numerical value of the hydration enthalpy as the attraction to the water molecules is appreciably stronger. So, for example, Mg^{2+} has a numerical value of 1923 whereas K^+ is 351.

> **Test yourself**
>
> 7 Use Table 4.4 to explain the difference in the values of the enthalpy changes of hydration of:
>   a) I^- and Br^-
>   b) Mg^{2+} and Al^{3+}.

An enthalpy cycle can be constructed which relates the values of lattice enthalpy and enthalpy change of hydration to the enthalpy change of solution ($\Delta_{sol}H$).

When an ionic solid dissolves, it splits into its hydrated ions. The enthalpy change of solution is a combination of the energy required to break the lattice and the energy released as the ions are hydrated. This can be represented on an enthalpy diagram as shown in Figure 4.9 for magnesium sulfate.

> **Key term**
>
> The **enthalpy change of solution** ($\Delta_{sol}H$) is the enthalpy change that occurs when 1 mole of an ionic solid dissolves in water.

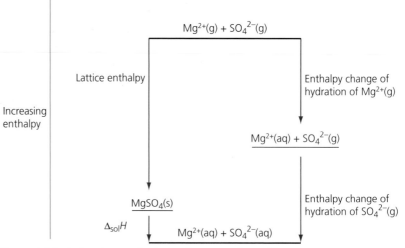

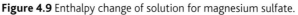

Figure 4.9 Enthalpy change of solution for magnesium sulfate.

The lattice enthalpy of magnesium sulfate is $-2833\,kJ\,mol^{-1}$.

The enthalpy of hydration of $Mg^{2+}(g) = -1923\,kJ\,mol^{-1}$ and the enthalpy of hydration of $SO_4^{2-}(g) = -1004\,kJ\,mol^{-1}$.

Using Hess' Law, the sum of the enthalpies of hydration = the lattice enthalpy + $\Delta_{sol}H$.

So $-1923 - 1004 = -2833 + \Delta_{sol}H$

$\Delta_{sol}H = -1923 - 1004 + 2833 = -94\,kJ\,mol^{-1}$

Enthalpy changes of solution are not always exothermic and do not always lead to values of ΔH that are negative. The enthalpy cycle for a solution of sodium chloride illustrates this (Figure 4.10).

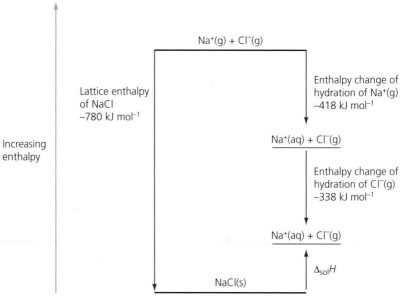

Figure 4.10 Enthalpy change of solution for sodium chloride.

The cycle above gives a value of $\Delta_{sol}H$ for sodium chloride of $+24\,kJ\,mol^{-1}$.

Entropy changes

The enthalpy change of solution for sodium chloride is positive, yet sodium chloride dissolves readily in water. This suggests enthalpies alone are not a reliable guide to the feasibility of a reaction. There are, in fact, many examples of endothermic reactions that occur readily. For example, the reaction between citric acid and sodium hydrogencarbonate creates a mixture which fizzes vigorously, while steadily getting colder.

There is a further energy-related factor that must be considered in order to obtain a complete understanding of the energy changes involved in a chemical process.

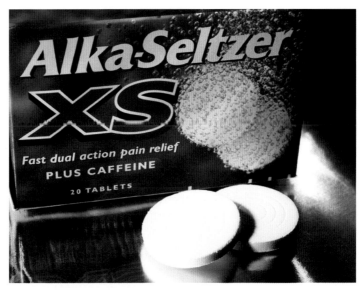

Figure 4.11 Alka-Seltzer tablets contain citric acid and sodium hydrogencarbonate. They fizz when added to water. The reaction of the acid and the hydrogencarbonate is an endothermic reaction.

When a reaction occurs, not all the energy is absorbed or released in a form that can be recognised immediately, such as heat. Some energy transfer is involved in the redistribution of the particles when the products are formed. The amount of this energy depends largely on the physical state of the substances and on the temperature. It is known as the **entropy** of the substance. It is given the symbol, S, or, under standard conditions, $S^\ominus$. The units of entropy are usually given in molar units as $J\,mol^{-1}K^{-1}$. It tells us the amount of energy (in $J\,mol^{-1}$) that the substance will require internally for every degree (in K) that the temperature rises. For example, the entropy for carbon dioxide is $213.6\,J\,mol^{-1}K^{-1}$. So, if there is a rise in the temperature of some carbon dioxide of $1\,K$, then an extra $213.6\,J\,mol^{-1}$ will be held internally by the carbon dioxide. Notice that the energy is measured in joules and not kilojoules as is usual for enthalpy.

The following points should be noted:

- A solid has a low value for its molar entropy because the movement of the particles, and hence the energy contained within the solid, is restricted to the vibrations of the particles. The more a solid is restricted by its structure, the lower the value of its molar entropy. For example, the rigid structure of diamond has an entropy value of $2.4 \, J \, mol^{-1} K^{-1}$. This means that for every increase in temperature of 1 K, $2.4 \, J \, mol^{-1}$ of energy is absorbed by its structure as increased vibrations and other distributions of energy. Atoms in a solid such as magnesium are less restricted in their movement and its entropy value is $32.7 \, J \, mol^{-1} K^{-1}$. So for magnesium an increase in temperature of 1 K has a more noticeable effect on the energy absorbed.

- The particles in liquids have a much greater freedom of movement and for every increase of 1 K they can disperse more energy than a solid. So water has a molar entropy of $69.9 \, J \, mol^{-1} K^{-1}$ while ice is only $38.0 \, J \, mol^{-1} K^{-1}$.

- Gases have predictably higher values for their molar entropies as they have the greatest freedom of movement. Steam has a molar entropy of $188.7 \, J \, mol^{-1} K^{-1}$. This is illustrated in Figure 4.12.

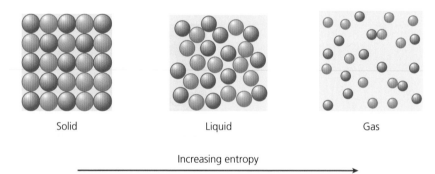

Solid   Liquid   Gas

Increasing entropy

Figure 4.12 The entropies of solids, liquids and gases.

Another way of looking at entropy is to note that the greater the disorder in the particles of a substance, the greater the value of its molar entropy because the energy can be dispersed more flexibly.

Entropy may seem to be a rather complicated concept, but it is quite easy to apply to a situation if the following points are remembered. Entropy always increases when there is a greater opportunity for energy to be spread out as a result of a change.

Therefore, entropy increases when:

- a solid becomes a liquid

- a liquid becomes a gas

- the temperature rises, even if there is no change in state

- a solid dissolves in a liquid to form a solution

- a reaction produces products with a greater degree of freedom of movement — for example, this could be because a gas is produced when a solid reacts, as in the decomposition of calcium carbonate:

$$CaCO_3(s) \rightarrow CaO(s) + CO_2(g)$$

or when a reaction produces more particles in the same state, as in the combustion of propane:

$$C_3H_8(g) + 5O_2(g) \rightarrow 3CO_2(g) + 4H_2O(g)$$

In the latter case, assuming the temperature is such that the H_2O is a gas, the reaction has produced seven gaseous molecules as products from the six gaseous reactants. (If the H_2O were liquid then the entropy would decrease as there would then be only 3 moles of gas formed, compared to 6 moles of gaseous reactants.)

Test yourself

9 Which substance in each of the following pairs is expected to have the higher standard molar entropy at 298 K?
   a) $Br_2(l)$, $Br_2(g)$
   b) $H_2O(s)$, $H_2O(l)$
   c) $NaCl(s)$, $NaCl(aq)$
   d) ethene, poly(ethene)
10 Predict whether the entropy of the system increases or decreases as a result of these changes:
   a) $KCl(s) \rightarrow KCl(aq)$
   b) $H_2O(l) \rightarrow H_2O(g)$
   c) $Mg(s) + Cl_2(g) \rightarrow MgCl_2(s)$
   d) $N_2O_4(g) \rightarrow 2NO_2(g)$
   e) $NaHCO_3(s) + HCl(aq) \rightarrow NaCl(aq) + H_2O(l) + CO_2(g)$

There are some significant differences between enthalpy and entropy.

We cannot determine the enthalpy of a substance – it is only possible to measure an **enthalpy change**. We must, therefore define a 'starting point'. This is done by defining the enthalpy of formation of an element as zero in its standard state at 25 °C and standard atmospheric pressure (100 kPa). This does not mean that the elements really have zero energy under these conditions; the definition enables values to be given to the enthalpy changes that take place.

However, with entropy it is possible to give a definite value to a substance, based on the assumption that when the temperature is at absolute zero the entropy is also zero. It is understood that at 0 K (−273 °C) the particles of a solid are completely unable to move. Entropy values are then the amount of energy added to particles for each 1 K rise in temperature.

The sign of ΔH and ΔS

Some care is needed when interpreting the signs of ΔH and ΔS – which may, at first glance, appear to be contradictory. In fact, this is not so as, in both cases, a positive sign is used if the conversion of reactants to products results in either the absorption of heat or a gain in entropy. Whereas a negative sign means heat has been lost (in other words the reaction is exothermic) or there has been in a reduction in entropy as a result of the reaction. Figure 4.13 illustrates these changes.

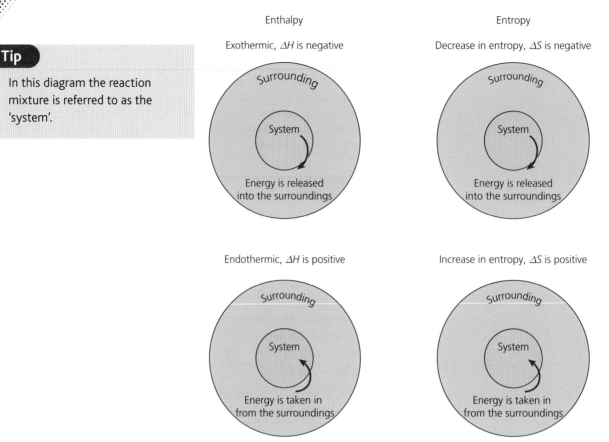

Figure 4.13 Enthalpies and entropies of a system and its surroundings.

Tip

In this diagram the reaction mixture is referred to as the 'system'.

Tip

The symbol 'Σ' means 'the sum of'.

Tip

Do not forget that, in the calculation of an entropy change, the elements in their standard states *do* have values. At 298 K the substance has the sum of all the entropy increases beyond their value of zero at 0 K.

Calculating entropy changes

You will have carried out calculations to determine ΔH for a reaction; calculations to determine ΔS are similar.

The change in entropy, ΔS, can be calculated using the formula:

$$\Delta S = \Sigma(\text{entropy of products}) - \Sigma(\text{entropy of reactants})$$

Example 1

Calculate the entropy change when 1 mol of graphite is oxidised to carbon dioxide under standard conditions.

$S^{\ominus}(\text{graphite}) = 5.7 \, \text{J mol}^{-1}\text{K}^{-1}$

$S^{\ominus}(O_2) = 102.5 \, \text{J mol}^{-1}\text{K}^{-1}$

$S^{\ominus}(CO_2) = 213.6 \, \text{J mol}^{-1}\text{K}^{-1}$

The equation is:

$$C(s) + O_2(g) \rightarrow CO_2(g)$$

Answer

$\Delta S^{\ominus} = \Sigma(\text{entropy of products}) - \Sigma(\text{entropy of reactants})$

$\Delta S^{\ominus} = (213.6) - (5.7 + 102.5) = 105.4 \, \text{J mol}^{-1}\text{K}^{-1}$

Example 2

Calculate the entropy change for the following reaction under standard conditions:

$$3O_2(g) \rightarrow 2O_3(g)$$
$$S^{\ominus}(O_3) = 237.7\,J\,mol^{-1}\,K^{-1}$$

$$S^{\ominus}(O_2) = 204.9\,J\,mol^{-1}\,K^{-1}$$

Answer

$$\Delta S^{\ominus} = \Sigma(\text{entropy of products}) - \Sigma(\text{entropy of reactants})$$

$$\Delta S^{\ominus} = (2 \times 237.7) - (3 \times 204.9) = -139.3\,J\,mol^{-1}\,K^{-1}$$

Test yourself

11 a) Calculate the entropy change for the synthesis of ammonia from nitrogen and hydrogen.

$$N_2(g) + 3H_2(g) \rightarrow 2NH_3(g)$$

$S^{\ominus}[NH_3(g)] = 192.4\,J\,mol^{-1}\,K^{-1}$; $S^{\ominus}[N_2(g)] = 191.6\,J\,mol^{-1}\,K^{-1}$;
$S^{\ominus}[H_2(g)] = 130.6\,J\,mol^{-1}\,K^{-1}$

b) Comment on the sign of the value that you obtained.

Free energy

The calculations shown above make it clear that, when a reaction takes place, there is always some change in entropy. There is also an enthalpy difference between the reactants and products because of the energy released or absorbed due to the rearrangements of bonds – in other words, an enthalpy change takes place. This release or absorption of energy inevitably affects the air surrounding the reaction mixture, because a change in temperature as a result of the reaction will also change the entropy of the air. For example, the molar entropy of nitrogen is $95.8\,J\,mol^{-1}\,K^{-1}$ and an increase of $10\,K$ would mean that $958\,J$ would be added to $1\,mol$ of nitrogen in the air. Of course, it is difficult to define the immediate surroundings of a reaction in an open laboratory and the temperature of the closest air would rise more than air which is further away.

The entropy change that occurs as a result of the reaction (known as the entropy change of the system) and the entropy change that occurs in the surroundings can be combined to give a total entropy change that has taken place as a result of the reaction

$$\Delta S_{total} = \Delta S_{system} + \Delta S_{surroundings}$$

This relationship is important because a key principle of chemistry, stated by the second law of thermodynamics, is that a reaction will only occur if ΔS_{total} is positive, so that the overall entropy increases as a result of the reaction.

In order to apply this important law to a reaction, an American, Willard Gibbs, devised an equation which is easier to use than the entropy equation above. He did this by introducing a quantity called free energy (more correctly referred to as the **Gibbs free energy**) which is given the symbol, ΔG. His equation is:

$\Delta G = \Delta H - T\Delta S$ (The ΔS in this equation is the entropy of the system.)

or under standard conditions,

$\Delta G^{\ominus} = \Delta H^{\ominus} - T\Delta S^{\ominus}$

In this equation:

- ΔH is the enthalpy of the reaction (or 'system')
- ΔS indicates the change in entropy when the reaction occurs
- both play a part in determining whether or not a reaction is feasible.

The importance of this equation is that it makes it possible to determine whether a reaction is feasible or not based on a full consideration of the energy changes involved.

The interpretation is straightforward:

- If ΔG is negative, the reaction is feasible.
- If ΔG is positive then, at the particular temperature chosen, the reaction is not feasible.

Any reaction will fit one of four possible scenarios.

1 **ΔH is negative and ΔS is positive.**

In these circumstances, the reaction is exothermic and the entropy has increased, so it will always be feasible.

$\Delta G = \Delta H - T\Delta S$

and therefore must be negative.

2 **ΔH is positive and ΔS is negative.**

Here the reaction can never be feasible, since the reaction is endothermic and the entropy has also decreased, so both the terms ΔH and $-T\Delta S$ are positive.

3 **ΔH is negative and ΔS is negative.**

This situation is more complicated. ΔH favours the reaction but $-T\Delta S$ is positive, thus resisting the change. The sign of ΔG depends on the relative size of the two terms. Although ΔS is likely to have a small value compared with ΔH, when ΔS is multiplied by the temperature in K it could be similar to ΔH. There is no simple answer other than to do the arithmetic. It should be noted that the higher the temperature, the more positive the value of $-T\Delta S$, so reactions in which both ΔH and ΔS are negative are more likely to be feasible at lower temperatures. Of course, low temperatures will probably mean a very slow rate of reaction and may also mean that liquids freeze, making a reaction impossible to achieve.

4 **ΔH is positive and ΔS is positive.**

The reaction will be opposed by ΔH but favoured by ΔS. The outcome can be determined only by establishing the relative sizes of the two terms in the expression for ΔG. The reaction will be favoured by increasing the temperature to make $T\Delta S$ more positive.

Tip

It is beyond the requirements of A Level to explain the reasoning that Gibbs applied to arrive at this equation, but it is essential that you can remember and use the equation $\Delta G^{\ominus} = \Delta H^{\ominus} - T\Delta S^{\ominus}$.

In this circumstance, a reaction that is not feasible under standard conditions of 298 K may be possible at a higher temperature – for example, the decomposition of a solid. Another example is the solubility of sodium chloride (page 68): ΔH has a value of +24 kJ mol^{-1} but the positive entropy value when multiplied by the temperature makes ΔG negative and so the process is feasible.

Calculating free energy changes

Free energy changes can be calculated in the same way as enthalpy or entropy changes.

$$\Delta G = \sum(\text{Gibbs free energy of products}) - \sum(\text{Gibbs free energy of reactants})$$

Example 3

Use the following standard free energies of formation to calculate the free energy change for the reaction:

$$C_2H_4(g) + HCl(g) \rightarrow C_2H_5Cl(g)$$

$\Delta_f G^{\ominus}(C_2H_4) = 68.1 \text{ kJ mol}^{-1}$

$\Delta_f G^{\ominus}(HCl) = -95.3 \text{ kJ mol}^{-1}$

$\Delta_f G^{\ominus}(C_2H_5Cl) = -59.4 \text{ kJ mol}^{-1}$

Answer

$\Delta G = \sum(\text{Gibbs free energy of products}) - \sum(\text{Gibbs free energy of reactants})$

$\Delta G = -59.4 - (68.1 - 95.3) = -59.4 + 27.2 = -32.2 \text{ kJ mol}^{-1}$

Test yourself

12 Calculate the free energy change for the following reaction:

$$C_6H_6O_6 \rightarrow 2C_2H_5OH + 2CO_2$$

when $\Delta_f G^{\ominus}(C_6H_6O_6) = -910.5 \text{ kJ mol}^{-1}$; $\Delta_f G^{\ominus}(C_2H_5OH) = -174.1 \text{ kJ mol}^{-1}$; $\Delta_f G^{\ominus}(CO_2) = -394.4 \text{ kJ mol}^{-1}$

Equilibrium

If ΔG is zero, then the system will be at equilibrium. Even if this does not occur under standard conditions, if both ΔH and ΔS are positive, then changing the temperature could allow the reaction to achieve equilibrium.

At equilibrium $\Delta G = 0$.

Since $\Delta G = \Delta H - T\Delta S$ this means that $\Delta H - T\Delta S = 0$

Therefore, $\Delta H = T\Delta S$

and the temperature, T, at which the reaction just reaches equilibrium is $\Delta H / \Delta S$.

A calculation to determine this temperature is illustrated in Example 4. An assumption is made that the values of ΔH and ΔS do not change as the temperature is changed. In fact, unless there is a change in state, entropies and enthalpies of formation do not change significantly and so calculations based on this assumption are normally valid.

Example 4

Use the data below to calculate the temperature at which the reaction

$$2NO_2(g) \rightarrow 2NO(g) + O_2(g)$$

reaches equilibrium.

$\Delta_f H^{\ominus}(NO) = 90.4 \text{ kJ mol}^{-1}$

$\Delta_f H^{\ominus}(NO_2) = 33.2 \text{ kJ mol}^{-1}$

$S^{\ominus}(NO) = 210.5 \text{ J mol}^{-1}\text{K}^{-1}$

$S^{\ominus}(NO_2) = 240.0 \text{ J mol}^{-1}\text{K}^{-1}$

$S^{\ominus}(O_2) = 204.9 \text{ J mol}^{-1}\text{K}^{-1}$

Answer

$\Delta H = \sum(\text{enthalpy of products}) - \sum(\text{enthalpy of reactants})$

ΔH for the reaction is $(2 \times 90.4) - (2 \times 33.2) = +114.4 \text{ kJ mol}^{-1}$

$\Delta S = \sum(\text{entropy of products}) - \sum(\text{entropy of reactants})$

$\Delta S = ((2 \times 210.5) + 204.9) - (2 \times 240.0) = +145.9 \text{ J mol}^{-1}\text{K}^{-1} =$
$+0.1459 \text{ kJ mol}^{-1}\text{K}^{-1}$

(Under standard conditions this gives $\Delta G = 114.4 - 298(0.1459) = 70.9 \text{ kJ mol}^{-1}$. So the reaction is not feasible.)

At equilibrium, $\Delta G = 0$

So, $T = \Delta H / \Delta S$

Therefore to achieve equilibrium, $T = 114.4/0.1459 = 784 \text{ K}$ or $511\,°C$

Tip

Remember that you must convert ΔS from J into kJ when calculating ΔG.

Test yourself

13 The values of the standard changes for the following reaction

$$4CuO(s) \rightarrow 2Cu_2O(s) + O_2(g)$$

are $\Delta H^{\ominus} = +287.4 \text{ kJ mol}^{-1}$ and $\Delta S^{\ominus} = +232.5 \text{ J mol}^{-1}\text{K}^{-1}$
   a) Calculate the standard free energy change for the reaction.
   b) Calculate the temperature in °C at which the reaction will just reach equilibrium.

Activity

The thermal stability of Group 2 carbonates

The decomposition of Group 2 metal carbonates is used on a large scale to make oxides such as magnesium and calcium oxides. The carbonates of Group 2 metals do not decompose at room temperature. They do decompose on heating:

$$MgCO_3(s) \rightarrow MgO(s) + CO_2(g)$$

$\Delta H^{\ominus} = +117\,kJ\,mol^{-1}; \Delta S^{\ominus} = +175\,J\,mol^{-1}\,K^{-1}$

$$BaCO_3(s) \rightarrow BaO(s) + CO_2(g)$$

$\Delta H^{\ominus} = +268\,kJ\,mol^{-1}; \Delta S^{\ominus} = +172\,J\,mol^{-1}\,K^{-1}$

1 Why does the entropy increase when a Group 2 carbonate decomposes?
2 a) Calculate the free energy change for the room temperature (298 K) decomposition of:
   i) magnesium carbonate
   ii) barium carbonate.
  b) Are these two compounds stable or unstable relative to decomposition into their oxide and carbon dioxide at room temperature?
3 Assuming that $\Delta H^{\ominus}$ and $\Delta S^{\ominus}$ for the reactions do not vary with temperature, estimate the temperatures at which the two decomposition reactions become feasible.
4 Moving down Group 2, do the metal carbonates become more stable or less stable relative to decomposition into the oxide and carbon dioxide?

Figure 4.14 Crystals of dolomite (foreground) and magnesite (background) from Brazil. Dolomite is a calcium–magnesium carbonate rock. Magnesite is pure magnesium carbonate. Large crystals of magnesite are rare.

Chemists seek to explain the trend in thermal stability of the Group 2 carbonates by analysing the energy changes.

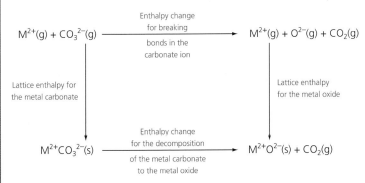

Figure 4.15 An energy cycle for the decomposition of the carbonate of a Group 2 metal, M.

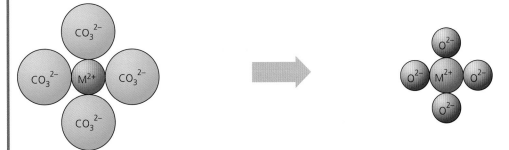

Figure 4.16 Decomposition of the Group 2 carbonate into its oxide.

5 Which of the enthalpy changes in Figure 4.15 are exothermic and which are endothermic?

6 Why is the lattice enthalpy of magnesium oxide more negative than the lattice enthalpy of barium oxide?

7 Why is the lattice enthalpy of magnesium oxide more negative than the lattice enthalpy of magnesium carbonate?

8 Why is the difference between the lattice enthalpies of the metal carbonates and oxides significant in explaining the trend in thermal stability of the Group 2 carbonates?

Practice questions

Multiple choice questions 1–10

1 Which one of the following equations represents the 3rd ionisation enthalpy of potassium ions?

A $K^{2+}(s) \rightarrow K^{3+}(s) + e^-$

B $K^{2+}(g) \rightarrow K^{3+}(g) + e^-$

C $K^{3+}(s) \rightarrow K^{4+}(s) + e^-$

D $K^{3+}(g) \rightarrow K^{4+}(g) + e^-$ (1)

2 Which one of the following equations represents the lattice enthalpy of 1 mol of sodium oxide?

A $2Na^+(g) + O^{2-}(g) \rightarrow Na_2O(g)$

B $Na^+(g) + \frac{1}{2}O^{2-}(g) \rightarrow \frac{1}{2}Na_2O(s)$

C $Na^+(s) + \frac{1}{2}O^{2-}(g) \rightarrow Na_2O(s)$

D $2Na^+(g) + O^{2-}(g) \rightarrow Na_2O(s)$ (1)

3 Which one of the following will have the highest numerical value (ignoring the sign) for its lattice enthalpy?

A NaCl

B KCl

C Na_2O

D K_2O (1)

4 Which one of the following would give the enthalpy of solution of potassium bromide?

A the lattice enthalpy of potassium bromide + the enthalpies of hydration of the ions of potassium and bromide

B the lattice enthalpy of potassium bromide − the enthalpies of hydration of the ions of potassium and bromide

C the enthalpies of hydration of the ions of potassium and bromide − the lattice enthalpy of potassium bromide

D − the enthalpies of hydration of the ions of potassium and bromide − the lattice enthalpy of potassium bromide. (1)

5 Which of the following describes the enthalpy of atomisation of chlorine ($\Delta_{at}H$) and the bond enthalpy of chlorine ($\Delta_{bond}H$)?

A $\Delta_{at}H = \Delta_{bond}H$

B $\Delta_{at}H = 2\Delta_{bond}H$

C $\Delta_{at}H = \frac{1}{2}\Delta_{bond}H$

D The two enthalpies are not directly related to each other. (1)

6 An endothermic reaction which has a negative value for its entropy change is

A never feasible

B always feasible

C more feasible at low temperatures

D more feasible at high temperatures. (1)

7 Which one of the following gives the shape of the graph that would be obtained if ΔG (y-axis) was plotted against temperature for the following reaction?

$$2C(s) + O_2(g) \rightarrow 2CO(g)$$

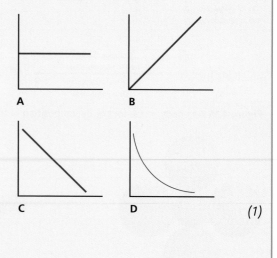

A B

C D (1)

Use the key below to answer questions 8, 9 and 10.

A	B	C	D
1, 2 & 3 correct	1 & 2 correct	2 & 3 correct	1 only correct

8 Which of the following have enthalpy values which are negative?
   1 ionisation enthalpies
   2 lattice enthalpies
   3 the formation of covalent bonds. *(1)*

9 The standard enthalpy of formation of ethene is endothermic and the standard enthalpy of formation of ethane is exothermic.
   Which of the following describe this reaction?

   $C_2H_4(g) + H_2(g) \rightarrow C_2H_6(g)$

   1 The entropy decreases as the reaction takes place.
   2 The reaction is exothermic.
   3 ΔG will be more negative at 100 °C than at 25 °C. *(1)*

10 Which of the following have values for ΔH and ΔS which are both negative?
   1 the formation of a crystal lattice from its gaseous ions
   2 the conversion of graphite into diamond
   3 the melting of ice *(1)*

11 The lattice enthalpy of LiF is $-1031\,kJ\,mol^{-1}$ and that of LiI is $-759\,kJ\,mol^{-1}$.
   Why is the lattice enthalpy of LiI less exothermic than the lattice enthalpy of LiF? *(2)*

12 Copy and complete the Born–Haber cycle for the formation of sodium chloride from its elements.

$Na^+(g) + Cl(g)$

$Na(g) + Cl(g)$ $Na^+(g) + Cl^-(g)$

$Na(g) + \frac{1}{2}Cl_2(g)$

$Na(s) + \frac{1}{2}Cl_2(g)$

$Na^+Cl^-(s)$

   a) Label each of the enthalpy changes.
   b) Use the data in the table to calculate the lattice enthalpy of sodium chloride.

	Enthalpy change/ kJ mol⁻¹
Enthalpy of atomisation of sodium	+108
1st ionisation enthalpy of sodium	+500
Enthalpy of atomisation of chlorine	+122
1st electron affinity of chlorine	−364
Enthalpy of formation of sodium chloride	−411

(8)

13 Use the data in the table to:
   a) construct a Born–Haber cycle for the formation of potassium oxide from its elements
   b) calculate the lattice enthalpy of potassium oxide.

	Enthalpy change / kJ mol⁻¹
Enthalpy of atomisation of potassium	+89.5
1st ionisation enthalpy of potassium	+420.0
Enthalpy of atomisation of oxygen	+249.4
1st electron affinity of oxygen	−141.4
2nd electron affinity of oxygen	+790.8
Enthalpy of formation of potassium oxide	−361.5

(9)

14 Silver fluoride, AgF, is a reasonably stable compound of silver, but gold fluoride, AuF, does not exist under normal conditions. It has been estimated that gold fluoride as AuF would be expected to have a lattice enthalpy of $-772.0\,kJ\,mol^{-1}$.
   a) Use the data in the table to construct Born–Haber cycles and calculate the enthalpies of formation of AgF and AuF.

	Enthalpy change/kJ mol⁻¹
Enthalpy of atomisation of silver	+286.2
1st ionisation enthalpy of silver	+730.0
Enthalpy of atomisation of fluorine	+79.1
1st electron affinity of fluorine	−332.6
Lattice enthalpy of silver(I) fluoride	−943.0
Enthalpy of atomisation of gold	+369.6
1st ionisation enthalpy of gold	+890.1
Lattice enthalpy of gold(I) fluoride	−772.0

   b) Comment on how your answers to part a) may explain why AgF is stable at room temperature but AuF is not. *(8)*

Ion	Li^+	Na^+	K^+	Mg^{2+}	Al^{3+}	Cl^-	Br^-	I^-	O^{2-}
Atomic Radius /nm	−545	−418	−351	−1923	−2537	−338	−304	−261	−937

a) How is the enthalpy change of hydration affected by increasing ionic charge?

b) List an appropriate series of ions and their enthalpy changes of hydration to illustrate your conclusion in part **a)**.

c) How is the enthalpy change of hydration affected by increasing ionic radius?

d) List an appropriate series of ions, their radii and their enthalpy changes of hydration to illustrate your conclusion in part **c)**. *(4)*

16 When calcium chloride dissolves in water, the process can be represented by the equation:

$$CaCl_2(s) + aq \rightarrow Ca^{2+}(aq) + 2Cl^-(aq)$$

The enthalpy change for this process is the enthalpy change of solution. Its value can be calculated from a Born–Haber cycle using the following data:

lattice enthalpy of calcium chloride = −2258 kJ mol^{-1}

enthalpy change of hydration of $Ca^{2+}(g)$ = −1653 kJ mol^{-1}

enthalpy change of hydration of $Cl^-(g)$ = −338 kJ mol^{-1}

a) Draw and label the Born–Haber cycle linking the enthalpy change of solution of calcium chloride with the enthalpy changes in the data above.

b) Use your Born–Haber cycle to calculate the enthalpy change of solution of calcium chloride. *(6)*

17 Use the data in the table to calculate the enthalpy of solution of:

a) silver chloride

b) silver iodide

	Enthalpy change/ kJ mol^{-1}
Enthalpy change of hydration of Ag^+	−510
Enthalpy change of hydration of Cl^-	−338
Enthalpy change of hydration of I^-	−261
Lattice enthalpy of AgCl	−905
Lattice enthalpy of AgI	−889

c) To what extent can your answers to parts **a)** and **b)** explain the solubilities of AgCl and AgI?

Solubilities: AgCl = 1.35 x 10^{-6} g per 100 cm^3 H$_2$O, AgI = 1.11 x 10^{-8} g per 100 cm^3 H$_2$O. *(6)*

18 For each of the following, predict whether the reaction will have a positive or negative value for the entropy change. Give a reason for your answer.

a) $H_2O(g) \rightarrow H_2O(s)$

b) $NaOH(s) \rightarrow NaOH(aq)$

c) $2Mg(s) + O_2(g) \rightarrow 2MgO(s)$

d) $2SO_2(g) + O_2(g) \rightarrow 2SO_3(g)$

e) $C_5H_{12}(g) + 8O_2(g) \rightarrow 5CO_2(g) + 6H_2O(g)$ *(5)*

19 Substances A, B and C are iodine, ammonia and methanol (but not necessarily in that order). Given the following entropies, identify which substance corresponds to which letter. Explain your answers.

A = 192.5 J mol^{-1} K^{-1}

B = 58.4 J mol^{-1} K^{-1}

C = 127.2 J mol^{-1} K^{-1} *(4)*

20 Calculate the entropy change when sodium reacts with oxygen.

$S^{\ominus}(Na) = 51.0$ J mol^{-1} K^{-1}

$S^{\ominus}(O_2) = 204.9$ J mol^{-1} K^{-1}

$S^{\ominus}(Na_2O) = 72.8$ J mol^{-1} K^{-1} *(3)*

21 Calculate the entropy change when 1 mol of ethane is burnt in excess oxygen.

$$C_2H_6(g) + 3\tfrac{1}{2}O_2(g) \rightarrow 2CO_2(g) + 3H_2O(l)$$

$S^{\ominus}(C_2H_6) = 229.5$ J mol^{-1} K^{-1}

$S^{\ominus}(O_2) = 204.9$ J mol^{-1} K^{-1}

$S^{\ominus}(CO_2) = 213.8$ J mol^{-1} K^{-1}

$S^{\ominus}(H_2O(l)) = 70.0$ J mol^{-1} K^{-1} *(4)*

22 a) Write a balanced equation for the catalytic reaction of ammonia with oxygen to form nitrogen monoxide, NO, and steam.

b) Calculate the entropy change of this reaction using the following standard entropy changes.

$S^\ominus(NH_3) = 192.3\,J\,mol^{-1}K^{-1}$

$S^\ominus(O_2) = 204.9\,J\,mol^{-1}K^{-1}$

$S^\ominus(NO) = 210.7\,J\,mol^{-1}K^{-1}$

$S^\ominus(H_2O(g)) = 188.7\,J\,mol^{-1}K^{-1}$

c) Comment on the value you obtained in part **b)**. *(5)*

23 Calculate the free energy change for each of the following reactions:

a) the addition of hydrogen to ethene

b) the addition of steam to ethene.

$\Delta_f G^\ominus(C_2H_4) = 68.1\,kJ\,mol^{-1}$

$\Delta_f G^\ominus(H_2) = 0\,kJ\,mol^{-1}$

$\Delta_f G^\ominus(H_2O(g)) = -228.6\,kJ\,mol^{-1}$

$\Delta_f G^\ominus(C_2H_6) = -32.8\,kJ\,mol^{-1}$

$\Delta_f G^\ominus(C_2H_5OH) = -174.9\,kJ\,mol^{-1}$ *(4)*

24 Explain whether or not an exothermic reaction which is not feasible at room temperature becomes feasible at a higher temperature if the entropy change of the reaction is negative? *(5)*

25 Consider the reaction of magnesium with oxygen:

$2Mg(s) + O_2(g) \rightarrow 2MgO(s)$

$\Delta H^\ominus = -602\,kJ\,mol^{-1}; \Delta S^\ominus = -217\,J\,mol^{-1}K^{-1}$

a) Why does the entropy of the system decrease?

b) Show why the reaction of magnesium with oxygen is feasible at 298 K despite the decrease in the entropy. *(5)*

26 a) Write a balanced equation for the synthesis of methanol from carbon monoxide and hydrogen.

b) Calculate ΔS and ΔH for the synthesis using the following data.

$\Delta_f H^\ominus(CH_3OH) = -239.1\,kJ\,mol^{-1}$

$\Delta_f H^\ominus(CO) = -110.5\,kJ\,mol^{-1}$

$S^\ominus(CH_3OH) = 239.7\,J\,mol^{-1}K^{-1}$

$S^\ominus(CO) = 197.6\,J\,mol^{-1}K^{-1}$

$S^\ominus(H_2) = 130.6\,J\,mol^{-1}K^{-1}$

c) Work out the temperature in °C at which the synthesis ceases to be feasible. *(6)*

27 At high temperature, it is possible to react carbon with steam to produce a mixture of carbon monoxide and hydrogen known as water gas. This is a useful fuel. The equation for the reaction is:

$C(s) + H_2O(g) \rightarrow CO(g) + H_2(g)$

Use the enthalpy and entropy values below to calculate the temperature in °C at which an equilibrium is established between the reactants and products.

$\Delta_f H^\ominus(H_2O(g)) = -241.8\,kJ\,mol^{-1}$

$\Delta_f H^\ominus(CO) = -110.5\,kJ\,mol^{-1}$

$S^\ominus(C) = 5.7\,J\,mol^{-1}K^{-1}$

$S^\ominus(H_2O(g)) = 188.7\,J\,mol^{-1}K^{-1}$

$S^\ominus(CO) = 197.9\,J\,mol^{-1}K^{-1}$

$S^\ominus(H_2) = 130.6\,J\,mol^{-1}K^{-1}$ *(4)*

Challenge

28 Answer the following questions about xenon fluorides.

a) In general, the noble gases do not form stable ionic compounds. However, chemists seeking to create a compound containing a noble gas cation thought that one possibility might be a fluoride, M^+F^-.

i) Why might the fluoride be a sensible initial choice?

ii) Explain why xenon would be the best noble gas to consider as a possibility.

b) Use the data in the table below to calculate the enthalpy change for the process:

	Enthalpy change/kJ mol⁻¹
1st ionisation enthalpy of xenon	+1170.0
Enthalpy of atomisation of fluorine	+79.1
1st electron affinity of fluorine	-332.6

c) Use the internet or a data book to find some values for the lattice energies of fluorides to explain why the formation of $Xe^+F^-(s)$ is unlikely to happen.

d) A solid compound, XeF_2, can be prepared by irradiating xenon and fluorine with UV. This compound sublimes at 114°C.

What two factors suggest that the bonding in this compound is covalent?

e) The covalent molecule XeF_4 has also been prepared.

Deduce the shape of the XeF_4 molecule.

f) XeF_4 reacts with water according to the equation:

$$6XeF_4 + 12H_2O \rightarrow 4Xe + 2XeO_3 + 24HF + 3O_2$$

Discuss whether or not this reaction is a redox reaction.

Justify your answer.

g) Explain the likely shape of the covalent molecule, XeO_3.

h) The mass spectrum of XeF_2 has significant peaks for the unfragmented molecular ion as detailed in the table.

Mass peak %	166	167	168	169	170	172	174
Occurrence	1.9	26.4	4.1	21.2	26.9	10.4	8.9

(The remaining 0.2% results from extremely small peaks at 162 and 164).

Calculate the relative molecular mass of the molecule.

i) Fluorine has only one isotope. What is the relative atomic mass of xenon? *(19)*

T/K	$\Delta_f G(CO)/kJ\,mol^{-1}$	$\Delta_f G(CO_2)/kJ\,mol^{-1}$	$\Delta_f G(Al_2O_3)/kJ\,mol^{-1}$	$\Delta_f G(Fe_2O_3)/kJ\,mol^{-1}$	$\Delta_f G(CuO)/kJ\,mol^{-1}$
298	−137.2	−394.5	−1582.4	−740.4	−127.7
500	−155.4	−395.1	−1519.2	−685.0	−109.1
750	−177.8	−395.9	−1440.9	−616.4	−86.0
1000	−200.3	−396.7	−1362.7	−547.9	−63.0
1250	−222.7	−397.5	−1284.4	−479.3	−39.9
1500	−245.1	−398.3	−1206.1	−410.7	−16.8
1750	−267.6	−399.1	−1127.9	−342.1	6.2
2000	−290.0	−399.9	−1049.6	−273.5	29.3
2250	−312.4	−400.7	−971.3	−204.9	52.4
2500	−334.9	−401.5	−893.1	−136.3	75.4
2750	−357.3	−402.3	−814.8	−67.7	98.5
3000	−379.8	−403.1	−736.6	0.8	121.6
3500	−424.6	−404.7	−580.0	138.0	167.7

29

Some metals are produced by the reduction of their oxides in a blast furnace using carbon (as coke) – for example, iron occurs naturally as an oxide ore. Other metals, for example zinc and copper, occur as sulfides that are roasted initially in air to form the oxide.

The oxide reduction takes place in a furnace. Care must be taken to provide an appropriate temperature to make the process effective. The reactions that occur within the furnace can be quite complex, particularly because carbon can be oxidised to either carbon monoxide or carbon dioxide depending on the air supply and the temperature.

The table above gives some details of the free energies of formation for some oxides at various temperatures.

a) Explain why, as the temperature is increased:
  i) $\Delta_f G(CO)$ becomes more negative
  ii) $\Delta_f G(CO_2)$ changes very little
  iii) $\Delta_f G$ for all three metal oxides becomes more positive.

b) Using the value of $\Delta_f G(CO)$ at 298 K show, by calculation, that the value at 2000 K is −290.0 kJ mol⁻¹. ($\Delta_f S(CO) = 89.8\,J\,mol^{-1}\,K^{-1}$)

c) Use appropriate values from the table to calculate the free energy change for the reaction
$CO + \frac{1}{2}O_2 \rightarrow CO_2$ at 500 K

d) The free energy change for the reaction
$CO + \frac{1}{2}O_2 \rightarrow CO_2$
at 750 K is −218.1 kJ mol⁻¹. Use this result and your result from part **c)** to calculate the values of ΔH and ΔS for this reaction. (Assume that ΔH and ΔS do not change with temperature.)

e) Estimate the temperature at which the reaction $CO + \frac{1}{2}O_2 \rightleftharpoons CO_2$ reaches equilibrium.

f) Copper metal was known to ancient civilisations and many artefacts made of bronze, which contains copper, have been found.
Suggest why copper could be produced so readily, while iron was almost unknown. (Some iron was obtained from meteorites but none was extracted from an ore.)

g) The blast furnaces used to produce iron operate at temperatures in excess of 1850 K. This is partly because iron melts at 1808 K and so the iron produced is molten and partly because the reduction of the oxide is only effective at temperatures in this region. Explain why the reduction of the oxide is effective only at temperatures in the region of 1850 K.

h) Aluminium is extracted by an electrolytic process. Explain why a method using reduction by carbon is not employed. How does this help to explain why aluminium was discovered only relatively recently (in 1808, by Humphry Davy)? *(25)*

Chapter 5

Redox and electrode potentials

Prior knowledge

In this chapter it is assumed that you can:
- assign an oxidation number to elements in compounds and ions
- understand and write ionic equations for:
  - precipitation reactions
  - the reactions of acids with oxides, hydroxides and carbonates
  - the reactions of hydrochloric acid or sulfuric acid with some metals
  - the reactions of halogens with halide ions
- recognise redox reactions as those involving electron transfer.

Test yourself on prior knowledge

1 Give the oxidation number of the element underlined in the following formulae:

a) $\underline{Cu}_2O$
b) $\underline{Fe}_2O_3$
c) $Ca\underline{I}_2$
d) $\underline{Cl}O_2$
e) $K\underline{Cl}O_3$
f) $H_2\underline{O}_2$
g) $K\underline{N}O_2$
h) $K\underline{Mn}O_4$

i) $K_2\underline{Cr}_2O_7$
j) $Na\underline{Br}O_3$
k) $\underline{Cu}Cl_4{}^{2-}$
l) $\underline{S}O_3{}^{2-}$
m) $\underline{N}O_2{}^{-}$
n) $\underline{Sn}Cl_6{}^{2-}$
o) $\underline{Zn}(OH)_4{}^{2-}$

2 Write ionic equations for the following reactions:

a) $BaCl_2(aq) + Na_2SO_4(aq) \rightarrow BaSO_4(s) + 2NaCl(aq)$
b) $AgNO_3(aq) + KI(aq) \rightarrow AgI(s) + KNO_3(aq)$
c) $Mg(NO_3)_2(aq) + 2NaOH(aq) \rightarrow Mg(OH)_2(s) + 2NaNO_3(aq)$
d) $Na_2CO_3(aq) + 2HCl(aq) \rightarrow 2NaCl(aq) + H_2O(l) + CO_2(g)$
e) $2KOH(aq) + H_2SO_4(aq) \rightarrow K_2SO_4(aq) + 2H_2O(l)$
f) $ZnO(s) + 2HCl(aq) \rightarrow ZnCl_2(aq) + H_2O(l)$
g) $Mg(s) + H_2SO_4(aq) \rightarrow MgSO_4(aq) + H_2(g)$
h) $Fe(s) + 2HCl(aq) \rightarrow FeCl_2(aq) + H_2(g)$
i) $Cl_2(g) + 2KI(aq) \rightarrow 2KCl(aq) + I_2(s)$

3 a) Explain why examples **2g), h)** and **i)** are described as redox reactions. In each case state what has been oxidised and what has been reduced.

b) The reaction of sodium with water forms sodium hydroxide and hydrogen. Explain why this is an example of a redox reaction.

c) Use the oxidation numbers and the equation below to show why this reaction is an example of a redox reaction in which the chlorine is both oxidised and reduced (a feature called disproportionation).

$$Cl_2(g) + 2NaOH(aq) \rightarrow NaCl(aq) + NaOCl(aq) + H_2O(l)$$

Ionic equations

Of all the types of reaction, redox reactions are perhaps the easiest to study. This is because it is possible to assess the energy change that occurs when electrons are transferred from one species to another during the reaction. This is done by allowing the electrons to flow through an external electrical circuit where the energy that they carry can be measured using a voltmeter (which measures energy in terms of joules per coulomb of electricity). The size of the voltage gives an indication of the energy difference between the reactants and products. First, however, you must become fluent at writing ionic equations which establish the species that are the source and those that are the recipient of the electrons.

An ionic equation is important because it indicates which particles have been involved in a reaction. The ions present that do not take part in the reaction are excluded from the equation.

To write ionic equations correctly, it is essential to remember the following points:

- Ionic substances that are solid do not have free-moving ions and, therefore, their ions cannot react independently. In an ionic equation their complete formulae must be given.

- Soluble compounds of metals and also strong acids in aqueous solution always split into their ions. These ions react independently of each other.

- Covalent compounds exist as complete molecules and are always shown as complete entities in ionic equations.

- The total charge on each side of the ionic equation **must** balance. It is important to balance both symbol **and** charge.

Redox reactions

You have already met some examples of redox reactions, but it is now necessary to consider a range of other examples. In most cases it is easier to work with ionic equations.

Example 1

Zinc and aqueous copper sulfate react to form aqueous zinc sulfate and copper metal:

$$Zn(s) + CuSO_4(aq) \rightarrow ZnSO_4(aq) + Cu(s)$$

Identify which species is reduced and which is oxidised.

Answer
The ionic equation is:

$$Zn(s) + Cu^{2+}(aq) + SO_4^{2-}(aq) \rightarrow Zn^{2+}(aq) + SO_4^{2-}(aq) + Cu(s)$$

The sulfate ions, $SO_4^{2-}(aq)$, are present on both sides of the equation and clearly do not take part in the reaction, so they are not included in the final ionic equation. This only shows the particles that have been involved in the reaction.

$$Zn(s) + Cu^{2+}(aq) \rightarrow Zn^{2+}(aq) + Cu(s)$$

The zinc is oxidised; each atom loses two electrons and becomes a zinc ion, Zn^{2+}. The electrons are taken up by a Cu^{2+} ion which is reduced to a copper atom, Cu.

The oxidation number of Zn is 0; it becomes +2 in the zinc ion. The oxidation number of copper in Cu^{2+} is +2; it becomes 0 in the metal.

Key terms

An **oxidising agent** accepts electrons from another reagent and during the course of the reaction is itself reduced.

A **reducing agent** gives electrons to another reagent and during the course of the reaction is itself oxidised.

Example 2

Aqueous iron(II) chloride reacts with chlorine to produce aqueous iron(III) chloride:

$$2FeCl_2(aq) + Cl_2(g) \rightarrow 2FeCl_3(aq)$$

Identify the **oxidising agent** and the **reducing agent**.

Answer

$$2Fe^{2+} + 4Cl^-(aq) + Cl_2(g) \rightarrow 2Fe^{3+} + 6Cl^-(aq)$$

There are four $Cl^-(aq)$ ions that do not take part in the reaction and the ionic equation is written as

$$2Fe^{2+}(aq) + Cl_2(g) \rightarrow 2Fe^{3+}(aq) + 2Cl^-(aq)$$

which shows only the particles that are involved in the reaction.

The $Fe^{2+}(aq)$ ion is oxidised to $Fe^{3+}(aq)$ and the $Cl_2(g)$ is reduced to $Cl^-(aq)$. Two electrons are transferred.

In terms of oxidation number, the oxidation number of iron in the Fe^{2+} ion is +2; it becomes +3 in Fe^{3+}. The oxidation number of chlorine in Cl_2 is 0; it becomes −1 in Cl^-.

Chlorine gas is the oxidising agent and Fe^{2+} is the reducing agent.

Example 3

Iodine and chlorine react together in a two-step reaction:

$$I_2(s) + Cl_2(g) \rightarrow 2ICl(l)$$
$$2ICl + 2Cl_2(g) \rightarrow I_2Cl_6(s)$$

Identify which species is reduced and which is oxidised.

Answer

These substances are all covalent, so ionic equations cannot be written.

Using oxidation numbers:

- In the first equation, the oxidation number of iodine starts at 0 and it is oxidised to +1; the oxidation number of chlorine starts at 0 and it is reduced to −1.

- In the second equation, iodine is oxidised further from +1 to +3; chlorine is again reduced from 0 to −1.

These examples are relatively straightforward, but they emphasise that, in order to recognise a redox reaction, it is sometimes helpful to think about the movement of electrons and sometimes easier to use oxidation numbers.

Test yourself

1 a) Write ionic equations for the following reactions:
   i) $Mg(s) + FeSO_4(aq) \rightarrow Fe(s) + MgSO_4(aq)$
   ii) $3Mg(s) + Fe_2(SO_4)_3(aq) \rightarrow 2Fe(s) + 3MgSO_4(aq)$
   iii) $Cl_2(g) + SnCl_2(aq) \rightarrow SnCl_4(aq)$
   iv) $2FeCl_3(aq) + SnCl_2(aq) \rightarrow SnCl_4(aq) + 2FeCl_2(aq)$
   b State what has been oxidised and what has been reduced in each of the examples above.

Activity

Redox reactions in the Space Shuttle

Unlike most vehicles on Earth, spacecraft must carry oxidising agents, as well as fuels. These mixtures of fuels plus oxidising agents are called propellants. In order to launch the Space Shuttle from ground level into the Earth's upper atmosphere, a solid propellant mixture of powdered aluminium and ammonium chlorate(VII), NH_4ClO_4, is used. The reaction involved is:

$$3Al(s) + 3NH_4ClO_4(s) \rightarrow Al_2O_3(s) + AlCl_3(s) + 3NO(g) + 6H_2O(g)$$

1 What is:
   a) the fuel
   b) the oxidising agent in the solid propellant?
2 What is the oxidation number of nitrogen in the NH_4^+ ion and what is the oxidation number of chlorine in the ClO_4^- ion?
3 a) Which elements are oxidised in the reaction?
   b) State the change in oxidation numbers of these elements.
4 Which element is reduced in the reaction, and what is the change in its oxidation number?
5 What actually propels the Space Shuttle?

Figure 5.1 The Space Shuttle *Endeavour* takes off on its journey to the International Space Station.

Propulsion of the Shuttle from the upper atmosphere into orbit is achieved using a mixture of liquid hydrogen and liquid oxygen. Once the hydrogen and oxygen are ignited, they continue to vaporise and burn continuously, producing a clean water-vapour exhaust.

▶▶▶ 87

6 Write an equation for the combustion of liquid hydrogen and oxygen during the second stage.

Once in orbit, a propulsion system is needed for manoeuvrability – one in which the fuel and oxidising agent ignite spontaneously on mixing and which can be readily started and stopped. In this stage, the fuel is liquid methylhydrazine (CH_3NHNH_2) with liquid dinitrogen tetroxide (N_2O_4) as the oxidiser. The reaction produces water vapour, carbon dioxide and nitrogen.

7 Write a balanced equation for the reaction between CH_3NHNH_2 and N_2O_4 in this stage.

8 Explain why CH_3NHNH_2 acts as the reducing agent and why N_2O_4 acts as the oxidising agent during the reaction.

Half-equations

In order to move on to more examples of redox reactions, the use of ionic half-equations is required. These make it clear which reagent is supplying electrons and which is receiving them.

Using the example of the reaction of zinc and aqueous copper sulfate,

$$Zn(s) + Cu^{2+}(aq) \rightarrow Zn^{2+}(aq) + Cu(s)$$

the half-equations are:

$$Zn(s) \rightarrow Zn^{2+}(aq) + 2e^-$$
$$Cu^{2+}(aq) + 2e^- \rightarrow Cu(s)$$

For Fe^{2+} and Cl_2,

$$2Fe^{2+}(aq) + Cl_2(g) \rightarrow 2Fe^{3+}(aq) + 2Cl^-(aq)$$

the half-equations are:

$$Fe^{2+}(aq) \rightarrow Fe^{3+}(aq) + e^-$$
$$2e^- + Cl_2(g) \rightarrow 2Cl^-(aq)$$

This reaction illustrates another use of half-equations, which is to provide a systematic method of constructing a balanced overall equation. In this case, it may seem to be an unnecessarily involved way of arriving at the answer, but it is often the best approach when the equations are more complex. The point to appreciate is that the number of electrons released by the reducing agent (Fe^{2+} in this example) must be the same as the number required by the oxidising agent (Cl_2). In the first half-equation Fe^{2+} supplies one electron as it is oxidised to Fe^{3+}, but each Cl_2 molecule requires two electrons to be reduced to two chloride ions. Therefore, it is necessary to have two Fe^{2+} ions for every one Cl_2 molecule.

The best way of seeing this clearly is to rebalance the half-equations until the number of electrons is the same. Here, this would be:

$$2Fe^{2+}(aq) \rightarrow 2Fe^{3+}(aq) + 2e^-$$
$$2e^- + Cl_2(g) \rightarrow 2Cl^-(aq)$$

The overall equation is then obtained by adding the two half-equations together, which eliminates the electrons.

$$2Fe^{2+}(aq) + Cl_2(g) \rightarrow 2Fe^{3+}(aq) + 2Cl^-(aq)$$

Note though that, as in Example 3 (page 86), this method is not appropriate when the reaction is wholly between covalent molecules.

Balancing charge in equations

When balancing ionic equations it is essential that charge, as well as symbols, balance. For example in the equation:

$$2Fe^{2+}(aq) + Cl_2(g) \rightarrow 2Fe^{3+}(aq) + 2Cl^-(aq)$$

The left-hand side of the equation has a total charge of +4 from the two Fe^{2+} ions.

The right-hand side of the equation also has a charge of +4. In this case, the two Fe^{3+} ions contribute a charge of +6 but the two Cl^- ions add a charge of −2, giving a net charge of +4.

An example of a more complex redox reaction will help to illustrate the ideas discussed.

Potassium manganate(VII) is an oxidising agent that readily converts $Fe^{2+}(aq)$ to $Fe^{3+}(aq)$. It is used in aqueous solution and must be acidified to provide a source of hydrogen ions, which form water with the oxygen in the MnO_4^-. During the oxidation, the acidified manganate(VII) ion, $MnO_4^-(aq)$, is reduced to $Mn^{2+}(aq)$ ions and H_2O. Potassium manganate(VII) in acidic solution is a powerful oxidising agent that can oxidise a range of inorganic ions as well as organic compounds.

So the basic process is:

$$MnO_4^-(aq) + H^+(aq) \rightarrow Mn^{2+}(aq) + H_2O(l)$$

However, this is not a balanced equation. It must be corrected to:

$$MnO_4^-(aq) + 8H^+(aq) \rightarrow Mn^{2+}(aq) + 4H_2O(l)$$

The equation must also be balanced for charge by the addition of electrons to the equation. As it stands, the total charge on the left-hand side is made up of 1− (from the MnO_4^-) and 8+ (from the $8H^+$) = 7+.

On the right-hand side, the charge is 2+ from the Mn^{2+} ion.

So $5e^-$ must be supplied, in order that the total charge on each side of the half-equation is 2+. This is achieved by including in the half-equation electrons which are supplied by the $Fe^{2+}(aq)$ ions in the reaction.

$$5e^- + MnO_4^-(aq) + 8H^+(aq) \rightarrow Mn^{2+}(aq) + 4H_2O(l)$$

| 5− | 1− | | 8+ | 2+ | 0 |

The half-equation for the oxidation of Fe^{2+} to Fe^{3+} is:

$$Fe^{2+}(aq) \rightarrow Fe^{3+}(aq) + e^-$$

The two half-equations can now be combined. $5Fe^{2+}$ are needed to supply the $5e^-$ required by each MnO_4^-:

$$5Fe^{2+}(aq) \rightarrow 5Fe^{3+}(aq) + 5e^-$$

Tip

This is an important reaction that can be performed as a titration.

Tip

The role of hydrogen ions in the reduction of anions containing oxygen is to convert the oxygen into water. (In a few rare cases, oxygen gas is produced.)

Tip

It is easy to make a mistake somewhere in the procedure. Making sure that the total charge on either side of the equation is the same is a useful check to confirm that the working has been completed correctly.

The overall equation is:

$$5Fe^{2+}(aq) + MnO_4^-(aq) + 8H^+(aq) \rightarrow 5Fe^{3+}(aq) + Mn^{2+}(aq) + 4H_2O(l)$$

10+   1−   8+   15+   2+

You should notice that the total charge on the left-hand side of the overall balanced equation must equal the total charge on the right-hand side. Here it does, as both sides have a net charge of 17+.

This can also be balanced using oxidation numbers. The oxidation number of the iron ion is increased from +2 to +3. The oxidation number of Mn in MnO_4^- is +7. During the reaction this is reduced to +2 in the Mn^{2+} ion. The relationship between the amounts of Fe^{2+} and MnO_4^- required can be deduced from the change in oxidation number. The oxidation number of manganese is reduced by 5 (+7 to +2). The reaction requires five times as much Fe^{2+} as MnO_4^- because the oxidation number of iron only increases by 1.

Ionic equations are the usual way of writing equations for redox reactions. There is no need to worry about the actual chemicals that supply the ions. For example, this reaction would work with any soluble iron(II) salt reacting with any soluble metal manganate(VII) in the presence of an acid.

In summary the steps necessary to produce an ionic equation for reactions in acid solution are as follows:

1 Write down the ions (atoms or molecules) to create the two half-equations for the reaction.

2 Balance the numbers of each atom in each half-equation.

3 If required, add H^+ and/or H_2O to balance the oxygen in the half-equations (be careful to check first that none of the oxygen is released as a gas).

4 Add up the total charge on either side of the equation and add electrons as necessary to make sure the charge is the same on both sides of each half-equation.

5 Multiply the quantities involved in each half-equation to make the number of electrons being supplied by one half-equation the same as the number being received by the other half-equation.

6 Add the two equations together, excluding the electrons.

7 Cancel out H^+ and H_2O if they appear on both sides of the equation.

8 Check the total charge on each side of the equation to make sure they are the same and you haven't made a mistake!

Tip

The procedure to generate equations may at first appear daunting but, **with practice**, it can become automatic and will eventually seem quite straightforward.

If the reaction is carried out in the presence of alkali (which is very much less usual) the same procedure is adopted but the ions added in step 3 are OH^- and H_2O.

For example iron(II) hydroxide can be oxidised using the oxygen in the air to form iron(III) hydroxide.

The half equations required are:

$$Fe(OH)_2(s) + OH^-(aq) \rightarrow Fe(OH)_3(s) + e^-$$

$$O_2(g) + 2H_2O(l) + 4e^- \rightarrow 4OH^-(aq)$$

and the overall equation is then obtained by multiplying the first equation by 4 and combining the two half-equations.

$$4Fe(OH)_2(s) + 4OH^-(aq) + O_2(g) + 2H_2O(l) \rightarrow 4Fe(OH)_3(s) + 4OH^-(aq)$$

or

$$4Fe(OH)_2(s) + O_2(g) + 2H_2O(l) \rightarrow 4Fe(OH)_3(s)$$

Unusually, the OH^- cancel out in the final equation. (This will probably not happen in other cases.)

Example 4

The reaction between potassium manganate(VII) and iodide ions can be summarised by two half-equations:

$$MnO_4^-(aq) + H^+(aq) \rightarrow Mn^{2+}(aq) + H_2O(l)$$
$$I^-(aq) \rightarrow I_2(s)$$

Deduce the overall equation for the reaction.

Answer

Steps 1, 2 and 3

Balancing the half-equations gives:

$$MnO_4^-(aq) + 8H^+(aq) \rightarrow Mn^{2+}(aq) + 4H_2O(l)$$
$$2I^-(aq) \rightarrow I_2(s)$$

Step 4

Adding electrons to make sure that the charges on either side of each half-equation are the same gives:

$$5e^- + MnO_4^-(aq) + 8H^+(aq) \rightarrow Mn^{2+}(aq) + 4H_2O(l)$$
$$2I^-(aq) \rightarrow I_2(s) + 2e^-$$

Step 5

To balance the electrons, the first half-equation must be multiplied by 2 and the second by 5, so that $10e^-$ are exchanged.

$$10e^- + 2MnO_4^-(aq) + 16H^+(aq) \rightarrow 2Mn^{2+}(aq) + 8H_2O(l)$$
$$10I^-(aq) \rightarrow 5I_2(s) + 10e^-$$

Steps 6 and 7

These are added together to give the overall balanced equation:

$$10I^-(aq) + 2MnO_4^-(aq) + 16H^+(aq) \rightarrow 5I_2(s) + 2Mn^{2+}(aq) + 8H_2O(l)$$

Step 8 The total charge on each side of the equation is 4+.

Tip

Don't forget to check the total charge on both sides of the equation.

Example 5

The reaction of potassium dichromate(VI) and tin(II) ions can be summarised in two half-equations:

$$Cr_2O_7^{2-}(aq) + H^+(aq) \rightarrow Cr^{3+}(aq) + H_2O(l)$$
$$Sn^{2+}(aq) \rightarrow Sn^{4+}(aq)$$

Deduce the overall equation for the reaction.

▶▶▶

Answer

Steps 1, 2 and 3

Balancing the half-equations gives:

$$Cr_2O_7^{2-}(aq) + 14H^+(aq) \rightarrow 2Cr^{3+}(aq) + 7H_2O(l)$$
$$Sn^{2+}(aq) \rightarrow Sn^{4+}(aq)$$

Step 4

Adding electrons to make sure that the charges on either side of each half-equation are the same gives:

$$6e^- + Cr_2O_7^{2-}(aq) + 14H^+(aq) \rightarrow 2Cr^{3+}(aq) + 7H_2O(l)$$
$$Sn^{2+}(aq) \rightarrow Sn^{4+}(aq) + 2e^-$$

Step 5

To balance the electrons the second half-equation must be multiplied by 3 so that $6e^-$ are released:

$$6e^- + Cr_2O_7^{2-}(aq) + 14H^+(aq) \rightarrow 2Cr^{3+}(aq) + 7H_2O(l)$$
$$3Sn^{2+}(aq) \rightarrow 3Sn^{4+}(aq) + 6e^-$$

Steps 6 and 7

These are added together to give the overall balanced equation:

$$3Sn^{2+}(aq) + Cr_2O_7^{2-}(aq) + 14H^+(aq) \rightarrow 3Sn^{4+}(aq) + 2Cr^{3+}(aq) + 7H_2O(l)$$

Step 8 The total charge on each side of the equation is 18+.

This agrees with the change in oxidation numbers. The oxidation number of tin has increased by 2 from +2 to +4; the oxidation number of chromium has decreased from +6 to +3. Since there are two chromium atoms in the dichromate, this is an overall reduction of 6, giving a ratio of $Sn^{2+}:Cr_2O_7^{2-}$ of 3:1.

Example 6

Iodine reacts with thiosulfate ions ($S_2O_3^{2-}$) to form iodide ions and tetrathionate ($S_4O_6^{2-}$) ions.

The two balanced half-equations are:

$$2S_2O_3^{2-}(aq) \rightarrow S_4O_6^{2-}(aq) + 2e^-$$
$$I_2(s) + 2e^- \rightarrow 2I^-(aq)$$

Deduce the overall equation for the reaction.

Answer

In this case no hydrogen ions are required so that the overall equation is:

$$2S_2O_3^{2-}(aq) + I_2(s) \rightarrow S_4O_6^{2-}(aq) + 2I^-(aq)$$

The total charge on each side of the equation is 4−.

Tip

The reaction may look obscure but, as will be seen in Chapter 6, it forms the basis of a useful titration method (see page 135).

Example 7

Sometimes a substance is oxidised to another species that contains more oxygen atoms. For example, SO_2 can be oxidised to SO_4^{2-} by MnO_4^- ions. In these cases the SO_2 uses water and generates hydrogen ions in the course of the reaction. The procedure to write the overall equation is the same.

Steps 1, 2, 3 and 4

$$SO_2(g) + 2H_2O(l) \rightarrow SO_4^{2-}(aq) + 4H^+(aq) + 2e^-$$

$$5e^- + 8H^+(aq) + MnO_4^-(aq) \rightarrow Mn^{2+}(aq) + 4H_2O(l)$$

Step 5

To balance the electrons the first half-equation must be multiplied by 5 and the second half-equation by 2 so that $10e^-$ are exchanged:

$$5SO_2(g) + 10H_2O(l) \rightarrow 5SO_4^{2-}(aq) + 20 H^+(aq) + 10e^-$$

$$10e^- + 16H^+(aq) + 2MnO_4^-(aq) \rightarrow 2Mn^{2+}(aq) + 8H_2O(l)$$

Step 6

When combined this gives:

$$5SO_2(g) + 10H_2O(l) + 16H^+(aq) + 2MnO_4^-(aq) \rightarrow 5SO_4^{2-}(aq) + 20 H^+(aq) + 2Mn^{2+}(aq) + 8H_2O(l)$$

Step 7

When H^+ and H_2O are cancelled this simplifies to:

$$5SO_2(g) + 2H_2O(l) + 2MnO_4^-(aq) \rightarrow 5SO_4^{2-}(aq) + 4H^+(aq) + 2Mn^{2+}(aq)$$

Step 8 The total charge on each side of the equation is 2−.

Test yourself

2 Use half-equations to write the overall ionic equation for each of the following reactions.
   a) $NO_3^-(aq)$ in acid solution reacts with $Cu(s)$ to form $Cu^{2+}(aq)$ and $NO_2(g)$.
   b) H_2O_2 in acid solution reacts with $Fe^{2+}(aq)$ to form $H_2O(l)$ and $Fe^{3+}(aq)$.
   c) $Cr_2O_7^{2-}(aq)$ in acid solution reacts with $Fe^{2+}(aq)$ to form $Cr^{3+}(aq)$ and $Fe^{3+}(aq)$.
   d) $MnO_4^-(aq)$ in acid solution reacts with $Br^-(aq)$ to form $Mn^{2+}(aq)$ and $Br_2(aq)$.

Electrode potentials

An advantage of redox reactions is that the half-equations of the reactions can be reproduced in electrical cells.

If an electrical circuit is set up as shown in Figure 5.2, an overall chemical reaction takes place, but the half-reactions are kept separate and the transfer of electrons takes place via the external circuit. The energy released by the transfer of electrons can be measured using a voltmeter.

The circuit in Figure 5.2 shows the reaction

$$Zn(s) + Cu^{2+}(aq) \rightarrow Zn^{2+}(aq) + Cu(s)$$

taking place.

At the zinc electrode, a metal atom loses two electrons to become an ion:

$$Zn(s) \rightarrow Zn^{2+}(aq) + 2e^-$$

The electrons travel around the external circuit through the voltmeter to the copper electrode. Once there, each pair of electrons is taken up by a copper ion and an atom of copper is formed:

$$Cu^{2+}(aq) + 2e^- \rightarrow Cu(s)$$

The voltmeter reading is a measure of the energy of the transferred electrons and, just as the size of a free energy change gives a measure of the energy difference between reactants and products, the voltmeter reading does the same for this reaction in terms of the number of joules per coulomb. The resultant measurement is called an electrode potential.

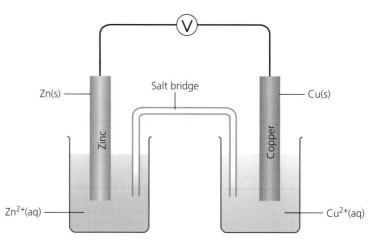

Figure 5.2 An electrochemical cell based on the reaction of zinc metal with aqueous copper ions.

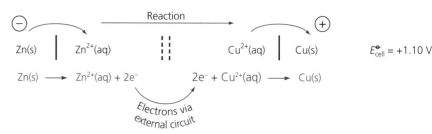

Figure 5.3 The direction of change in the electrochemical cell.

To measure an electrode potential, the voltage should be determined while the flow of current is minimal, otherwise some of the energy to be measured will be lost. To do this, a high resistance voltmeter is used.

The salt bridge contains a solution or gel of a substance that does not interfere with the reaction. Potassium chloride is commonly used, although this would not be appropriate if the chloride ion formed a precipitate with one of the cations. Precipitation would occur if, for example, a silver electrode and silver nitrate solution were used. In this case the salt bridge could contain potassium nitrate.

The purpose of the salt bridge needs some explanation. As electrons are released, zinc ions are formed in the solution in the container on the left, while copper ions are removed from the solution on the right (see Figure 5.3). The charge of all of the ions (cations and anions combined) in each container must remain zero. The only way for this to happen and to keep the electron flow continuing, is to transfer excess ions between the containers. This is accomplished by movement of these ions into, and eventually through, the salt bridge. So zinc ions move into the salt bridge and are balanced by the anion moving from the copper salt solution. The external circuit allows movement of electrons from one half-cell to the other whilst the salt bridge allows movement of ions from one to the other.

The standard hydrogen electrode

Any redox reaction can be made to take place in a circuit of this type and so a wide variety of electrode potentials can be measured. However, to organise and make the information useful, it is necessary to measure the voltages against a standard electrode. The electrode chosen for this purpose is the hydrogen electrode, which is illustrated in Figure 5.4. It is used as the standard because it is relatively easy to control its purity and reproducibility. The standard conditions used for the electrode are 298 K with the hydrogen being bubbled over the electrode (which is made of platinum) at the standard pressure of 100 kPa.

The standard half-cell is connected via an external circuit and through a salt bridge to the experimental half-cell. The voltage measured is the **electrode potential** of the cell compared with the half-reaction:

$$2H^+(aq) + 2e^- \rightarrow H_2(g)$$

which is given the arbitrary value of zero.

If the concentrations used in the cells are 1 mol dm^{-3}, the temperature is 298 K and all gases are at a pressure of 100 kPa, the voltage measured is the **standard electrode potential**, $E^{\ominus}$.

The standard electrode potential for Cu(s) is represented as:

$$Cu^{2+}(aq) + 2e^- \rightleftharpoons Cu(s) \quad E^{\ominus} = +0.34\,V$$

oxidised form   reduced form

This serves to emphasise that standard electrode potentials represent reduction.

When the copper half-cell is connected, the electrons flow from the hydrogen to the copper, equivalent to the reaction:

$$Cu^{2+}(aq) + H_2(g) \rightarrow Cu(s) + 2H^+(aq)$$

If the standard hydrogen electrode were connected to a zinc half-cell (Figure 5.5), the voltmeter would read −0.76 volts. In the case of zinc, electrons flow from the zinc to the hydrogen electrode, equivalent to the reaction:

$$Zn(s) + 2H^+(aq) \rightarrow Zn^{2+}(aq) + H_2(g)$$

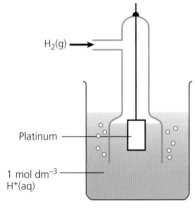

$H_2(g)$

Platinum

1 mol dm^{-3}
$H^+(aq)$

Figure 5.4 A standard hydrogen electrode. To provide a better surface for the hydrogen, the platinum electrode is usually coated with very finely divided platinum, known as 'platinum black'.

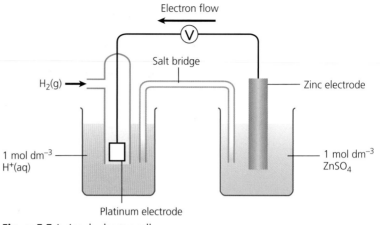

Figure 5.5 A zinc–hydrogen cell.

A table can be produced to include all processes in which half-reactions occur with the loss or gain of electrons. The results are called standard electrode potentials or standard reduction potentials and some examples are shown in Table 5.1.

Table 5.1 Standard cell potential values at 298 K.

Electrode reaction	$E^{\ominus}/V$
$Li^+(aq) + e^- \rightleftharpoons Li(s)$	−3.03
$K^+(aq) + e^- \rightleftharpoons K(s)$	−2.92
$Ba^{2+}(aq) + 2e^- \rightleftharpoons Ba(s)$	−2.90
$Ca^{2+}(aq) + 2e^- \rightleftharpoons Ca(s)$	−2.87
$Na^+(aq) + e^- \rightleftharpoons Na(s)$	−2.71
$Mg^{2+}(aq) + 2e^- \rightleftharpoons Mg(s)$	−2.37
$Al^{3+}(aq) + 3e^- \rightleftharpoons Al(s)$	−1.66
$Zn^{2+}(aq) + 2e^- \rightleftharpoons Zn(s)$	−0.76
$Fe^{2+}(aq) + 2e^- \rightleftharpoons Fe(s)$	−0.44
$Cr^{3+}(aq) + e^- \rightleftharpoons Cr^{2+}(aq)$	−0.41
$V^{3+}(aq) + e^- \rightleftharpoons V^{2+}(aq)$	−0.26
$Sn^{2+}(aq) + 2e^- \rightleftharpoons Sn(s)$	−0.14
$Pb^{2+}(aq) + 2e^- \rightleftharpoons Pb(s)$	−0.13
$H^+(aq) + e^- \rightleftharpoons \frac{1}{2}H_2(g)$	0
$S(s) + 2H^+(aq) + 2e^- \rightleftharpoons H_2S(g)$	+0.14
$Sn^{4+}(aq) + 2e^- \rightleftharpoons Sn^{2+}(aq)$	+0.15
$Cu^{2+}(aq) + e^- \rightleftharpoons Cu^+(aq)$	+0.15
$Cu^{2+}(aq) + 2e^- \rightleftharpoons Cu(s)$	+0.34

$Cu^+(aq) + e^- \rightleftharpoons Cu(s)$	+0.52
$I_2(s) + 2e^- \rightleftharpoons 2I^-(aq)$	+0.54
$MnO_4^-(aq) + e^- \rightleftharpoons MnO_4^{2-}(aq)$	+0.56
$MnO_4^{2-}(aq) + 2H_2O(l) + 2e^- \rightleftharpoons MnO_2(s) + 4OH^-(aq)$	+0.59
$O_2(g) + 2H^+(aq) + 2e^- \rightleftharpoons H_2O_2(aq)$	+0.68
$Fe^{3+}(aq) + e^- \rightleftharpoons Fe^{2+}(aq)$	+0.77
$Ag^+(aq) + e^- \rightleftharpoons Ag(s)$	+0.80
$Br_2(l) + 2e^- \rightleftharpoons 2Br^-(aq)$	+1.09
$O_2(g) + 4H^+(aq) + 4e^- \rightleftharpoons 2H_2O(l)$	+1.23
$Cr_2O_7^{2-}(aq) + 14H^+(aq) + 6e^- \rightleftharpoons 2Cr^{3+}(aq) + 7H_2O(l)$	+1.33
$Cl_2(g) + 2e^- \rightleftharpoons 2Cl^-(aq)$	+1.36
$MnO_4^-(aq) + 8H^+(aq) + 5e^- \rightleftharpoons Mn^{2+}(aq) + 4H_2O(l)$	+1.51
$Pb^{4+}(aq) + 2e^- \rightleftharpoons Pb^{2+}(aq)$	+1.69
$H_2O_2(aq) + 2H^+(aq) + 2e^- \rightleftharpoons 2H_2O(l)$	+1.77
$F_2(g) + 2e^- \rightleftharpoons 2F^-(aq)$	+2.87

If a cell uses hydrogen and another gas, then the two half-cells look similar (Figure 5.6).

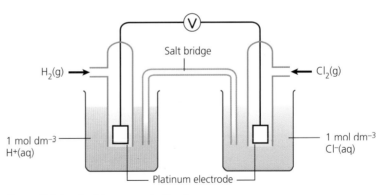

Figure 5.6 A cell with two gas half-cells.

If the other reaction involves the transfer of electrons between two ions, then the solution in the half-cell must contain both ions and the electrode should be platinum (Figure 5.7).

For a *standard* electrode potential to be measured the concentration of all the ions involved should, in theory, be $1.0\,mol\,dm^{-3}$. In practice, it is rare that substances are sufficiently soluble to achieve this. However, if all the ions are at the same concentration in $mol\,dm^{-3}$ ('equimolar') then the electrode potential has the same value irrespective of the concentration. For example, the standard electrode potential of the redox half-reaction $Fe^{3+}(aq) + e^- \rightleftharpoons Fe^{2+}(aq)$ can be measured with both the ions, $Fe^{3+}(aq)$ and $Fe^{2+}(aq)$ at a concentration of $0.1\,mol\,dm^{-3}$.

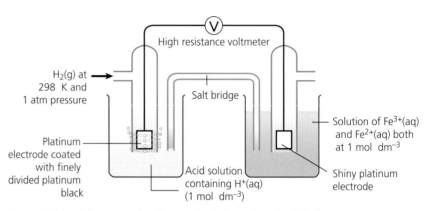

Figure 5.7 A cell for measuring the standard electrode potential of $Fe^{3+}(aq) + e^- \rightleftharpoons Fe^{2+}(aq)$.

Cell potentials

Electrode potentials can be used to predict the voltage or cell potential of any combination of half-cells.

Using the example of a zinc electrode combined with a copper electrode, the zinc electrode provides the electrons and the copper ion receives them.

The standard electrode potential of $Zn^{2+}(aq) + 2e^- \rightleftharpoons Zn(s)$ is $-0.76\,V$.

This tells us that, relative to the standard hydrogen electrode, the energy required to convert $Zn^{2+}(aq)$ into $Zn(s)$ is measured as $-0.76V$.

But, in this cell, $Zn(s)$ is converted into $Zn^{2+}(aq)$ as electrons are transferred to the $Cu^{2+}(aq)$. Thus, the energy transferred will be $+0.76\,V$.

The standard electrode potential of $Cu^{2+}(aq) + 2e^- \rightleftharpoons Cu(s)$ is $+0.34\,V$.

Therefore, the overall cell potential is $+0.76 + 0.34$, which is $+1.10\,V$.

Standard cell potentials can always be calculated by combining the two half-cell standard electrode potentials, since each standard electrode potential is measured relative to the standard hydrogen electrode. Remember that the sign convention is important.

A useful method of obtaining the overall cell potential is to look at the electrode potentials of the two electrodes making up the cell, regarding the most positive electrode potential as being the 'right-hand' electrode and the least positive as being the 'left-hand' electrode. It then follows that:

$$E^{\ominus}_{cell} = E^{\ominus}_{(right\text{-}hand\ electrode)} - E^{\ominus}_{(left\text{-}hand\ electrode)}$$

The following examples should make this procedure clear.

Example 8

Use the following standard electrode potentials to calculate the standard cell potential for a combination of chromium and cadmium electrodes.

In which direction will the electrons flow in the external circuit?

Reaction	$E^{\ominus}$/V	
$Cr^{3+}(aq) + 3e^{-} \rightleftharpoons Cr(s)$	−0.74	left-hand electrode (as it is the more negative)
$Cd^{2+}(aq) + 2e^{-} \rightleftharpoons Cd(s)$	−0.40	right-hand electrode

Answer

Electrons must be supplied by one half-reaction and received by the other half-reaction. The overall cell potential must be positive. This can only be achieved if the electrons are supplied by chromium to the cadmium ions

$$E^{\ominus}_{cell} = E^{\circ}_{(right\text{-}hand\ electrode)} - E^{\circ}_{(left\text{-}hand\ electrode)}$$

which gives $E^{\ominus}_{cell} = (-0.40V) - (-0.74V) = +0.34V$

Example 9

Use the following standard electrode potentials to calculate the standard cell potential for a combination of copper and silver electrodes.

In which direction will the electrons flow in the external circuit?

Reaction	$E^{\ominus}$/V	
$Cu^{2+}(aq) + 2e^{-} \rightleftharpoons Cu(s)$	+0.34	left-hand electrode (as it is the less positive)
$Ag^{+}(aq) + e^{-} \rightleftharpoons Ag(s)$	+0.80	right-hand electrode

▶▶▶

Answer

Electrons must be supplied by one half-reaction and received by the other half-reaction. The overall cell potential must be positive. This can only be achieved if the electrons are supplied by copper to the silver ions.

$$E^{\ominus}{}_{cell} = E^{\ominus}{}_{(right\text{-}hand\ electrode)} - E^{\ominus}{}_{(left\text{-}hand\ electrode)}$$

which gives $E^{\ominus}{}_{cell} = (+0.80V) - (0.34V) = +0.46V$

When you look at the electrode potentials in Example 2 and notice that they are both positive, don't just add them together. Both processes release energy, but within a cell, one process must provide electrons and the other process must receive them.

Also, do not be tempted to adjust the cell potentials to match the number of electrons transferred. It is true that an atom of copper supplies two electrons and that a silver ion only requires one electron to form a silver atom. However, this simply means that two atoms of silver are deposited on the electrode for each atom that is lost from the copper electrode. Do **not** multiply the silver half-cell potential by 2.

To obtain the overall potential, just add or subtract the electrode potentials of the half-reactions as usual.

Tip

The reason that electrode potentials can be treated in this simple way is because a volt is a unit of energy (joule) carried per coulomb of electricity. So for each half-cell, the energy is for a fixed number of electrons (the coulomb) compared with the same fixed number of electrons from the other half-cell.

Example 10

Use the following standard electrode potentials to predict the standard cell potential for a combination of a standard half-cell containing a platinum electrode in contact with $Fe^{2+}(aq)$ and $Fe^{3+}(aq)$, with a half-cell consisting of a Pb electrode and $Pb^{2+}(aq)$.

In which direction will the electrons flow in the external circuit?

Reaction	$E^{\ominus}/V$	
$Fe^{3+}(aq) + e^- \rightleftharpoons Fe^{2+}(aq)$	+0.77	right-hand electrode
$Pb^{2+}(aq) + 2e^- \rightleftharpoons Pb(s)$	−0.13	left-hand electrode

Answer

Electrons must be supplied by one half-reaction and consumed in the other half-reaction. The overall cell potential must be positive. This can only be achieved if the electrons are supplied by lead to the $Fe^{3+}(aq)$ ions.

$$E^{\ominus}{}_{cell} = E^{\ominus}{}_{(right\text{-}hand\ electrode)} - E^{\ominus}{}_{(left\text{-}hand\ electrode)}$$

which gives $E^{\ominus}{}_{cell} = (+0.77V) - (-0.13V) = +0.90V$

If you are given a list of electrode potentials, you will notice that the cell potential is always the difference between the two electrode potentials, taking into account the sign. Although this provides a quick method of obtaining the potential, it is better (for reasons that will become clear when you have considered the next section on the feasibility of reactions) to consider the chemical processes that are taking place.

3 Use the standard electrode potentials in Table 5.1 (pages 96 and 97) to:

   a) calculate the standard cell potential for the combination of electrodes shown below

   b) deduce an ionic equation for the overall reaction in each case.

   i) $Mg/Mg^{2+}(aq)$ and $Zn/Zn^{2+}(aq)$

   ii) $Cl_2/Cl^-(aq)$ and Cu/Cu^{2+}

   iii) $Sn^{2+}(aq)/Sn^{4+}(aq)$ and $I_2(s)/I^-(aq)$

   iv) $Sn^{2+}(aq)/Sn^{4+}(aq)$ and $H^+/MnO_4^-(aq)/Mn^{2+}(aq)$

   v) $Cl^-(aq)/Cl_2(g)$ and $H^+/MnO_4^-(aq)/Mn^{2+}(aq)$

Using electrode potentials to predict the feasibility of a reaction

The reactions that take place in electrical cells are equivalent to the redox reactions that can be performed in the laboratory. By using electrode potentials, the energy change that takes place when a redox reaction occurs can be established and it is possible to decide whether a chemical reaction is feasible or not.

To illustrate this, we will predict whether $Fe^{3+}(aq)$ can be reduced to $Fe^{2+}(aq)$ under standard conditions using:

- $I^-(aq)$

- $Br^-(aq)$.

The relevant standard electrode potentials are shown in Table 5.2.

Table 5.2

Reaction	$E^\ominus$/V
$I_2(aq) + 2e^- \rightleftharpoons 2I^-(aq)$	+0.54
$Fe^{3+}(aq) + e^- \rightleftharpoons Fe^{2+}(aq)$	+0.77
$Br_2(aq) + 2e^- \rightleftharpoons 2Br^-(aq)$	+1.09

With $I^-(aq)$:

Is the reaction $2Fe^{3+}(aq) + 2I^-(aq) \rightleftharpoons 2Fe^{2+}(aq) + I_2(aq)$ feasible?

The electrode potentials for the half-reactions are:

   $Fe^{3+}(aq) + e^- \rightleftharpoons Fe^{2+}(aq)$ $E^\ominus = +0.77\,V$

   $2I^-(aq) \rightleftharpoons I_2(aq) + 2e^-$ the potential would then be $-0.54\,V$ (the reverse of the standard electrode potential)

Adding these together, gives a cell potential of $+0.77 - 0.54 = +0.23\,V$.

The positive answer indicates that the reaction is feasible.

With $Br^-(aq)$:

Is the reaction $2Fe^{3+}(aq) + 2Br^-(aq) \rightleftharpoons 2Fe^{2+}(aq) + Br_2(aq)$ feasible?

The electrode potentials for the half-reactions are:

$$Fe^{3+}(aq) + e^- \rightleftharpoons Fe^{2+}(aq) \qquad E^\ominus = +0.77\,V$$

$$2Br^-(aq) \rightleftharpoons Br_2(aq) + 2e^- \qquad \text{the potential would then be } -1.09\,V \text{ (the reverse of the standard electrode potential)}$$

Adding these together gives an overall potential of $+0.77 - 1.09 = -0.32\,V$.

The negative answer indicates that the reaction is not feasible.

(In fact, you may be able to see that $Br_2(aq)$ will be able to oxidise $Fe^{2+}(aq)$ to $Fe^{3+}(aq)$.)

Notice that the more positive a standard redox potential is, the stronger the oxidising power of the reagent on the left-hand side of the equation. Using Table 5.2, $Br_2(aq)$ is the strongest oxidising agent, followed by $Fe^{3+}(aq)$ and then $I_2(aq)$. This is why $Fe^{3+}(aq)$ is able to oxidise $I^-(aq)$ but not $Br^-(aq)$. Table 5.1 can be used to deduce which are the strongest oxidising agents. For example, $Cl_2(g)$ is a good oxidising agent ($E^\ominus = +1.36V$) but H_2O_2 (aq) ($E^\ominus = +1.77V$) is better.

Example 11

Under standard conditions, can acidified aqueous potassium manganate(VII) ions oxidise $Br^-(aq)$ to $Br_2(aq)$?

The standard electrode potentials are given in the table.

Reaction	$E^\ominus/V$
$Br_2(aq) + 2e^- \rightleftharpoons 2Br^-(aq)$	+1.09
$MnO_4^-(aq) + 8H^+(aq) + 5e^- \rightleftharpoons Mn^{2+}(aq) + 4H_2O(l)$	+1.51

Answer

The process $2Br^-(aq) \rightarrow Br_2(aq) + 2e^-$ is the reverse of that given by the $E^\ominus$ value, and therefore has a value of $-1.09V$. The reduction of the manganate(VII) has an $E^\ominus$ of $+1.51V$.

Adding these together gives a cell potential of $-1.09 + 1.51 = +0.42V$.

The positive cell potential indicates that the reaction is feasible.

Example 12

Under standard conditions, can aqueous Ag^+ ions oxidise $Fe^{2+}(aq)$ ions to $Fe^{3+}(aq)$ ions?

The standard electrode potentials are shown in the table.

Reaction	$E^\ominus/V$
$Ag^+(aq) + e^- \rightleftharpoons Ag(s)$	+0.80
$Fe^{3+}(aq) + e^- \rightleftharpoons Fe^{2+}(aq)$	+0.77

Answer

The process $Fe^{2+}(aq) \rightarrow Fe^{3+}(aq) + e^-$ is the reverse of that given by the $E^\ominus$ value, and therefore has a value of $-0.77V$. The reduction of the $Ag^+(aq)$ to $Ag(s)$ has an $E^\ominus$ of $+0.80V$.

Adding these together gives an overall value of $+0.03V$.

The positive cell potential indicates that the reaction is feasible.

Example 12 illustrates a potential trap that must be avoided. It is tempting to suggest that the low value of the electrode potential ($+0.03\,V$) will mean that the reaction will be slow. However, it is not possible to deduce this. As with the value of ΔG, the overall potential indicates the relative energies of the reactants and products. It gives **no** information about the rate of the reaction. Whether a reaction occurs quickly or slowly depends on the activation energy of the reaction and activation energy is not related to the energy difference between the reactants and products. A clear distinction must be made between thermodynamic stability (which relates to the relative energies of the reactants and products) and kinetic stability (which refers to how fast the conversion will take place).

Test yourself

4 Use Table 5.1 to decide whether under standard conditions:
 a) $Sn^{2+}(aq)$ can reduce $Fe^{3+}(aq)$ ions to $Fe^{2+}(aq)$ ions
 b) $Zn(s)$ will react with $Sn^{2+}(aq)$
 c) acidified $Cr_2O_7^{2-}(aq)$ can oxidise chloride ions to $Cl_2(g)$.
5 Use Table 5.1 to arrange the following sets of molecules/ions in order of decreasing strength as oxidising agents in acid solution:
 Br_2, Cl_2, H_2O_2, O_2.

The effect of concentration on the feasibility of reactions

The predictions made so far have been based on the use of standard electrode potentials, which specify particular conditions of temperature, pressure and concentration. Normal laboratory conditions of temperature and pressure are usually quite close to standard conditions. However, concentration is an important factor – reactions are not often carried out using concentrations of $1\,mol\,dm^{-3}$ and, in any case, if a reaction does occur, the concentrations change as the reaction takes place.

For concentrations usually used in the laboratory, the values of electrode potentials do not change significantly. A reduction in concentration from $1\,mol\,dm^{-3}$ to $0.1\,mol\,dm^{-3}$ only changes the electrode potential of a half-reaction by $0.06\,V$ or less. Therefore, predictions based on standard electrode potentials are nearly always correct for the normal concentrations used in the laboratory.

Using an argument based around le Chatelier's principle, it is possible to decide qualitatively whether a change in concentration will increase or decrease the value of the electrode potential. Electrode potentials refer to equilibria and the position of the equilibrium varies in accordance with le Chatelier's principle. A change in concentration will, therefore, cause a change in the balance of equilibrium and hence the value of the standard potential.

For example, if the equilibrium:

$$Fe^{3+}(aq) + e^- \rightleftharpoons Fe^{2+}(aq) \quad \text{for which } E^\circ = +0.77 \text{ V}$$

has a reduced concentration of $Fe^{2+}(aq)$, this will encourage a shift in the equilibrium position from left to right, which will cause the electrode potential to increase. If there is a reduced concentration of $Fe^{3+}(aq)$ then the equilibrium position will move to the left and the value of $E^\ominus$ will decrease. In both cases, it would require a large change in concentration for the difference to be apparent.

Storage cells and fuel cells

Storage cells

Storage cells are used commonly in appliances to supply electricity. There is a wide range available as manufacturers strive to provide improved reliability and a longer life. The principles discussed above can be used to anticipate the voltage that would be expected from such cells.

For obvious reasons, it is better if a battery that is going to be sold to the general public does not contain liquid. The cells usually contain pastes that surround the electrodes. Some examples will help you appreciate the relative complexity of the reactions taking place.

An 'alkaline battery' has a positive electrode made from graphite and manganese(IV) oxide and a negative electrode made of either zinc or nickel-plated steel (Figure 5.8).

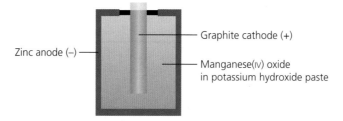

Zinc anode (−)

Graphite cathode (+)

Manganese(IV) oxide in potassium hydroxide paste

Figure 5.8 An alkaline battery.

The half-cell that provides the electrons to the external circuit is the negative electrode; the half-cell that receives them is the positive electrode. The paste or solution is the electrolyte.

The reactions that take place are:

at the negative electrode

$$Zn + 2OH^- \rightarrow ZnO + H_2O + 2e^-$$

at the positive electrode

$$2MnO_2 + H_2O + 2e^- \rightarrow Mn_2O_3 + 2OH^-$$

The electrolyte is potassium hydroxide.

The overall equation taking place is found by combining the two half-reactions:

$$Zn + 2MnO_2 \rightarrow ZnO + Mn_2O_3$$

Both OH⁻ and H_2O can be eliminated from the equation because their concentrations should remain constant. (In practice, some loss does occur.) The overall voltage is about 1.5 V and is a combination of the electrode potentials of the two half-reactions.

Another useful battery is the rechargeable nickel–cadmium (Ni–Cd) cell. While it is supplying electricity, these reactions taking place are:

at the negative electrode

$$Cd + 2OH^- \rightarrow Cd(OH)_2 + 2e^-$$

at the positive electrode

$$2NiO(OH) + 2H_2O + 2e^- \rightarrow 2Ni(OH)_2 + 2OH^-$$

The overall reaction is:

$$Cd + 2NiO(OH) + 2H_2O \rightarrow Cd(OH)_2 + 2Ni(OH)_2$$

The battery can be recharged by applying an external voltage that reverses the reactions shown above. A disadvantage of this type of battery is that cadmium is toxic and care needs to be taken when disposing of the batteries. A voltage of around 1.2 V can be obtained during use.

Many other storage cells are manufactured and the intense research into the development of batteries for use in vehicles has led to a number of different constructions. The lead–acid accumulator is still employed widely (Figure 5.9).

However, its considerable weight is a disadvantage. Alternatives such as a sodium–sulfur cell are much lighter, but can be expensive and more risky in use.

Modern mobile phones and laptop computers use lithium batteries. The advantages of electrodes based on lithium are that the metal has a low density, so cells based on lithium electrodes can be relatively light. Also, lithium is very reactive, which means that the electrode potential of a lithium half-cell is relatively high and each cell has a large e.m.f.

Figure 5.9 A car battery.

The difficulty to overcome is that lithium is so reactive that it readily combines with oxygen in the air, forming a layer of non-conducting oxide on the surface of the metal. The metal also reacts rapidly with water. Research workers have solved these technical problems by developing electrodes with lithium atoms and ions inserted into the crystal lattices of other materials. In addition, the electrolyte is a polymeric material rather than an aqueous solution. However a disadvantage of lithium cells is that lithium is toxic and its high reactivity means there is a potential fire risk.

Fuel cells

Perhaps more promising than storage cells as a source of energy, are fuel cells. A fuel cell produces electrical power from the chemical reaction of a fuel (for example hydrogen, hydrocarbons or alcohols) with oxygen (Figure 5.10). The fuel cell operates like a conventional storage cell, except that the fuels are supplied externally. The cell, therefore, operates more or less indefinitely, so long as the fuel supply is maintained.

The hydrogen–oxygen fuel cell is used widely and illustrates the principles behind fuel cells in general.

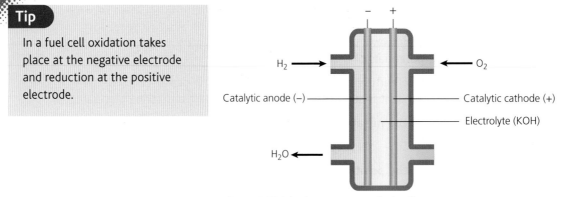

Figure 5.10 A hydrogen–oxygen fuel cell.

Tip

In a fuel cell oxidation takes place at the negative electrode and reduction at the positive electrode.

Tip

You will not be asked to draw a diagram of a fuel cell in an exam. It is the principles of how a fuel cell operates that are required.

The electrodes are made of a material such as a titanium sponge coated in platinum. The electrolyte is an acid or alkaline membrane that allows ions to move from one compartment of the cell to the other. (In other words, it acts like a salt bridge, see page 95.)

In alkaline solution, hydrogen reacts with hydroxide ions at the negative electrode to form water while, at the positive electrode, oxygen reacts with water to form hydroxide ions:

$$H_2(g) + 2OH^-(aq) \rightleftharpoons 2H_2O(l) + 2e^- \quad \text{potential is } +0.83\,V$$

$$\tfrac{1}{2}O_2(g) + H_2O(l) + 2e^- \rightleftharpoons 2OH^-(aq) \quad E^\ominus = +0.40\,V$$

In an acidic solution, hydrogen is converted to hydrogen ions at the negative electrode while, at the positive electrode, oxygen reacts with hydrogen ions to make water:

$$H_2(g) \rightleftharpoons 2H^+(aq) + 2e^-$$

$$\tfrac{1}{2}O_2(g) + 2H^+(aq) + 2e^- \rightleftharpoons H_2O(l) \quad E^\ominus = +1.23\,V$$

The overall reaction in both cases is the same and is equivalent to the burning of the fuel – in this case, hydrogen.

$$H_2(g) + \tfrac{1}{2}O_2(g) \rightarrow H_2O(l)$$

and the voltage produced is 1.23 V.

Other fuel cells can be constructed using different 'fuels'. For example, a methanol fuel cell uses methanol instead of hydrogen. The overall reaction that is occurring within the cell is again equivalent to burning the methanol, so that the overall equation is:

$$CH_3OH + 1\tfrac{1}{2}O_2 \rightarrow CO_2 + 2H_2O$$

The processes that take place at the electrodes mirror this equation. In an acidic medium:

at the positive electrode

$$\tfrac{1}{2}O_2 + 2H^+ + 2e^- \rightarrow H_2O$$

at the negative electrode

$$CH_3OH + H_2O \rightarrow CO_2 + 6H^+ + 6e^-$$

Tip

Notice that these equations are constructed in just the same way as other redox equations, using H^+ and H_2O.

And combining these two equations so that the supply of electrons is balanced gives:

$$1\tfrac{1}{2}O_2 + 6H^+ + CH_3OH + H_2O \rightarrow 3H_2O + CO_2 + 6H^+$$

Cancelling the $6H^+$ and simplifying then gives the equation for the combustion of methanol.

$$CH_3OH + 1\tfrac{1}{2}O_2 \rightarrow CO_2 + 2H_2O$$

Test yourself

6 A commonly used cell has a negative terminal made of zinc. The positive terminal is a carbon (graphite) rod where ammonium ions from the ammonium chloride electrolyte are converted to ammonia and hydrogen as the cell delivers a current. The cell can produce a potential difference of 1.50 V.
  a) Write a half-equation for the reaction at the zinc terminal.
  b) Write a half-equation for the reaction at the carbon (graphite) terminal.
  c) Assuming that the electrode potential for the $Zn^{2+}(aq)/Zn$ half-cell is -0.76 V, calculate the electrode potential of the other half-cell.
7 Construct the equations for processes that take place at the electrodes in a fuel cell when the fuel used is methane.

Practice questions

Multiple choice questions 1–10

1 Which one of the following gives the correctly balanced equation for the reaction between $Br^-(aq)$ and $BrO_3^-(aq)$ in the presence of hydrogen ions at room temperature to form bromine and water?
  A $Br^-(aq) + BrO_3^-(aq) + 6H^+(aq) \rightarrow Br_2(aq) + 3H_2O(l)$
  B $2Br^-(aq) + BrO_3^-(aq) + 6H^+(aq) \rightarrow \tfrac{3}{2}Br_2(aq) + 3H_2O(l)$
  C $3Br^-(aq) + BrO_3^-(aq) + 6H^+(aq) \rightarrow 2Br_2(aq) + 3H_2O(l)$
  D $5Br^-(aq) + BrO_3^-(aq) + 6H^+(aq) \rightarrow 3Br_2(aq) + 3H_2O(l)$ *(1)*

2 When 1 mol of $NO_3^-(aq)$ is reduced to form $N_2O(g)$, which of the following must be supplied?
  A 4 mol of hydrogen ions and 3 mol of electrons
  B 5 mol of hydrogen ions and 4 mol of electrons
  C 10 mol of hydrogen ions and 8 mol of electrons
  D 12 mol of hydrogen ions and 10 mol of electrons *(1)*

3 These are ionic equations for reactions of VO_4^{3-} (aq):
  Reaction 1
  $VO_4^{3-}(aq) + 4H^+(aq) \rightarrow VO_2^+(aq) + 2H_2O(l)$
  Reaction 2
  $VO_4^{3-}(aq) + 4S^{2-}(aq) + 4H_2O(l) \rightarrow VS_4^{3-}(aq) + 8OH^-(aq)$
  Which one of the following statements describes the behaviour of VO_4^{3-} (aq) in these two reactions?
  A It is not oxidised or reduced in either reaction.
  B It is reduced in both reactions.
  C It is oxidised in reaction 1 and reduced in reaction 2.
  D It is reduced in reaction 2 but is neither oxidised nor reduced in reaction 1. *(1)*

4 When connected to a standard hydrogen electrode, the standard electrode potential for the process $Fe^{3+}(aq) + e^- \rightleftharpoons Fe^{2+}(aq)$ can be measured using a half-cell containing:
  A an iron electrode in a solution containing $0.1\,mol\,dm^{-3}$ iron(II) sulfate and $0.1\,mol\,dm^{-3}$ iron(III) sulfate

B an iron electrode in a solution containing $0.5\,mol\,dm^{-3}$ iron(II) sulfate and $0.25\,mol\,dm^{-3}$ iron(III) sulfate

C a platinum electrode in a solution containing $0.1\,mol\,dm^{-3}$ iron(II) sulfate and $0.1\,mol\,dm^{-3}$ iron(III) sulfate

D a platinum electrode in a solution containing $0.04\,mol\,dm^{-3}$ iron(II) sulfate and $0.02\,mol\,dm^{-3}$ iron(III) sulfate *(1)*

Use the following standard electrode potentials to answer questions 5, 6 and 7.

Right-hand electrode	$E^{\ominus}/V$
$Fe^{3+}(aq) + e^- \rightleftharpoons Fe^{2+}(aq)$	+0.77
$Fe^{3+}(aq) + 3e^- \rightleftharpoons Fe(s)$	−0.04
$Fe^{2+}(aq) + 2e^- \rightleftharpoons Fe(s)$	−0.44
$Cu^+(aq) + e^- \rightleftharpoons Cu(s)$	+0.52
$Cu^{2+}(aq) + 2e^- \rightleftharpoons Cu(s)$	+0.34
$Cu^{2+}(aq) + e^- \rightleftharpoons Cu^+(aq)$	+0.15

5 Which one of the following is the strongest reducing agent?

A $Fe(s)$

B $Fe^{2+}(aq)$

C $Cu^{2+}(aq)$

D $Cu(s)$ *(1)*

6 Which one of the following will be the standard cell potential of a cell made from $Cu(s)/Cu^{2+}(aq)$ connected to $Fe(s)/Fe^{3+}(aq)$?

A +0.30V

B +0.38V

C +0.94V

D +1.10V *(1)*

7 A solution of a metal ion $X^{2+}(aq)$ does not react with Cu to produce X, but does give a precipitate of the metal $X(s)$ when reacted with Fe. Which one of the following could be the standard electrode potential for $X^{2+}(aq)/X(s)$?

A −0.85V

B −0.61V

C −0.39V

D −0.25V *(1)*

Use the key below to answer questions 8, 9 and 10.

A	B	C	D
1, 2 & 3 correct	1 & 2 correct	2 & 3 correct	1 only correct

8 In a standard cell being used to measure the electrode potential of zinc in contact with a $1.00\,mol\,dm^{-3}$ solution of zinc sulfate compared to standard hydrogen half-cell in contact with $1\,mol\,dm^{-3}$ hydrochloric acid, which of the following is a function of the salt bridge?

1 To transfer electrons from the zinc to the hydrogen half-cell.

2 To allow chloride ions to move across the bridge to the zinc half-cell.

3 To allow zinc ions to move across the bridge to the hydrogen half-cell. *(1)*

9 Which of the following could be used as the salt bridge when determining the electrode potential of $Ag^+(aq)/Ag(s)$?

1 aqueous potassium nitrate

2 aqueous potassium sulfate

3 aqueous potassium chloride *(1)*

10

Right-hand electrode	$E^{\ominus}/V$
$Fe^{3+}(aq) + 3e^- \rightleftharpoons Fe(s)$	−0.04
$Fe^{2+}(aq) + 2e^- \rightleftharpoons Fe(s)$	−0.44
$Cu^+(aq) + e^- \rightleftharpoons Cu(s)$	+0.52
$Cu^{2+}(aq) + 2e^- \rightleftharpoons Cu(s)$	+0.34

For this question use the table of electrode potentials above and the following standard electrode potential:

$Co^{2+}(aq) + 2e^- \rightleftharpoons Co(s)\ E^{\ominus} = -0.28V$

Which of the following would react with $1.00\,mol\,dm^{-3}$ aqueous Co^{2+} under standard conditions?

1 $Fe(s)$

2 $Cu(s)$

3 $Fe^{3+}(aq)$ *(1)*

11 Write ionic equations for each of the following reactions:

a) aqueous potassium carbonate with nitric acid

b) solid calcium carbonate with hydrochloric acid

c) the precipitation of calcium carbonate from aqueous calcium chloride and aqueous sodium carbonate

d) the neutralisation of aqueous calcium hydroxide by hydrochloric acid

e) the precipitation of copper(II) hydroxide from aqueous copper nitrate and aqueous potassium hydroxide

f) zinc oxide with nitric acid. *(4)*

12 Give the oxidation number of the element in **bold** in each of the following.

a) **H**$_2$
b) K**Cl**O$_3$
c) Na$_3$**P**O$_4$
d) Mg**S**O$_4$
e) Ca(**Cl**O)$_2$
f) K$_2$**Cr**$_2$O$_7$
g) I**Cl**$_3$
h) Al(**N**O$_3$)$_3$
i) **S**O$_3^{2-}$
j) **Cl**O$_4^-$
k) **N**O$_2^-$
l) **V**O$_3^-$
m) **V**O^{2+}
n) **Al**(OH)$_4^-$
o) **Al**F$_6^{3-}$
p) **S**$_2$O$_3^{2-}$ (16)

13 Complete each of the following pairs of half-equations and construct an overall equation assuming the reaction occurs.

a) Fe^{3+}(aq) → Fe^{2+}(aq)
   I$^-$(aq) → I$_2$(s)

b) MnO$_4^-$(aq) + H$^+$(aq) → Mn^{2+}(aq) + H$_2$O(l)
   V^{2+}(aq) → V^{3+}(aq)

c) MnO$_4^-$(aq) + H$^+$(aq) → Mn^{2+}(aq) + H$_2$O(l)
   Sn^{2+}(aq) → Sn^{4+}(aq)

d) MnO$_4^-$(aq) + H$^+$(aq) → Mn^{2+}(aq) + H$_2$O(l)
   V^{2+}(aq) + H$_2$O → VO$_3^-$(aq) + H$^+$(aq)

e) Cr$_2$O$_7^{2-}$(aq) + H$^+$(aq) → Cr^{3+}(aq) + H$_2$O(l)
   Br$^-$(aq) → Br$_2$(aq)

f) Cr$_2$O$_7^{2-}$(aq) + H$^+$(aq) → Cr^{3+}(aq) + H$_2$O(l)
   SO$_2$(aq) + H$_2$O(l) → SO$_4^{2-}$(aq) + H$^+$(aq)

g) NO$_3^-$(aq) → NO(g)
   Cu(s) → Cu^{2+}(aq)   (16)

14 a) Use Table 5.1 of electrode potentials on pages 96 and 97 to write the half-equations involved in the reaction between hydrogen peroxide and iron(II)sulfate in acid solution.

b) Write a full redox equation for the reaction involved.

c) Describe the change you would see in the solution when the reaction occurs.

d) How many moles of iron(II) sulfate react with 1 mole of H$_2$O$_2$?   (4)

15 a) Write the half-equations involved when dichromate(VI) ions in acid solution react with sulfite ions, SO$_3^{2-}$, to form chromium(III) ions and sulfate ions, SO$_4^{2-}$.

b) Write a full, balanced redox equation for the reaction.

c) Describe the change you would see in the solution when the reaction occurs.

d) How many moles of sulfite react with one mole of dichromate?   (4)

16 Use the standard electrode potential data supplied to calculate the cell potential and write the overall equation for the reaction which tends to happen for each of the following:

a) V^{3+}(aq) + e$^-$ ⇌ V^{2+}(aq);
   Zn^{2+}(aq) + 2e$^-$ ⇌ Zn(s)

b) Br$_2$(aq) + 2e$^-$ ⇌ 2Br$^-$(aq);
   I$_2$(aq) + 2e$^-$ ⇌ 2I$^-$(aq)   (4)

V^{3+}(aq) → V^{2+}(aq) + e$^-$	−0.26V
Zn^{2+}(aq) + 2e$^-$ ⇌ Zn(s)	−0.76V
Br$_2$(aq) + 2e$^-$ ⇌ 2Br$^-$(aq)	+1.09V
I$_2$(aq) + 2e$^-$ ⇌ 2I$^-$(aq)	+0.54V

17 Use the standard electrode potential data in Table 5.1 on pages 96 and 97 to arrange the following sets of metals in order of decreasing strength as reducing agents:

a) Ca, K, Li, Mg, Na
b) Cu, Fe, Pb, Sn, Zn.   (4)

18 Use the standard electrode potential data below to arrange the following sets of molecules/ions in order of decreasing strength as oxidising agents in acid solution:

Cr$_2$O$_7^{2-}$, Fe^{3+}, H$_2$O$_2$, MnO$_4^-$

Cr$_2$O$_7^{2-}$(aq) + 14H$^+$(aq) + 6e$^-$ ⇌ 2Cr^{3+}(aq) + 7H$_2$O(l)	+1.33V
Fe^{3+}(aq) + e$^-$ ⇌ Fe^{2+}(aq)	+0.77V
H$_2$O$_2$(aq) + 2H$^+$(aq) ⇌ 2H$_2$O(l)	+1.77V
MnO$_4^-$(aq) + 8H$^+$(aq) + 5e$^-$ ⇌ Mn^{2+}(aq) + 4H$_2$O(l)	+1.51V

                (2)

19 Use the standard electrode potentials in the table below to calculate the cell potential for each of the following pairs of half-cells:

a) Mg(s)/Mg^{2+}(aq) and Zn(s)/Zn^{2+}(aq)
b) Sn^{4+}(aq)/Sn^{2+}(aq) and Fe^{3+}(aq)/Fe^{2+}(aq)
c) I$_2$(aq)/2I$^-$(aq) and Br$_2$(aq)/2Br$^-$(aq)
d) Zn(s)/Zn^{2+}(aq) and I$_2$(aq)/2I$^-$(aq)
e) Sn^{4+}(aq)/Sn^{2+}(aq) and Br$_2$(aq)/2Br$^-$(aq)   (5)

Reaction	$E^{\ominus}/V$
Mg^{2+}(aq) + 2e$^-$ ⇌ Mg(s)	−2.37
Zn^{2+}(aq) + 2e$^-$ ⇌ Zn(s)	−0.76
Sn^{4+}(aq) + 2e$^-$ ⇌ Sn^{2+}(aq)	+0.15
I$_2$(aq) + 2e$^-$ ⇌ 2I$^-$(aq)	+0.54
Fe^{3+}(aq) + e$^-$ ⇌ Fe^{2+}(aq)	+0.77
Br$_2$(aq) + 2e$^-$ ⇌ 2Br$^-$(aq)	+1.09

20 Dental amalgam contains about 40% mercury (Hg) combined with an alloy that is made largely of silver and tin. (Small amounts of copper and zinc are also present.)

A number of redox reactions are possible with the amalgam as one electrode and saliva in the mouth as the electrolyte.

Two examples are:

$$Ag^+ + e^- \rightleftharpoons Ag/Hg \text{ (amalgam)} \quad E^\ominus = +0.85V$$

$$Sn^{2+} + 2e^- \rightleftharpoons Sn/Hg \text{ (amalgam)} \quad E^\ominus = -0.13V$$

If a piece of aluminium foil is bitten by teeth containing an amalgam, an unpleasant sharp pain is experienced. This happens because a temporary cell is set up between the amalgam and the aluminium.

$$Al^{3+}(aq) + 3e^- \rightleftharpoons Al(s) \quad E^\ominus = -1.66V$$

Describe what happens in the cell and why it results in a pain being felt. *(3)*

21 Use the standard electrode potentials in the table below to decide whether these reagents will react together under standard conditions:

a) $Co(s) + Pb^{2+}(aq)$
b) $Pb(s) + Ni^{2+}(aq)$
c) $I_2(aq) + Ni^{2+}(aq)$
d) $Cl^-(aq) + MnO_4^-(aq) + H^+(aq)$
e) $I^-(aq) + Cr_2O_7^{2-}(aq) + H^+(aq)$
f) $Cl^-(aq) + Cr_2O_7^{2-}(aq) + H^+(aq)$
g) $Cr^{3+}(aq) + H_2O(l) + MnO_4^-(aq) + H^+(aq)$ *(13)*

Reaction	$E^\ominus/V$
$Co^{2+}(aq) + 2e^- \rightleftharpoons Co(s)$	−0.28
$Ni^{2+}(aq) + 2e^- \rightleftharpoons Ni(s)$	−0.25
$Pb^{2+}(aq) + 2e^- \rightleftharpoons Pb(s)$	−0.13
$I_2(aq) + 2e^- \rightleftharpoons 2I^-(aq)$	+0.54
$Fe^{3+}(aq) + e^- \rightleftharpoons Fe^{2+}(aq)$	+ 0.77
$Cr_2O_7^{2-}(aq) + 14H^+(aq) + 6e^- \rightleftharpoons$ $2Cr^{3+}(aq) + 7H_2O(l)$	+1.33
$Cl_2(aq) + 2e^- \rightleftharpoons 2Cl^-(aq)$	+1.36
$MnO_4^-(aq) + 8H^+(aq) + 5e^- \rightleftharpoons$ $Mn^{2+}(aq) + 4H_2O(l)$	+1.51

22 Use the standard electrode potentials in the table below to answer the questions that follow.

A	$Fe^{3+}(aq) + e^- \rightarrow Fe^{2+}(aq)$	$E^\ominus = +0.77V$
B	$Cu^{2+}(aq) + 2e^- \rightarrow Cu(s)$	$E^\ominus = +0.34V$
C	$2H^+(aq) + e^- \rightarrow H_2(g)$	$E^\ominus = 0V$
D	$O_2(g) + 4H^+(aq) + 4e^- \rightarrow H_2O(l)$	$E^\ominus = 0.037V$

a) An electrochemical cell was arranged using reactions A and B.
   i) Write half-equations for the reactions which occur in each half-cell when a current flows.
   Say which half-equation involves oxidation and which involves reduction.
   ii) Calculate the change in oxidation number of the oxidised and reduced elements in each half-cell.
   iii) Determine the cell potential of the cell.
b) Fuel cells using reactions C and D are increasingly being used to generate electricity.
   i) Construct an overall equation for the cell reaction and show your working.
   ii) From which half-cell do electrons flow into the external circuit?
   iii) State two advantages and two disadvantages of using fuel cells based on C and D to generate energy rather than using fossil fuels. *(12)*

23 Using the following electrode potentials:

$Fe^{3+}(aq) + e^- \rightarrow Fe^{2+}(aq)$ $\quad$ +0.77V
$2I^-(aq) \rightarrow I_2(aq) + 2e^-$ $\quad$ −0.54V
$2Cl^-(aq) \rightarrow Cl_2(aq) + 2e^-$ $\quad$ −1.36V

explain why iron(III) chloride is a stable compound but iron(III) iodide cannot be made. Consider the relevant electrode potentials and explain why iron(III) iodide cannot be formed. *(4)*

24 A cell that can be used for heart pacemakers contains a negative electrode made of zinc and a positive electrode made of platinum. In the cell, zinc ions are discharged and oxygen in the blood reacts with water to form hydroxide ions.
a) Write half-equations for the reactions taking place at the two electrodes.
b) Write an overall equation for the reaction in the cell.
c) The standard electrode potential for Zn^{2+}/Zn is −0.76V and that for the reaction of oxygen is +0.40V.

What does this suggest the overall voltage of the cell would be if it were operated under standard conditions?

d) Suggest two reasons why the actual operating cell voltage is likely to be different from that in your answer to part c)? *(5)*

25 Lead–sulfuric acid accumulators are used widely as electrical power supplies in vehicles. They do, however, have the disadvantages of being heavy and the soft lead metal being damaged easily. More expensive cells have been developed that are lighter, stronger and, in some cases, more reliable. An example is a zinc–silver cell in which concentrated potassium hydroxide is the electrolyte.

a) When supplying current, the chemical change at the negative electrode of a lead–sulfuric acid accumulator is:

$$Pb + HSO_4^- \rightarrow PbSO_4 + H^+$$

Balance this half-equation by adding electrons.

b) The overall equation in the accumulator is:

$$Pb + PbO_2 + 2H^+ + 2HSO_4^- \rightarrow 2PbSO_4 + 2H_2O$$

Give the half-equation for the reaction at the positive electrode.

c) The overall reaction taking place in the zinc–silver cell with a potassium hydroxide electrolyte is:

$$Ag_2O + Zn + H_2O \rightarrow 2Ag + Zn(OH)_2$$

Suggest half-equations for the reactions at the negative and positive electrodes within the cell. *(5)*

26 A fuel cell developed by NASA in 1980 used an arrangement similar to the hydrogen–oxygen fuel cell described in this chapter (page 106). However, instead of pumping gases into the negative and positive electrode compartments, solutions were run from external storage tanks over inert electrodes within each compartment. One tank supplied $Cr^{2+}(aq)$ and the other tank supplied $Fe^{3+}(aq)$.

In such a cell, a redox reaction occurs and, by continuously supplying fresh solutions, the cell is able to run for an indefinite length of time. Electrode potentials that may be relevant are given in the table.

Reaction	$E^{\ominus}/V$
$Cr^{3+}(aq) + e^- \rightleftharpoons Cr^{2+}(aq)$	−0.41
$Cr^{2+}(aq) + 2e^- \rightleftharpoons Cr(s)$	−0.74
$Fe^{3+}(aq) + e^- \rightleftharpoons Fe^{2+}(aq)$	+0.77
$Fe^{3+}(aq) + 3e^- \rightleftharpoons Fe(s)$	−0.037

a) Use the electrode potentials in the table to suggest the overall reaction. State the voltage produced by the cell.

b) The solutions are supplied as chlorides and the membrane of the cell allows the passage of chloride ions through it.
Explain why it is essential for the membrane to allow the passage of chloride ions. *(5)*

27 Construct the equations for processes that take place at the positive and negative electrodes in a fuel cell when the fuel used is ethanol. *(2)*

Challenge

28 The health of most plants requires good circulation of air through the soil in which they grow. This is, in part, because a good supply of oxygen maintains ions in the soil in the fully oxidised form that may be necessary to support plant growth.

$E^{\ominus}$ for the half-reaction
$O_2(g) + 4H^+(aq) + 4e^- \rightleftharpoons 2H_2O(l)$ is 1.23V. However, the conditions in the soil are considerably different from standard conditions and, as a result, the electrode potential falls to about 0.80V.

a) In what ways do the conditions in soil differ from standard conditions?

b) Explain, qualitatively, why the different conditions lead to a lower value for the electrode potential.

c) Iron is usually present in soils in the form of Fe^{3+} ions. However, if the oxygen circulation is restricted because the soil becomes compacted or waterlogged, the iron is found as Fe^{2+} and the soil takes on a characteristic green–grey colouration.
Explain why this occurs.

d) In a landfill site, rubbish is compacted heavily and circulation of air is almost impossible. Under these circumstances, sulfate ions are reduced to sulfide and the foul-smelling, toxic gas hydrogen sulfide, H_2S, can be released.

Organic matter, such as CH_3COOH, in the waste is also reduced with the result that methane is formed. Hydrogen ions are required for these reductions and they are provided by surrounding decomposing materials.

Write balanced half-equations for the reductions:

i) $H^+ + SO_4^{2-} \rightarrow H_2S + H_2O$

ii) $H^+ + CH_3COOH \rightarrow CH_4 + H_2O$
(to illustrate the reduction of organic matter) *(9)*

29

Figure 5.11 A bombardier beetle (×27).

The remarkable bombardier beetle (Figure 5.11) has a unique method of defending itself against attack. The beetle stores two separated solutions:
- a 25% solution (by mass) of hydrogen peroxide, H_2O_2
- a 10% solution (by mass) of hydroquinone (Figure 5.12).

Figure 5.12 Hydroquinone.

When the beetle is threatened, the solutions mix, oxygen is generated and an explosion takes place! The surroundings are sprayed with an unpleasant cocktail of chemicals and steam. The explosion is caused by redox reactions that proceed via an intermediate complex called quinhydrone and then result in the formation of quinone as the final product:

Figure 5.13 Quinone.

The mixture that is sprayed contains both quinhydrone and quinone.

Oxygen, which comes from the catalytic decomposition of some of the hydrogen peroxide, provides the propulsion for the explosion.

The half-equations and standard electrode potentials are:

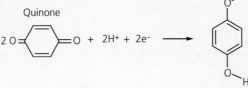

Quinone

$2 C_6H_4O_2 + 2H^+ + 2e^- \longrightarrow C_{12}H_{10}O_4$

$E^{\ominus} = +0.75\,V$
Unstable intermediate

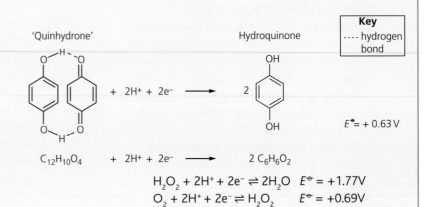

'Quinhydrone'   Hydroquinone

$E^{\ominus} = +0.63\,V$

$C_{12}H_{10}O_4 + 2H^+ + 2e^- \longrightarrow 2 C_6H_6O_2$

$H_2O_2 + 2H^+ + 2e^- \rightleftharpoons 2H_2O \quad E^{\ominus} = +1.77V$

$O_2 + 2H^+ + 2e^- \rightleftharpoons H_2O_2 \quad E^{\ominus} = +0.69V$

Key
---- hydrogen bond

a) It is not entirely clear how the catalytic breakdown of hydrogen peroxide occurs. The likely answer is that it is catalysed by an enzyme, but it is possible that hydrogen peroxide undergoes an autocatalytic process in which one redox reaction of hydrogen peroxide promotes the other.
Use the two hydrogen peroxide half-equations to explain how autocatalysis might be possible.

b) i) A further possibility is that a metal ion such as Fe^{3+} initiates the hydrogen peroxide decomposition. Use the half-equation $Fe^{3+}(aq) + e^- \rightarrow Fe^{2+}(aq)$ ($E^{\ominus} = +0.77V$) to explain how this could be possible.

 ii) Suggest why Fe^{3+} might effectively be a catalyst for this process.

c) Calculate the concentration in $mol\,dm^{-3}$ of:
 i) a 10% solution of hydroquinone (assume this is 10 g of hydroquinone in $100\,cm^3$ of solution)
 ii) a 25% solution of hydrogen peroxide (assume this is 25 g of hydrogen peroxide in $100\,cm^3$ of solution).

d) i) If just $0.1\,cm^3$ of hydrogen peroxide were converted to oxygen using the redox reaction with $Fe^{3+}(aq)$ ions, what volume of oxygen would be produced at 100°C (the temperature at which the beetle expels liquids at the point of the explosion).
(Assume that 1 mol of gas occupies a volume of $30.1\,dm^3$ at 100°C.)

 ii) How does this volume of oxygen differ from the volume of oxygen that would be collected at 100°C by heating $0.1\,cm^3$ of hydrogen peroxide?

e) Use the electrode potentials to explain how hydroquinone and quinone can both be formed as part of the beetle's spray. *(13)*

30 Using the data below, explain why $E^{\ominus}$ for the half-reaction $Fe^3(aq) + 3e^- \rightarrow Fe(s)$ is $-0.037V$?
$E^{\ominus}$ for the half-reaction

$Fe^{3+}(aq) + e^- \rightarrow Fe^{2+}(aq)$ is $+0.77V$

$E^{\ominus}$ for the half-reaction

$Fe^{2+}(aq) + 2e^- + Fe(s)$ is $-0.44V$ *(4)*

Chapter 6

Transition elements and qualitative analysis

Prior knowledge

In this chapter it is assumed that you are familiar with:

- structure of the atom and electronic configurations
- dative covalent bonding and shapes of molecules and ions
- tests for anions
- oxidation numbers and ionic half-equations.

You should be familiar with atomic number and mass number, and how they can be used to determine the number of sub-atomic particles within an atom or ion. You should be aware that the electrons within an atom are arranged in shells which can be sub-divided into subshells. You should be able to use the electronic sub-shell notation, $1s^2 2s^2 2p^6$... etc. up to and including $_{36}Kr$.

Dative covalent bonding, like covalent bonding, involves a shared pair of electrons but, with dative covalent bonding, one atom provides both of the shared pair of electrons. Shapes of simple covalent molecules, and ions, can be predicted using the electron-pair repulsion theory which states that all electron pairs repel and the shape is determined by the number, and type, of electrons pairs around the central ion.

You should also be familiar with simple chemical tests for identifying some common anions, including the halides, sulfate and carbonate ions.

Test yourself on prior knowledge

1 Copy and complete the table below for each atom or ion.

		Number of			Electron configuration (use sub-shell notation)
		protons	neutrons	electrons	
a)	$_{9}^{19}F$				
b)	$_{13}^{27}Al$				
c)	$_{35}^{79}Br^-$				
d)	$_{34}^{79}Se^{2-}$				

2 Predict the shape and the bond angles in:
  a) NH_3 and NH_4^+
  b) H_2O and H_3O^+.
3 Describe a simple chemical test for:
  a) the carbonate ion, CO_3^{2-}
  b) the halide ions, Cl^-, Br^- and I^-.
  c) State how aqueous ammonia is used to distinguish between the halide ions in part b).

4 Determine the oxidation state of the element underlined in each of:

a) $\underline{Cr}_2O_3$

b) $\underline{V}_2O_5$

c) $\underline{S}O_3^{2-}$

d) $\underline{S}O_4^{2-}$

5 Construct ionic half-equations for each of the following:

a) $Fe^{2+} \rightarrow Fe^{3+}$

b) $NO_3^- / H^+ \rightarrow NO$

c) $MnO_4^- / H^+ \rightarrow Mn^{2+}$

d) $S_2O_3^{2-} / H^+ \rightarrow S$

The electron configuration of the d-block elements

As the shells of electrons around the nuclei of atoms get further from the nucleus, they become closer in energy. So the difference in energy between the second and third shells is less than that between the first and second. When the fourth shell is reached, there is, in fact, an overlap between the orbitals of highest energy in the third shell (the 3d-orbitals) and that of lowest energy in the fourth shell (the 4s-orbital). The 3d sub-shell is, on average, closer to the nucleus than the 4s sub-shell, but at a higher energy level. So, once the 3s and 3p sub-shells are filled, the next electrons go into the 4s sub-shell because it occupies a lower energy level than the 3d sub-shell, even though the 3d sub-shell is closer to the nucleus. This means that potassium and calcium have electrons in the 4s sub-shell but none in the 3d sub-shell.

- $_{19}$K has the electronic configuration $1s^2 2s^2 2p^6 3s^2 3p^6 4s^1$.

- $_{20}$Ca has the electronic configuration $1s^2 2s^2 2p^6 3s^2 3p^6 4s^2$.

After the 4s sub-shell is filled, the 3d sub-shell is filled in the elements from scandium to zinc. $_{21}$Sc has the electronic configuration $1s^2 2s^2 2p^6 3s^2 3p^6 3d^1 4s^2$.

Table 6.1 The electron configurations from potassium to zinc in Period 4. [Ar] represents the electron configuration of argon as $1s^2 2s^2 2p^6 3s^2 3p^6$.

Element	Symbol	Electron structure	
		spdf notation	Electrons-in-boxes notation
potassium	K	[Ar]4s^1	[Ar] (3d empty) 4s ↑
calcium	Ca	[Ar]4s^2	[Ar] (3d empty) 4s ↑↓
scandium	Sc	[Ar]3d^{1}4s^2	[Ar] ↑ 4s ↑↓
titanium	Ti	[Ar]3d^{2}4s^2	[Ar] ↑ ↑ 4s ↑↓
vanadium	V	[Ar]3d^{3}4s^2	[Ar] ↑ ↑ ↑ 4s ↑↓
chromium	Cr	[Ar]3d^{5}4s^1	[Ar] ↑ ↑ ↑ ↑ ↑ 4s ↑
manganese	Mn	[Ar]3d^{5}4s^2	[Ar] ↑ ↑ ↑ ↑ ↑ 4s ↑↓
iron	Fe	[Ar]3d^{6}4s^2	[Ar] ↑↓ ↑ ↑ ↑ ↑ 4s ↑↓
cobalt	Co	[Ar]3d^{7}4s^2	[Ar] ↑↓ ↑↓ ↑ ↑ ↑ 4s ↑↓
nickel	Ni	[Ar]3d^{8}4s^2	[Ar] ↑↓ ↑↓ ↑↓ ↑ ↑ 4s ↑↓
copper	Cu	[Ar]3d^{10}4s^1	[Ar] ↑↓ ↑↓ ↑↓ ↑↓ ↑↓ 4s ↑
zinc	Zn	[Ar]3d^{10}4s^2	[Ar] ↑↓ ↑↓ ↑↓ ↑↓ ↑↓ 4s ↑↓

From Sc to Mn a single electron occupies one of the 3d-orbitals. From Fe to Zn the electrons pair up and fill the 3d-orbitals. The electron configurations of chromium and copper do not fit the general pattern. The explanation of these irregularities lies in the stability associated with half-filled and filled sub-shells. So, the electron structure of chromium, $[Ar]3d^5 4s^1$, with half-filled 3d and 4s sub-shells is more stable than the electron structure $[Ar]3d^4 4s^2$. Similarly, the electron structure of copper, $[Ar]3d^{10} 4s^1$, with a filled 3d sub-shell and a half-filled 4s sub-shell, is more stable than $[Ar]3d^9 4s^2$.

Along the series of d-block elements from scandium to zinc, the number of protons in the nucleus increases by one from one element to the next. The added electrons go into an inner d sub-shell. The outer electrons are always in the 4s sub-shell, and this means that the changes in chemical properties across the series are much less marked than the big changes across a series of p-block elements such as aluminium to argon. This is clearly illustrated by the variation in first ionisation energies as shown in the figure below:

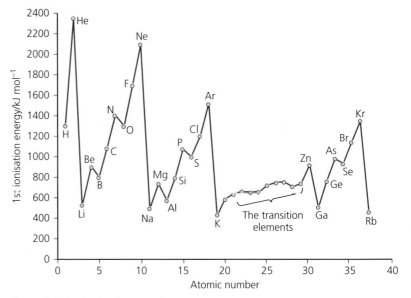

Figure 6.1 First ionisation energies.

The elements Sc to Zn all react by losing electrons and, since their ionisation energies are similar, you would expect their chemistry to be similar. In addition, the chemistry of an element is determined, to a large extent, by its outer shell electrons – because they are the first to get involved in reactions. All the d-block elements from Sc to Zn have their outer electrons in the 4s sub-shell, so they are similar in many ways.

When transition metals form their ions, electrons are lost initially from the 4s sub-shell and not the 3d sub-shell. This may seem somewhat illogical because, prior to holding any electrons, the 4s level is more stable than the 3d level. But, once the 3d sub-shell is occupied by electrons, being closer to the nucleus, these 3d electrons repel the 4s electrons to a higher energy level. The 4s electrons are, in fact, pushed to an energy level higher than those occupying the 3d sub-shell. So, when transition metals form ions they lose electrons from the 4s orbital before the 3d level. Because of this, it is usual to write the ground state of the transition metals with the 3d electrons listed before the 4s. For example, iron is normally written as $[Ar]3d^6 4s^2$ rather than $[Ar]4s^2 3d^6$. This further emphasises the fact that transition metals have similar chemical properties, which are dictated by the behaviour of the 4s electrons in their outer shells.

> **Example 1**
>
> What is the electron configuration of the Mn^{2+} ion?
>
> **Answer**
>
> A manganese atom has the structure $[Ar]3d^54s^2$.
>
> When it forms an ion the 4s electrons are lost, so the ion has an electron configuration $[Ar]3d^5$.

> **Test yourself**
>
> 1 Write the full electron configurations of the:
>   a) Sc atom and Sc^{3+} ion
>   b) Cr atom and Cr^{3+} ion
>   c) Zn atom and Zn^{2+} ion.
> 2 a) Write the electron structure of the Fe atom, Fe^{2+} ion and Fe^{3+} ion.
>   b) Which ion, Fe^{2+} or Fe^{3+}, would you expect to be most stable? Explain your answer.

Transition elements

> **Key term**
>
> A transition element is a d-block element that forms an ion with an incomplete d sub-shell.
> The above definition of a transition element excludes scandium and zinc.

It is not unusual for the d-block elements to be referred to as the transition elements, and the simplest and neatest way to define the transition elements might be to say that they are the elements in the d-block of the periodic table. However, this is not strictly correct because the elements scandium and zinc do not show many of the properties that are listed below as being characteristic of transition elements. This is because most of these properties depend on the presence of ions that contain d-orbitals which are not completely filled. There are clear differences between Sc and Zn, when compared to the elements Ti to Cu:

● Scandium and zinc have only one oxidation state in their compounds (scandium +3, zinc +2), whereas the elements from titanium to copper have two or more.

● The compounds of scandium and zinc are usually white, unlike other elements in the d-block which are generally coloured.

● Scandium and zinc, and their compounds, show little of the catalytic activity that is a property of other elements in the d-block.

When a scandium atom forms an ion, it does so by losing three electrons to make Sc^{3+}. This has the electron configuration $1s^22s^22p^63s^23p^6$, which has no d-electrons. (The configuration is the same as that of argon.)

A zinc ion, Zn^{2+}, does have d-electrons; the configuration $1s^22s^22p^63s^23p^6$ $3d^{10}$ but the d-orbitals are completely filled.

Typical properties of transition elements

Transition metals share a number of common properties. They:

● are metals, and are good conductors, have high melting points and high boiling points

- form compounds in more than one oxidation state
- form coloured compounds
- act as catalysts, either as the elements or as their compounds
- form a variety of complex ions.

Conductivity, melting points and boiling points

As solids, atoms pack together closely and, because of the ease with which electrons can move from orbital to orbital, the transition elements are all metals that conduct electricity well. Their atoms have relatively small radii and pack together closely, which means that there are strong bonds between the metal ions and the delocalised electrons. This contributes to their high density, and high melting and boiling points.

Activity

Studying the ionisation energies of transition metals

In the Year 1 book it was explained that ionisation energies provide us with information about the electronic structures of the main group elements. Ionisation energies can also be used to provide us with detail of the electronic structures of the d-block elements.

Look carefully at Figure 6.2 which shows graphs of the first, second and third ionisation energies of the elements from scandium to zinc.

1 Write the electronic structure of the following atoms and ions, using [Ar] for the electronic structure of argon:
   a) Zn
   b) Cu^+
   c) Zn^+
   d) Cr^+
   e) Mn^+.

2 Write an equation for:
   a) the second ionisation energy of chromium
   b) the third ionisation energy of iron.

3 Explain the general trend in ionisation energies as atomic number increases.

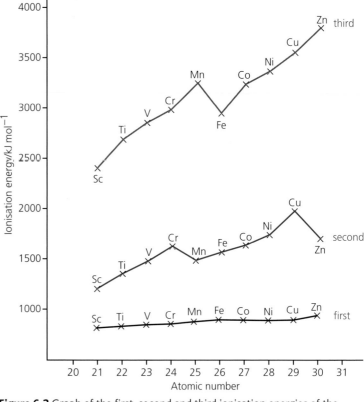

Figure 6.2 Graph of the first, second and third ionisation energies of the elements from scandium to zinc in the periodic table.

4 Why is the first ionisation energy of zinc significantly higher than that of any of the other elements?

5 Why is the second ionisation energy of every element higher than its first ionisation energy?

6 By referring to electron structures, explain why the second ionisation energy of both chromium and copper is higher than that of the next element.

7 By referring to electron structures, explain why the third ionisation energy of manganese is higher than the third ionisation energy of iron.

Variable oxidation states

The transition elements all exhibit two or more oxidation states in their compounds. Most of the transition elements have an outer shell electron configuration of $4s^2$ and they readily lose the two outer electrons and have an oxidation state +2. The reasons for the existence of multiple oxidation states are quite complicated, but can be explained partly by the similarity of the sizes of their atomic radii and their ionisation energies.

You do not need to know all of the variations in oxidation states but it is worth recalling that:

● all transition elements can form an ion of oxidation state +2

● the maximum oxidation state cannot exceed the total number of 3d and 4s electrons in the electron configuration of the atom.

Some oxidation states are listed in Table 6.2.

Table 6.2 Common oxidation states of transition elements.

Ti	V	Cr	Mn	Fe	Co	Ni	Cu
+4, +3, +2	+5,+4, +3, +2	+6, +3, +2	+7, +6, +4, +3, +2	+3, +2	+3, +2	+2, +3	+2, +1

Test yourself

3 Give examples of compounds, other than oxides, of:
   a) chromium in the +3 and +6 states
   b) manganese in the +2 and +7 states
   c) iron in the +2 and +3 states
   d) copper in the +1 and +2 states.
4 Explain why the oxidation number of vanadium in VO_2^+ ions is +5.
5 a) Why do you think that copper is the only transition metal to have +1 as a main oxidation state?
   b) Which other transition metal might you expect to form some compounds in the +1 state? Explain your answer.

The formation of coloured ions

A particularly noticeable feature of the transition elements is the characteristic colours of their aqueous ions. Aqueous copper ions are blue, aqueous chromium(III) ions are green, aqueous dichromate(VI) ions are orange and aqueous manganate(VII) ions are purple; the list could be extended. One of the most attractive and effective demonstrations of the range of oxidation states in a transition element can be shown by shaking a solution of ammonium vanadate(V), NH_4VO_3, in dilute sulfuric acid with zinc. The VO_3^- reacts with the H^+ ions in the sulfuric acid to form VO_2^+ ions, resulting in a bright yellow solution. The zinc then reduces the vanadium ions, through various colour changes, finally resulting in the formation a violet solution. The colour changes are shown in Figure 6.3.

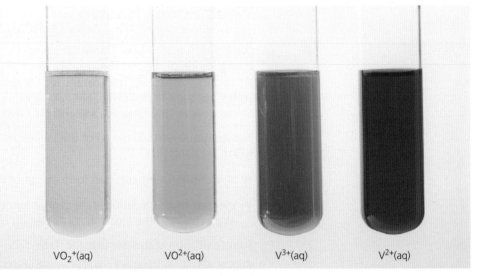

VO$_2^+$(aq)   VO^{2+}(aq)   V^{3+}(aq)   V^{2+}(aq)

Figure 6.3 The colours of the oxidation states of vanadium.

The reasons why the ions are coloured is complex and it is sufficient to appreciate that two essential characteristics are required:

● a central ion containing incompletely filled d-orbitals

● the central ion has ions or molecules attached to it.

Water is a common attachment and it is this that makes aqueous copper(II) salts appear blue and aqueous iron(II) salts appear light green.

Test yourself

6 Deduce the oxidation state of the vanadium in each of the following ions:
   a) i) VO$_3^-$ iii) V^{3+}
       ii) VO^{2+} iv) V^{2+}.
   b) Construct ionic equations for each of the following changes that occur in acid solution:
       i)   VO$_3^- \rightarrow$ VO$_2^+$
       ii) VO$_2^+ \rightarrow$ VO^{2+}
       iii) VO$^{2+} \rightarrow$ V^{3+}.
   c) Suggest why the half equation for VO$_3^- \rightarrow$ VO$_2^+$ contains no electrons.
7 Write the electron configurations for each of the following pairs:
   a) Cu$^+$ and Cu^{2+}
   b) Ti^{3+} and Ti^{4+}.
   c) For each pair above, predict which ion in the pair is likely to be colourless.
   d) TiO$_2$ is used as a paint pigment. Predict the colour of the paint. Justify your answer.

Use as catalysts

Transition metals are often used as **heterogeneous catalysts** in reactions, because they can use their d-orbitals to adsorb other molecules or ions onto the metal's surface and, in doing so, weaken the internal bonds of the adsorbed species. This lowers the activation energy for the reaction and allows it to proceed. The products escape from the surface of the catalyst (desorb), which remains unchanged, and more reactants can then be adsorbed onto the surface, and so the reaction continues.

There are many examples of transition metals as catalysts including the use, in industry, of:

- iron as a catalyst in the Haber process to produce ammonia
- platinum, palladium and rhodium in the catalytic converters of car exhausts
- nickel in the hydrogenation of alkenes.

In the laboratory, an example is the use of aqueous copper ions to encourage the reaction of zinc with acids or manganese(IV) oxide to promote the release of oxygen from hydrogen peroxide:

$$2H_2O_2(aq) \rightarrow 2H_2O(l) + O_2(g)$$

Transition metal ions may also act as **homogeneous catalysts**.

You do not need to know an example, but you should appreciate that, in these cases, the catalyst ions take part in the reaction and are then regenerated. The transition metal ion is usually converted to a different oxidation state during the course of the reaction, and it is the relative ease with which this takes place that makes them so useful as catalysts.

Test yourself

8 The redox reaction between peroxodisulfate ions, $S_2O_8^{2-}(aq)$, and iodide ions, $I^-(aq)$, is very slow and can be monitored by the change in colour as $I_2(aq)$ is produced. The overall reaction is:

$$2I^-(aq) + S_2O_8^{2-}(aq) \rightarrow 2SO_4^{2-}(aq) + I_2(aq)$$

a) In the overall equation, deduce the changes in oxidation state and identify the oxidising agent.
b) If $Fe^{3+}(aq)$ ions are added, the reaction is much faster. It is believed that, initially, the $Fe^{3+}(aq)$ ions are reduced to $Fe^{2+}(aq)$ by the $I^-(aq)$ and are then oxidised back to $Fe^{3+}(aq)$ by the $S_2O_8^{2-}(aq)$.
Construct ionic half-equations, including state symbols, and use these to construct overall equations for the two steps in which the $Fe^{3+}(aq)$ is first reduced and then oxidised.
c) Justify that $Fe^{3+}(aq)$
   i) is a catalyst in the overall reaction
   ii) is a homogeneous catalyst.

Ligands and the formation of complex ions

In Chapter 4 it was shown how water encourages ionic compounds to dissolve by hydrating their ions. The lone pairs of electrons on the oxygen atom of a molecule of water are attracted strongly to positive ions in the ionic lattice, binding with them firmly enough to allow the lattice to break down. This can happen with any ion that possesses lone pairs of electrons.

Key terms

A **ligand** is a molecule or anion that has at least one lone pair of electrons and bonds to a central metal ion by a dative or **co-ordinate bond**.

A **co-ordinate bond** is a dative covalent bond formed between the ligand and the central metal ion.

A **complex ion** is an ion in which a number of molecules or anions are bound to a central cation by co-ordinate bonds.

The ion making the attachment via a lone pair of electrons is called a **ligand**. The resulting ion is called a **complex ion**.

An example of a complex ion with water as a ligand is $[Cu(H_2O)_6]^{2+}$ (Figure 6.4).

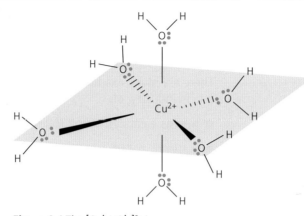

Figure 6.4 The $[Cu(H_2O)_6]^{2+}$ ion.

A range of complex ions can be formed with a variety of central cations and a range of surrounding ligands. Ammonia is an effective ligand and, for example, forms the complex $[Cr(NH_3)_6]^{3+}$ (Figure 6.5).

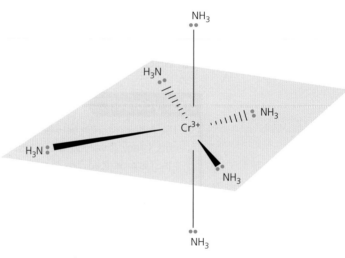

Figure 6.5 The $[Cr(NH_3)_6]^{3+}$ ion.

The word used for the ammonia ligands is 'ammine' with two letter 'm's in the name. The $[Cr(NH_3)_6]^{3+}$ ion is called a hexaammine chromium(III) ion. Do not confuse this with amine, which indicates a $-NH_2$ group.

Anions such as chloride ions, Cl^-, can also behave as ligands and can form complex ions such as $[CuCl_4]^{2-}$.

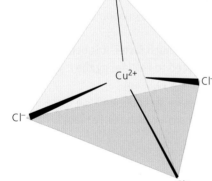

Figure 6.6 The $[CuCl_4]^{2-}$ ion.

Key term

The **co-ordination number** of a metal ion is the number of co-ordinate bonds to the metal ion from the surrounding ligands.

The overall charge of a complex ion is always the sum of the individual charges of the ions and molecules from which the complex is made. So, in this case, the net charge is 2- because the complex ion contains a Cu^{2+} central anion and four surrounding Cl^- cations (Figure 6.6).

The Cl^- ion has, in fact, four separate lone pairs of electrons and one of those is used to form the co-ordinate bond. The arrows represent the sharing of a lone pair of electrons.

Of the three complex ions shown above, the first two have a **co-ordination number** of 6 and the last one a co-ordination number of 4.

Shapes of complex ions

The shape of a complex ion depends on the number of ligands around the central metal ion (the co-ordination number of the central metal ion).

In complex ions with co-ordination number:

- **6** – the shape is octahedral and the bond angles are all 90°

- **4** – the shape is sometimes tetrahedral, bond angle 109° 5' but there are also a number of complex ions with a co-ordination number of 4 that are square planar, bond angle 90° .

- **2** – the shape is linear with bond angle 180°.

The structures are illustrated in Figure 6.7.

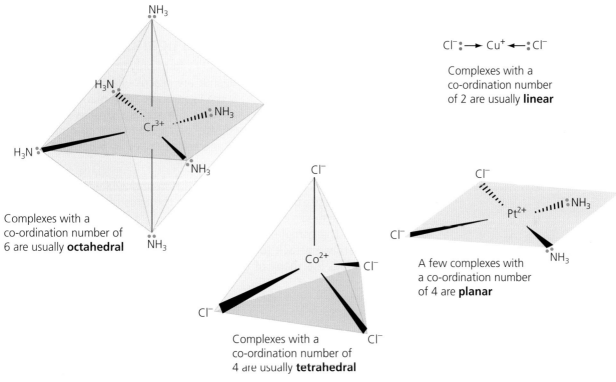

Figure 6.7 The shapes of complex ions.

An octahedral complex is said to show six-fold co-ordination; square planar and tetrahedral complexes show four-fold co-ordination.

Most complexes are formed by ions. However, it is also possible to form a complex using the uncharged metal atom. For example, nickel, iron and chromium all form a complex with carbon monoxide. The three complexes are illustrated in Figure 6.8.

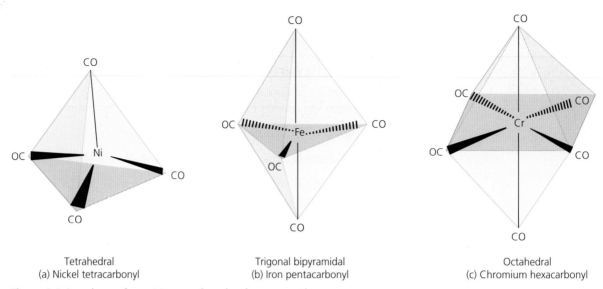

Tetrahedral
(a) Nickel tetracarbonyl

Trigonal bipyramidal
(b) Iron pentacarbonyl

Octahedral
(c) Chromium hexacarbonyl

Figure 6.8 Complexes of transition metals and carbon monoxide.

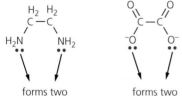

forms two
co-ordinate bonds

forms two
co-ordinate bonds

Figure 6.9 Bidentate ligands.

Monodentate and bidentate ligands

Most ligands, H_2O, NH_3, Cl^-, OH^- and CN^-, form only one co-ordinate bond with the central metal ion. These ligands are described as **monodentate** which literally means 'one tooth'. Some ligands can form more than one co-ordinate bond with the same central metal ion. Ligands such as 1,2-diaminoethane and the ethanedioate ion can form two co-ordinate bonds with the same central cation and are described as **bidentate** ('two-toothed', Figure 6.9).

The bidentate ligand 1,2-diaminoethane (often abbreviated in this context to 'en') uses the nitrogen atoms at the ends of the molecule, which have a lone pair of electrons, to bind onto a central cation at two points. In the complex ion shown in Figure 6.10, three molecules of 1,2-diaminoethane attach to the nickel ion. The overall co-ordination number is 6 and the shape of the complex is octahedral with bond angles 90°.

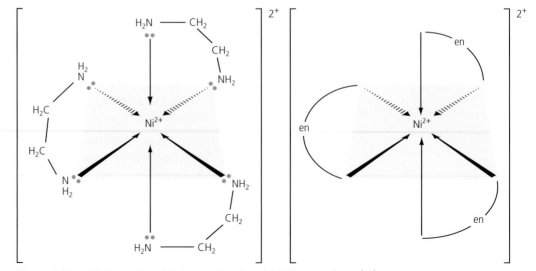

Figure 6.10 A nickel complex with three molecules of 1,2 diaminoethane (en).

Figure 6.11 The walrus, the tiger and the crab could perhaps all be described as 'bidentate'.

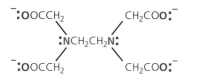

Figure 6.12 The EDTA^{4-} ion.

Polydentate ligands (those with several points of attachment) also exist. One of the most commonly encountered is EDTA, which has six points of attachment – it is hexadentate. The structure of EDTA is shown in Figure 6.12.

Each of the lone pairs is able to form a coordinate (dative) bond and is therefore able to attach in six places.

You will not be expected to draw ions like [Pb(EDTA)]$^{2-}$ but it gives an indication of the variety of complexes that can be formed.

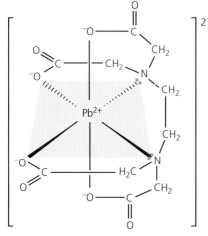

Figure 6.13 The complex ion formed by EDTA with a Pb^{2+} ion. The ligand can fold itself around metal ions, such as Pb^{2+}, so that four oxygen atoms and two nitrogen atoms form co-ordinate bonds to the metal ion. This is the ion formed when EDTA is used to treat lead poisoning. The EDTA forms such a stable complex with Pb^{2+} ions that it can be excreted through the kidneys.

> ### Test yourself
>
> 9 Draw dot-and-cross diagrams of each of the following and explain whether or not they can behave as a ligand in a complex ion.
>   a) NH_3
>   b) CH_4
>   c) CN^-
>   d) CO
>
> 10 Predict the likely shape and bond angle in each of the following. Draw the complex ion, clearly show the lone pairs of electrons that form the co-ordinate bonds. State the oxidation state of central cation in each complex.
>   a) $[Ag(CN)_2]^-$
>   b) $[Co(NH_3)_5Cl]^{2+}$
>   c) $[NiCl_4]^{2-}$
>   d) $[Fe(CN)_6]^{3-}$

Stereoisomerism of complex ions

Isomerism is commonplace in organic compounds. However, as a result of the spatial positions around a central cation, four and six co-ordinated complex ions can also have stereoisomers. You will recall that stereoisomerism defines *compounds that have the same molecular formula*

and same structural formula, but have a different arrangement in space.
Stereoisomerism falls into one of two categories:

● *cis–trans* (*E/Z*) isomerism

● optical isomerism.

Cis–trans (E/Z) isomerism

The square planar structure of $[Ni(NH_3)_2Cl_2]$ and the octahedral structure of $[Co(NH_3)_4Cl_2]^+$ give rise to *cis–trans* (*E/Z*) isomerism (Figure 6.14).

Diamminedichloronickel(II)   Tetraamminedichlorocobalt(III)

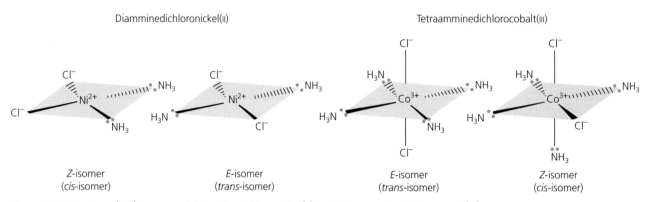

Z-isomer   E-isomer   E-isomer   Z-isomer
(cis-isomer)   (trans-isomer)   (trans-isomer)   (cis-isomer)

Figure 6.14 *Cis–trans* (*E/Z*) isomers of diamminedichloronickel(II) and tetraamminedichlorocobalt(III).

In Figure 6.14, consider the bond angle between the two chloride ligands and the central metal ion.

● If the Cl^-–(metal ion)–Cl^- bond angle is 90°, a *cis* (or *Z*) isomer forms.

● If the Cl^-–(metal ion)–Cl^- bond angle is 180°, a *trans* (or *E*) isomer results.

Optical isomerism

Octahedral complexes can also form optical isomers which are non-superimposable mirror images of each other. Optical isomerism can only occur with a multi-dentate ligand for example 1,2-diaminoethane, $H_2NCH_2CH_2NH_2$. Each of the NH_2 groups has a lone pair of electrons and they can approach the central metal ion in a vertical plane or a horizontal plane (Figure 6.15).

These two ligands are in the vertical plane, but twisted at 90° to each other.

This ligand is in the horizontal plane.

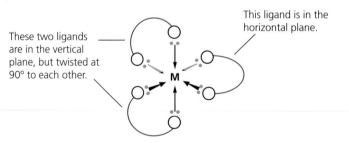

Figure 6.15 The shape of a complex ion showing M surrounded by three molecules of a bidentate ligand.

This gives rise to two possible isomers (Figure 6.16).

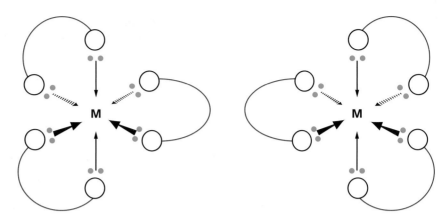

Figure 6.16 Optical isomers of a complex formed by three molecules of a bidentate ligand.

An example of a complex ion with optical isomers is $[Ni(H_2NCH_2CH_2NH_2)_3]^{2+}$, shown in Figure 6.17

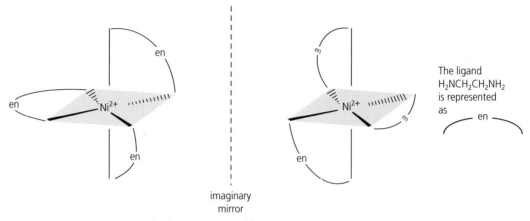

The ligand $H_2NCH_2CH_2NH_2$ is represented as

en

imaginary mirror

Figure 6.17 Optical isomers of $[Ni(H_2NCH_2CH_2NH_2)_3]^{2+}$.

As is the case with organic molecules, it is the asymmetry of the structure that leads to optical isomerism as the two molecules cannot be superimposed and are therefore spatially not the same.

Complex ions have various uses. A particularly interesting example is the *cis* (Z) form of the molecule $[PtCl_2(NH_3)_2]$. (The platinum is present as platinum(II), so the complex has no overall charge.) The structure of $[PtCl_2(NH_3)_2]$ is shown in Figure 6.18.

This compound is known as *cis*-platin. Used during chemotherapy as an anti-cancer drug, it is a colourless liquid that is usually administered by a drip into a vein. It works by binding onto the DNA of cancerous cells and preventing their division.

The importance of the exact shape and structure of the molecule is emphasised by the fact that the *trans* isomer is ineffective. Great care has to be taken when using chemotherapy to treat patients as the compounds can lead to unpleasant side effects and a balance has to be struck between the advantages of the treatment and the disadvantages.

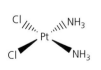

Figure 6.18 *Cis*-platin.

Cis-platin

This activity is most suitable for those studying biology. Others may find some parts need a certain amount of research to complete.

The neutral complex, $PtCl_2(NH_3)_2$, in which Cl^- ions and NH_3 molecules act as ligands, has two isomers. These isomers have different melting points and different chemical properties. One isomer, called *cis*-platin, is used in the treatment of certain cancers – the other isomer is ineffective against cancer.

Patients are given an intravenous injection of *cis*-platin which circulates all around the body, including the cancerous area. *Cis*-platin diffuses relatively easily through the tumour cell membrane because it has no overall charge, like the cell membrane. Its action is shown in Figure 6.19.

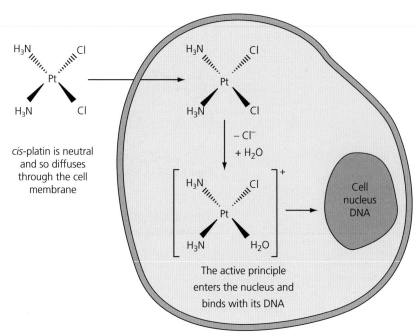

cis-platin is neutral and so diffuses through the cell membrane

The active principle enters the nucleus and binds with its DNA

Figure 6.19 The action of *cis*-platin.

Once inside the cell, *cis*-platin reacts by exchanging one of its chloride ions for a molecule of water forming $[Pt(NH_3)_2(Cl)(H_2O)]^+$ which is the 'active principle'. This positively charged ion then enters the cell nucleus where it readily bonds with two sites on the DNA. Binding involves co-ordinate bonding from the nitrogen or oxygen atoms in the bases of DNA to the platinum ion.

The *cis*-platin binding changes the overall structure of the DNA helix, pulling it out of shape and shortening the helical turn. The badly shaped DNA can no longer replicate and divide to form new cells, although the affected cells continue to grow. Eventually the cells die and, if enough of the cancerous cells absorb *cis*-platin, the tumour is destroyed. Unfortunately, *cis*-platin is not a miracle cure without risks or drawbacks. It is toxic, resulting in unpleasant side-effects, and can cause kidney failure. However, clinical trials have led to the discovery of other platinum complexes which cause fewer problems and are already being used as anti-cancer drugs.

1 Why is it possible to conclude that *cis*-platin has a square planar rather than a tetrahedral structure?
2 What type of isomerism do *cis*-platin and its isomer show?
3 **a)** What is the oxidation number of platinum in *cis*-platin?
   b) Write the systematic name of *cis*-platin.
   c) Draw the structure of *cis*-platin.
4 Why does *cis*-platin diffuse easily through the membrane of cells?
5 What is meant by the term 'active principle' applied to $[Pt(NH_3)_2(Cl)(H_2O)]^+$?
6 When $[Pt(NH_3)_2(Cl)(H_2O)]^+$ has formed inside the cell, it cannot diffuse out through the cell membrane. Why is this?
7 Why is a cell with *cis*-platin binding to DNA unable to replicate?

8 Why is the binding to *cis*-platin from nitrogen and oxygen atoms rather than from carbon and hydrogen atoms in the bases of DNA?

9 Why is *cis*-platin more likely to affect cancerous cells than normal cells?

10 Why is any anti-cancer drug which acts like *cis*-platin likely to have undesirable side-effects?

Ligand substitution

The ligands that are attached to an ion can sometimes be exchanged for another ligand in a **ligand substitution** reaction. This occurs when the substituting ligand has a stronger attachment to the metal than does the original ligand.

Two colourful illustrations of this occur with the blue aqueous copper ion, $[Cu(H_2O)_6]^{2+}$. On adding concentrated ammonia solution, a ligand substitution occurs. Four of the water ligands are replaced by ammonia ligands, which creates an intensely dark blue solution of the ion $[Cu(NH_3)_4(H_2O)_2]^{2+}$. Both ions are octahedral.

$$[Cu(H_2O)_6]^{2+} + 4NH_3 \rightleftharpoons [Cu(NH_3)_4(H_2O)_2]^{2+} + 4H_2O$$

The aqueous chromium ion $[Cr(H_2O)_6]^{3+}$ can also undergo a ligand exchange with aqueous ammonia. A precipitate of dark green chromium hydroxide is first obtained, but this dissolves to give a solution (which can appear to be green or almost violet depending on the anion present) and contains the ion $[Cr(NH_3)_6]^{3+}$ ion.

Many other ligand substitutions occur and complex ions are found widely in nature. Some ions are transported through plants as complexes, but a particularly interesting example is the transport of oxygen in blood by haemoglobin.

Haemoglobin and oxygen transportation

Ligand substitution is important in the transport of oxygen by haemoglobin. The active part of haemoglobin is the Fe^{2+} ion which forms co-ordinate bonds to four nitrogen atoms in the haem ligand and a further two dative bonds to nitrogen atoms in the globin. The bond between the central Fe^{2+} ion and one of the nitrogen atoms in the globin is weak and can readily be substituted by oxygen.

$$haemoglobin(aq) + O_2(g) \rightleftharpoons haemoglobin–O_2(aq)$$

(oxyhaemoglobin)

When air is breathed in, oxygen is transferred into the bloodstream and binds to the sixth position. It is then transported through the body. The binding of the oxygen molecule is weak, which allows it to be removed to fuel oxidation reactions within the body. When the oxygen is released, it is replaced by a water ligand which, in turn, will be replaced by another oxygen. This is an efficient system, but it can be disrupted if another ligand were to bind so strongly at the sixth position that oxygen could not replace it. One such ligand is carbon monoxide. When a person inhales carbon monoxide, it occupies this sixth site and the molecule of haemoglobin can no longer transport oxygen. This results initially in the person feeling drowsy and then losing consciousness. Ultimately, the poisoning can be fatal. When carbon monoxide occupies the sixth position the haemoglobin has a distinctive cherry-red colour, which is a tell-tale sign of its presence.

11 Draw the following ions and for each complex determine:
   a) the oxidation state of the central ion
   b) whether or not there are any stereoisomers. If there are, state the type of stereoisomerism and draw the isomers.
      i) $[FeF_4(OH)_2]^{3-}$
      ii) $[Co(NH_3)_5Cl]^{2+}$
      iii) $[Cr(H_2O)_4Cl_2]^+$
      iv) $[Cr(ox)_3]^{3-}$ where 'ox' is

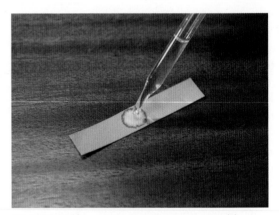

12 Write equations for the ligand substitution reactions which occur when:
   a) hexaaquacobalt(II) ions react with ammonia molecules to form hexaamminecobalt(II) ions
   b) hexaamminecobalt(II) ions react with chloride ions to form tetrachlorocobaltate(II) ions
   c) hexaaquairon(II) ions react with cyanide ions to form hexacyanoferrate(II) ions.

13 Explain the following changes with the help of equations.
   a) Adding ammonia solution to a pale blue solution of hydrated copper(II) ions produces a pale blue precipitate of the hydrated hydroxide.
   b) On adding more ammonia solution, the precipitate dissolves to give a deep blue solution.

Figure 6.20 Filter paper soaked in pink cobalt(II) chloride solution and dried in an oven until it is blue can be used to test for the presence of water.

14 A dilute solution of cobalt(II) chloride is pink because it contains hydrated cobalt(II) ions. The solution turns blue on adding concentrated hydrochloric acid with the formation of tetrachlorocobaltate(II) ions.
   a) Write an equation for the reaction which occurs when concentrated HCl is added to dilute cobalt chloride solution and indicate the colour of relevant species.
   b) Explain the chemical basis for the test illustrated in Figure 6.20.

Precipitation reactions

Precipitation reactions are not confined to transition metal ions. However, many transition metals undergo reactions that produce characteristic coloured precipitates, which can be useful in identification of the ion (Table 6.3). The addition of aqueous hydroxide ions, OH⁻(aq), (which can be obtained from an alkali such as sodium hydroxide or from aqueous ammonia) to aqueous transition metal ions often results in a coloured precipitate. The precipitation reaction can be summarised as:

$$M^{n+}(aq) + nOH^-(aq) \rightarrow M(OH)_n(s) \text{ where } M^{n+} \text{ is the transition metal ion}$$

Table 6.3

Ion	Reaction	Colour of precipitate	Comments
$Cu^{2+}(aq)$	$Cu^{2+}(aq) + 2OH^-(aq) \rightarrow Cu(OH)_2(s)$	light blue	see note 1 below
$Fe^{2+}(aq)$	$Fe^{2+}(aq) + 2OH^-(aq) \rightarrow Fe(OH)_2(s)$	green	turns brown*
$Fe^{3+}(aq)$	$Fe^{3+}(aq) + 3OH^-(aq) \rightarrow Fe(OH)_3(s)$	rust	
$Mn^{2+}(aq)$	$Mn^{2+}(aq) + 2OH^-(aq) \rightarrow Mn(OH)_2(s)$	pale pink almost white	turns brown**
$Cr^{3+}(aq)$	$Cr^{3+}(aq) + 3OH^-(aq) \rightarrow Cr(OH)_3(s)$	dark green	see note 2 below
Either $NaOH(aq)$ or dilute $NH_3(aq) + H_2O \rightarrow NH_4^+(aq) + OH(aq)$ are the source of the $OH^-(aq)$ ions			

*$Fe(OH)_2(s)$ slowly oxidises to $Fe(OH)_3(s)$, which is rust coloured.

**$Mn(OH)_2(s)$ readily oxidises to $Mn_2O_3(s)$, which is brown.

Note 1

As explained in the section on ligand exchange, if aqueous ammonia is used in excess, the blue precipitate of $Cu(OH)_2(s)$ dissolves to form a much darker blue solution due to the formation of $[Cu(NH_3)_4(H_2O)_2]^{2+}$.

Note 2

As with $Cu(OH)_2(s)$, the precipitate of $Cr(OH)_3$ will also dissolve if aqueous ammonia is added in excess. The solution obtained contains the ion, $[Cr(NH_3)_6]^{3+}$.

$Cr^{3+}(aq)$ also reacts with aqueous sodium hydroxide and a precipitate of chromium(III) hydroxide. $Cr(OH)_3$ is formed but a further reaction occurs if excess aqueous sodium hydroxide is added and a green solution containing the ion $[Cr(OH)_6]^{3-}$ is obtained.

Like all hydroxides, $Cr(OH)_3$ can also react with acids. For example, the addition of dilute HCl will result in the formation of a solution of chromium(III) chloride, $CrCl_3$. So $Cr(OH)_3$ has the property of being able to react with both acids and aqueous sodium hydroxide and, where this occurs, the hydroxide is referred to as an amphoteric hydroxide. It is not unique to chromium(III) hydroxide and, for example, zinc hydroxide and aluminium hydroxide are both amphoteric.

Precipitations and ligand exchange reactions commonly occur and the above examples should serve to illustrate the processes that take place.

Redox reactions and titrations

Redox reactions

It has already been mentioned that a feature of transition metals is their ability to exist in more than one oxidation state (page 120). It is not surprising, therefore, that their redox chemistry is extensive. The importance of an understanding of electrode potentials in deciding on the feasibility of these processes was considered in Chapter 5. The principles established there can be used to predict the feasibility of the redox reactions of transition metals.

Interconversions between Fe^{2+} and Fe^{3+}

The electrode potential $Fe^{3+}(aq) + e^- \rightarrow Fe^{2+}(aq)$ $E^\ominus = +0.77\,V$ governs the redox behaviour of iron ions. Any oxidising agent with an $E^\ominus$ value greater than $0.77\,V$ will be able to oxidise $Fe^{2+}(aq)$ to $Fe^{3+}(aq)$. A typical example is acidified $MnO_4^-(aq)$,

$$MnO_4^-(aq) + 8H^+(aq) + 5e^- \rightarrow Mn^{2+}(aq) + 4H_2O(l)$$

which has an $E^\ominus$ value equal to $+1.51\,V$ so it will react with $Fe^{2+}(aq)$. The overall equation is:

$$MnO_4^-\,(aq) + 8H^+\,(aq) + 5Fe^{2+}(aq) \rightarrow 5Fe^{3+}(aq) + Mn^{2+}\,(aq) + 4H_2O(l)$$

and this has an overall cell potential of $+0.74\,V$, so the reaction is feasible. When the reaction is carried out, the light green solution containing $Fe^{2+}(aq)$ is oxidised to a yellow solution containing $Fe^{3+}(aq)$, while the deep purple solution of $MnO_4^-(aq)$ is reduced to the very pale pink solution containing $Mn^{2+}(aq)$. In reality, it is the colour change of the $MnO_4^-(aq)$ that is most conspicuous. Because of its easily observed colour change, $MnO_4^-(aq)$ is used in redox titrations (see page 133).

A reducing agent which will convert $Fe^{3+}(aq)$ into $Fe^{2+}(aq)$ is $I^-(aq)$. This was discussed in Chapter 5 but is summarised here.

The electrode potentials for the half-reactions are:

$$Fe^{3+}(aq) + e^- \rightleftharpoons Fe^{2+}(aq)\ E^\ominus = +0.77\,V$$

$$2I^-(aq) \rightleftharpoons I_2(aq) + 2e^-\quad E^\ominus = -0.54\,V \text{ (the reverse of the standard electrode potential)}$$

Adding these together, gives a cell potential of $+0.77 - 0.54 = +0.23\,V$.

The positive value indicates that the reaction is feasible.

The colourless solution containing $I^-(aq)$ ions will become brown (the colour of iodine in the solution) as the $Fe^{3+}(aq)$ is reduced to $Fe^{2+}(aq)$.

Interconversions between Cr^{3+} and $Cr_2O_7^{2-}$

The relevant electrode potential is

$$Cr_2O_7^{2-}(aq) + 14H^+ + 6e^- \rightleftharpoons 2Cr^{3+}(aq) + 7H_2O\ E^\ominus = +1.33\,V$$

The electrode potential is large and acidified $Cr_2O_7^{2-}(aq)$ will be able to oxidise many substances. For example, zinc has an electrode potential for the process $Zn^{2+}(aq) + 2e^- \rightleftharpoons Zn(s)$ of $-0.76\,V$. So the following reaction is feasible:

$$3Zn(s) + Cr_2O_7^{2-}(aq) + 14H^+ \rightleftharpoons 3Zn^{2+}(aq) + 2Cr^{3+}(aq) + 7H_2O$$

This has an overall cell potential of $+0.57\,V$ and, as the zinc dissolves, you would observe the orange dichromate ion changing colour to form the green aqueous Cr^{3+} ion.

Oxidation of $Cr^{3+}(aq)$ is quite difficult because the oxidising agent must have an electrode potential that exceeds $+1.33\,V$. But a reagent which is able to achieve this is hydrogen peroxide which reacts with $Cr^{3+}(aq)$ in alkaline solution to form the yellow chromate(VI) ion $CrO_4^{2-}(aq)$.

The relevant half-equations must be constructed using $OH^-(aq)$ rather than $H^+(aq)$:

$$Cr^{3+}(aq) + 8OH^-(aq) \rightleftharpoons CrO_4^{2-}(aq) + 4H_2O(l) + 3e^-$$

$$H_2O_2(aq) + 2e^- \rightleftharpoons 2OH^-(aq)$$

and these combine to give an overall equation:

$$2Cr^{3+}(aq) + 3H_2O_2(aq) + 10OH^-\,(aq) \rightleftharpoons 2CrO_4^{2-}(aq) + 8H_2O(l)$$

The CrO_4^{2-}(aq) can be converted into $Cr_2O_7^{2-}$(aq) simply by adding excess hydrogen ions:

$$2CrO_4^{2-}(aq) + 2H^+(aq) \rightleftharpoons Cr_2O_7^{2-}(aq) + H_2O(l)$$

Cu^+ and Cu^{2+}

Cu^{2+}(aq) can be reduced to Cu^+(aq) ions using I^-.

$$Cu^{2+}(aq) + e^- \rightleftharpoons Cu^+(aq) \quad E^\ominus = +0.15\,V$$

$$I_2(aq) + 2e^- \rightleftharpoons 2I^-(aq) \quad E^\ominus = +0.54\,V$$

The overall cell potential would therefore be $-0.39\,V$ and this suggests the reaction will not be feasible. However, remember these are standard electrode potentials and equilibrium reactions. In this case, the exceptionally low solubility of copper(I) iodide causes the reaction to take place. As the reagents are mixed, a very tiny amount of Cu^+ is formed but it is sufficient to begin the precipitation of copper(I) iodide (CuI), which then drags the equilibria to produce more I^- and the process continues until the reaction is complete. In fact, the reaction is so efficient that it can be used as a basis for a titration, which is considered in the next section. The grey/white precipitate of copper(I) iodide is observed, but it is coloured brown by the iodine produced.

Unless copper(I) compounds are very insoluble, the Cu^+ ion readily disproportionates (is both oxidised and reduced) to form copper metal and copper(II) ions. For example, if a copper(I) salt is left in contact with a dilute acid such as sulfuric acid, a blue colour is soon observed as the Cu^+(aq) ion forms Cu^{2+}(aq) and a precipitate of Cu(s) is obtained. The electrode potentials explain why.

$$Cu^{2+}(aq) + e^- \rightleftharpoons Cu^+(aq) \quad E^\ominus = +0.15\,V$$

$$Cu^+(aq) + e^- \rightleftharpoons Cu(s) \quad E^\ominus = +0.34\,V$$

Therefore, the cell potential for the reaction $2Cu^+(aq) \rightleftharpoons Cu^{2+}(aq) + Cu(s)$ will be $+0.34 - (+0.15) = +0.19\,V$.

Since this cell potential is positive, the reaction will be feasible.

Other redox reactions of transition metal ions can be deduced using electrode potentials in a similar way.

Redox titrations

There is often a colour change when a transition element changes oxidation state, so it is possible to make use of this in redox titrations. As has been explained, acidified potassium manganate(VII) is particularly useful as it is able to oxidise a wide range of ions and, in doing so, has a distinct colour change from an intense purple colour to the pale pink of Mn^{2+}(aq) ions.

$$5e^- + MnO_4^-(aq) + 8H^+(aq) \rightarrow Mn^{2+}(aq) + 4H_2O(l)$$

Excess H^+(aq) ions are required and these are usually supplied by adding dilute sulfuric acid.

A redox titration is possible if the reagent used to reduce MnO_4^-(aq) is either colourless or only lightly coloured. An example is a solution containing pale green iron(II) ions. On oxidation, these are converted to yellow iron(III) ions.

Example 2

Two iron tablets (total mass 1.30 g) containing iron(II) sulfate were dissolved in dilute sulfuric acid and made up to $100.0\,cm^3$. $10.0\,cm^3$ of this solution required $12.00\,cm^3$ of a standard solution of $0.00500\,mol\,dm^{-3}$ $KMnO_4$ to produce a faint pink colour. What is the percentage of iron in the iron tablets?

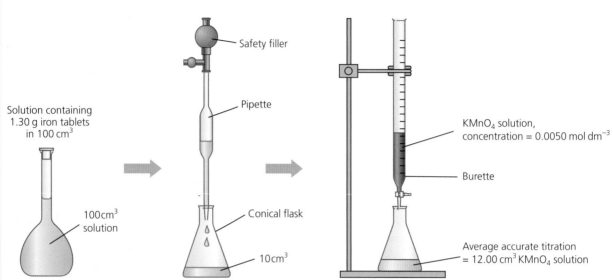

Figure 6.21 Finding the percentage of iron in iron tablets.

Answer

In any calculation it is essential that you:

- use the two half equations to deduce the mole ratio between the $MnO_4^-(aq)$ and the $Fe^{2+}(aq)$

- calculate the amount in moles of $MnO_4^-(aq)$ that reacts in the titration

- deduce the amount in moles of $Fe^{2+}(aq)$ that was pipetted in the titration

- work out the total amount in moles of $Fe^{2+}(aq)$ in the original solution

- calculate the percentage of iron in the tablets.

$$MnO_4^- + 8H^+ + 5e^- \longrightarrow Mn^{2+} + 4H_2O$$

$$\underline{Fe^{2+} \longrightarrow Fe^{3+} + e^- \qquad \times 5}$$

$$MnO_4^- + 8H^+ + 5Fe^{2+} \longrightarrow Mn^{2+} + 5Fe^{3+} + 4H_2O$$

amount in moles of:

$$MnO_4^- \text{ reacting} = \frac{12.00}{1000}\,dm^3 \times 0.0050\,mol\,dm^{-3} = 6.0 \times 10^{-5}\,mol$$

$$Fe^{2+} \text{ in the }10\,cm^3 \text{ pipette} = 6.0 \times 10^{-5} \times 5 = 3.0 \times 10^{-4}\,mol$$

$$\text{mass of } Fe^{2+} \text{ in the two tablets} = 3.0 \times 10^{-3} \times 55.8\,g\,mol^{-1} = 0.1674\,g$$

The two tablets had a mass of 1.30 g so the percentage iron in the tablets is

$$\frac{0.1674\,g}{1.30\,g} \times 100 = 12.9\%$$

Several other titrations involving manganate(VII) ions are possible and they all follow a similar procedure.

Although not exclusively related to transition metals, there is another titration that is employed widely. It is based on the redox reaction between iodine and thiosulfate ions. The equation for this reaction was considered in Example 6 on page 92.

$$2S_2O_3^{2-} (aq) + I_2(s) \rightarrow S_4O_6^{2-} (aq) + 2I^- (aq) \qquad \text{equation (1)}$$

This titration is not usually used directly to determine the concentration of an iodine solution. Rather, it allows the determination of the concentration of a reagent that generates iodine as a result of a reaction.

An example is the determination of the concentration of a copper(II) sulfate solution. A known volume of copper(II) sulfate is reacted with excess potassium iodide:

$$2Cu^{2+}(aq) + 4I^-(aq) \rightarrow 2CuI(s) + I_2(s) \qquad \text{equation (2)}$$

The iodine produced is then titrated against a solution of sodium thiosulfate of known concentration.

At the start of the titration, the solution appears brown–purple because of the presence of the iodine. As the titration proceeds, this colour fades to yellow and the end point is reached when the solution is colourless. In practice, this colour change is quite difficult to see, mainly because of the presence of the copper(I) iodide precipitate. To help, some starch solution is added as the end point is approached. This gives rise to a dark-blue coloration that disappears sharply at the end point.

Calculations based on this titration may look rather daunting, but the equations provide information:

- equation (2) — 2 mol of Cu^{2+} react to produce 1 mol of I_2

- equation (1) — 1 mol of I_2 reacts with 2 mol of $S_2O_3^{2-}$

It follows therefore that for every 1 mole of Cu^{2+}, 1 mole of $S_2O_3^{2-}$ is required. So the amount (in moles) of thiosulfate used is equivalent to the amount (in moles) of copper(II) ions present originally.

Example 3

25.0 cm³ of a solution of copper(II) sulfate reacts with excess potassium iodide solution to produce an amount of iodine that reacts exactly with 27.15 cm³ of a 0.100 mol dm⁻³ solution of sodium thiosulfate. Calculate the concentration, in mol dm⁻³, of the copper(II) sulfate solution.

Answer
Amount (in moles) of $S_2O_3^{2-}$ used in the titration = (27.15/1000) × 0.100 = 2.715×10^{-3} mol

Amount (in moles) of copper sulfate in 25.00 cm³ = 2.715×10^{-3} mol

Concentration (in mol dm⁻³) of copper(II) sulfate = (1000/25.0) × 2.715 × 10^{-3} = 0.109 mol dm⁻³ (calculator value = 0.1086)

Iodine is produced readily in a number of reactions, so this type of titration has wide application.

Figure 6.22 Supermarket products that contain bleach.

The active reagent in household bleaches is sodium chlorate(I), NaClO. To increase the cleaning power of these bleaches, manufacturers usually add detergents, and to improve their smell they add perfumes. Sodium chlorate(I) is a strong oxidising agent which bleaches by oxidising coloured materials to colourless or white substances.

The half-equation when sodium chlorate(I) acts as an oxidising agent is:

$$ClO^-(aq) + 2H^+(aq) + 2e^- \rightarrow Cl^-(aq) + H_2O$$

A student was asked to determine the concentration of sodium chlorate(I) in a supermarket bleach.

100.0 cm³ of the bleach was added to a graduated flask and made up to a volume of 1000 cm³. 10.0 cm³ of the diluted solution was then pipetted into a conical flask, followed by the addition of excess potassium iodide.

The iodine produced was finally titrated with 0.100 mol dm⁻³ sodium thiosulfate solution giving an average accurate titration of 26.60 cm³.

1 Write a half-equation for the oxidation of iodide ions to iodine.
2 Write a balanced equation for the reaction of chlorate(I) ions with iodide ions in acid solution to form iodine, chloride ions and water.
3 Write a balanced equation for the reaction of iodine with thiosulfate ions during the titration.
4 Using your answers to questions **2** and **3**, calculate the number of moles of thiosulfate that react with the iodine produced by 1 mole of chlorate(I) ions.

5 Calculate the number of moles of thiosulfate in the average accurate titration, and hence the number of moles of sodium chlorate(I) in 10.0 cm³ of the diluted bleach.

6 Calculate the mass of sodium chlorate(I) in 100 cm³ of undiluted bleach. (Na = 23.0, Cl = 35.5, O = 16.0)

7 What precautions should the student take to ensure that the result is accurate?

Test yourself

15 All the iron in 1.34 g of some iron ore was dissolved in acid and any iron(III) ions were reduced to iron(II) ions. The solution was then titrated with 0.0200 mol dm⁻³ potassium manganate(VII) solution. The titre was 26.75 cm³. Calculate the percentage by mass of iron in the ore. (Fe = 55.8)

16 0.275 g of an alloy containing copper was dissolved in nitric acid, and then diluted with water, producing a solution of copper(II) nitrate. An excess of potassium iodide was then added. The copper(II) ions reacted with the iodide ions to form a precipitate of copper(I) iodide and iodine. In a titration, the iodine reacted with 22.50 cm³ of 0.140 mol dm⁻³ sodium thiosulfate solution. (Cu = 63.5)

   a) Write equations for:
      i) the reaction of copper(II) ions with iodide ions to form copper(I) iodide and iodine
      ii) the reaction of iodine with sodium thiosulfate during the titration.
   b) Calculate the percentage by mass of copper in the alloy.

Qualitative analysis

During your A Level course you have met several reactions which can be used to identify the presence of particular ions. The chemistry of the reactions is covered in the relevant sections of the book but Table 6.4 gives a basic summary of the identification tests that you need to know.

Table 6.4

Ion	Identification test
CO_3^{2-}	Add a dilute acid, bubbling occurs and the gas given off turns limewater milky. Carbon dioxide precipitates calcium carbonate when bubbled into aqueous calcium hydroxide (limewater). $2H^+(aq) + CO_3^{2-}(aq) \rightarrow H_2O(l) + CO_2(g)$
Cl^-	Dissolve the suspected chloride and add aqueous silver nitrate. A white precipitate forms which dissolves if shaken with dilute aqueous ammonia. This is due to the formation of the $[Ag(NH_3)_2]^+$ complex ion. $Ag^+(aq) + Cl^-(aq) \rightleftharpoons AgCl(s)$
Br^-	Dissolve the suspected bromide and add aqueous silver nitrate. A cream precipitate forms which dissolves if shaken with concentrated aqueous ammonia. Silver bromide is less soluble than silver chloride and only supplies a few Ag^+ ions. Therefore, the ammonia must be concentrated to allow the complex $[Ag(NH_3)_2]^+$ to form. $Ag^+(aq) + Br^-(aq) \rightleftharpoons AgBr(s)$
I^-	Dissolve the suspected iodide and add aqueous silver nitrate. A yellow precipitate forms which will not dissolve in aqueous ammonia. Silver iodide is the least soluble of the silver halides and does not provide sufficient silver ions to allow ammonia to form $[Ag(NH_3)_2]^+$. $Ag^+(aq) + I^-(aq) \rightleftharpoons AgI(s)$
SO_4^{2-}	Dissolve the suspected sulfate and add aqueous barium nitrate. A white precipitate of barium sulfate forms which is insoluble in dilute nitric acid. $Ba^{2+}(aq) + SO_4^{2-}(aq) \rightarrow BaSO_4(s)$
NH_4^+	Warm the suspected ammonium compound with dilute aqueous sodium hydroxide. The smell of ammonia is apparent but can be tested with moist red litmus paper which will turn blue as aqueous ammonia is alkaline. $NH_3(g) + H_2O(l) \rightleftharpoons NH_4^+(aq) + OH^-(aq)$
Cu^{2+}	Dissolve the suspected compound and add dilute aqueous ammonia. A light blue precipitate forms which forms a deep blue solution containing the ion $[Cu(NH_3)_4(H_2O)_2]^{2+}$ when excess ammonia is added. $Cu^{2+}(aq) + 2OH^-(aq) \rightarrow Cu(OH)_2(s)$
Fe^{2+}	Dissolve the suspected compound and add aqueous sodium hydroxide. A green precipitate forms. $Fe^{2+}(aq) + 2OH^-(aq) \rightarrow Fe(OH)_2(s)$
Fe^{3+}	Dissolve the suspected compound and add aqueous sodium hydroxide. A rust-brown precipitate forms. $Fe^{3+}(aq) + 3OH^-(aq) \rightarrow Fe(OH)_3(s)$
Mn^{2+}	Dissolve the suspected compound and add aqueous sodium hydroxide. A very pale pink precipitate forms. $Mn^{2+}(aq) + 2OH^-(aq) \rightarrow Mn(OH)_2(s)$
Cr^{3+}	Dissolve the suspected compound and add aqueous sodium hydroxide. A green precipitate forms which dissolves when excess sodium hydroxide is added. Chromium compounds will also give a green precipitate with dilute aqueous ammonia which dissolves when excess ammonia is added (see Note 2 on page 131). $Cr^{3+}(aq) + 3OH^-(aq) \rightarrow Cr(OH)_3(s)$

Practice questions

Multiple choice questions 1–10

1 Which one of the following has the same number of unpaired electrons as sulfur?

A scandium   B vanadium

C cobalt   D nickel   *(1)*

2 Which of the following pairs of atoms do **not** have the same number of 4s electrons?

A K and Cu

B Ca and Ti

C Cr and Mn

D Ni and Zn   *(1)*

3 $10\,cm^3$ of a $0.1\,mol\,dm^{-3}$ solution of the sulfate of a metal M exactly reacts with $20\,cm^3$ of a $0.15\,mol\,dm^{-3}$ solution of sodium hydroxide to produce a precipitate of the hydroxide of M. The precipitate does not dissolve when excess sodium hydroxide is added.

Which one of the following could be metal M?

A manganese

B iron

C chromium

D copper   *(1)*

4 Cu^{2+} ions can form a complex ion with Cl^- ions and also with NH_3.

These complex ions have:

A different shapes but the same colour

B different shapes and different colours

C the same shape and the same colour

D the same shape but a different colour.   *(1)*

5 Cr^{3+} forms a complex ion with both ammonia and also with hydroxide ions.

These complex ions have:

A a different charge but the same coordination number

B a different charge but a different coordination number

C the same charge but the same coordination number

D the same charge but a different coordination number.   *(1)*

6 $25\,cm^3$ of $0.02\,mol\,dm^{-3}$ $KMnO_4$ reacts exactly with $25\,cm^3$ of a $0.05\,mol\,dm^{-3}$ solution containing $X^{2+}(aq)$ ions.

Which one of the following is the oxidation number of the ion formed by X as a result of the reaction?

A III   B IV

C V   D VI   *(1)*

7 Use the following $E^{\ominus}$ values to answer the question below.

$$VO^{2+}(aq) + 2H^+(aq) + e^- \rightleftharpoons V^{3+}(aq) \quad + H_2O(l)$$
$$E^{\ominus} = +0.34V$$

$$Fe^{3+}(aq) + e^- \rightleftharpoons Fe^{2+}(aq) \quad\quad E^{\ominus} = +0.77V$$

$$Sn^{4+}(aq) + 2e^- \rightleftharpoons Sn^{2+}(aq) \quad\quad E^{\ominus} = +0.15V$$

Which one of the following would be the best choice to convert $VO^{2+}(aq)$ into $V^{3+}(aq)$?

A $Fe^{3+}(aq)$

B $Fe^{2+}(aq)$

C $Sn^{4+}(aq)$

D $Sn^{2+}(aq)$ *(1)*

Use the key below to answer questions 8, 9 and 10.

A	B	C	D
1, 2 & 3 correct	1 & 2 correct	2 & 3 correct	1 only correct

8 Which of the following would first form a precipitate and then a solution when added to aqueous manganese(II) chloride?

1 $AgNO_3(aq)$ followed by excess $NH_3(aq)$

2 a few drops of $NH_3(aq)$ followed by an excess of $NH_3(aq)$

3 a few drops of $NaOH(aq)$ followed by an excess of $NaOH(aq)$ *(1)*

9 When $KCN(aq)$ is added in excess to $Ni^{2+}(aq)$ ions a square planar complex ion is formed.

Which of the following have the same charge as the nickel cyanide complex ion?

1 a sulfide ion

2 the complex ion formed between Cu^{2+} ions and Cl^- ions

3 the octahedral complex ion formed between $Fe^{3+}(aq)$ ions and $CN^-(aq)$ ions.   *(1)*

10 Which of the following are always true for square planar complexes of an ion M^{2+}?

1 $[MWZ]$, where W and Z are bidentate ligands, will have optical isomers.

2 $[MX_3Y]$ will not have cis/trans isomers.

3 $[MX_2Y_2]$ will have cis/trans isomers.   *(1)*

11 This question concerns the chemistry of transition metals.

a) Define and explain the terms:

    i) transition metal

    ii) oxidation number

    iii) complex ion.   *(8)*

b) Give one example (including an equation and observation) of the following reactions that occur with a transition metal ion:
- precipitation
- ligand substitution
- redox reaction. (9)

12 Copy the formula for each of the following complex ions and deduce their net charge, if any.
a) $[Zn(NH_3)_4(H_2O)_2]$
b) $[Fe(CN)_6]$ containing Fe(II)
c) $[Co(NH_3)_5Cl]$ containing Co(III)
d) $[Co(C_2O_4)_3]$ formed from Co^{2+} ions and $K_2C_2O_4$
e) $[Cr(CH_3COO)_2(H_2O)_2]$ containing Cr(III) (5)

13 The reaction scheme below involves various compounds of copper.

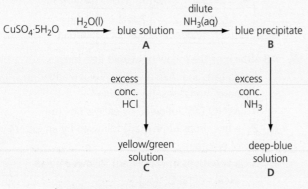

$CuSO_4 \cdot 5H_2O$ →$^{H_2O(l)}$ blue solution **A** →$^{dilute\ NH_3(aq)}$ blue precipitate **B**

A → excess conc. HCl → yellow/green solution **C**

B → excess conc. NH$_3$ → deep-blue solution **D**

a) Write the formulae of the species responsible for the colour in each of the products A to D.
b) Describe and explain, with an equation, what you would see when solution C is diluted with excess water.
c) Describe and explain, with an equation, what you would see when dilute hydrochloric acid is added to solution D. (13)

14 Explain the following:
a) When 1,2-diaminoethane is added to light-blue aqueous copper(II) sulfate solution, the colour of solution intensifies to dark blue.
b) When hydrochloric acid is added to the dark blue solution formed as a result of the experiment in part **a)**, the colour returns to light blue. (7)

15 Suggest explanations for each of the following:
a) When concentrated hydrochloric acid is added to aqueous cobalt(II) chloride solution, the colour of the solution changes from pink to blue.

b) If water is added to some of the blue solution formed in part **a)**, the colour changes back to pink.
c) If aqueous silver nitrate is added to some of the blue solution formed in part **a)**, the solution changes to pink again and a precipitate is formed.
d) If propanone is added carefully to the pink aqueous solution of cobalt(II) chloride, the propanone floats on the surface of the water. At the junction of the two liquids a blue layer is observed. (6)

16 Salts that contain a complex cation with an anion that is not a complex can be crystallised. For example, cobalt forms a salt, A, of formula $[Co(NH_3)_6]Cl_3$, which is orange in colour. It is possible to prepare isomers of this cobalt complex that retain the same *number* of chloride ions but which have a chloride ion substituting one of the ammonia molecules as the ligand. One example is salt, B, which has the formula $[Co(NH_3)_5(Cl)]Cl_2$.
a) If equal volumes and concentrations of salts A and B are reacted separately with excess aqueous silver nitrate, salt A produces a precipitate that has a mass 1.5 times greater than the precipitate produced by salt B. Explain why.
b) Two different salts, C and D, can be prepared that contain Co(III) coordinated with four ammonia ligands. Each salt also contains three Cl$^-$ ions. When the experiment in part **a)** is repeated separately with C and D, each produces a mass of precipitate that is half the mass obtained from compound B. Suggest structures for C and D and explain your answer. (7)

17 A solution is made that contains the VO^{2+}(aq) ion. When 25.0 cm^3 of this solution is titrated against 0.0150 mol dm^{-3} MnO$_4^-$(aq) ions in the presence of excess sulfuric acid, it is found that 23.30 cm^3 is required to reach the end point.
The half-equation for the oxidation of the VO^{2+}(aq) ions is:

$$VO^{2+}(aq) + 2H_2O(l) \rightarrow VO_3^-(aq) + 4H^+(aq) + e^-$$

Calculate the concentration (in mol dm^{-3}) of the solution containing VO^{2+}(aq). (5)

18 A general purpose solder contains antimony, lead and tin. When reacted with an acid, the solder dissolves forming a solution containing $Sb^{3+}(aq)$, $Pb^{2+}(aq)$ and $Sn^{2+}(aq)$. Neither $Sb^{3+}(aq)$ nor $Pb^{2+}(aq)$ reacts with dichromate $(Cr_2O_7^{2-})$ ions. However, $Sn^{2+}(aq)$ can be oxidised to $Sn^{4+}(aq)$ by an acidified solution of potassium dichromate(VI). In an experiment, 10.00 g of solder is dissolved in acid to make 1.00 dm³ of solution. When 25.0 cm³ of this solution is titrated against an acidified potassium dichromate solution of concentration 0.0175 mol dm⁻³, 20.00 cm³ of the dichromate is required to reach the end point.
The half-equation for the reduction of $Cr_2O_7^{2-}(aq)$ is:

$$Cr_2O_7^{2-}(aq) + 14H^+(aq) + 6e^- \rightarrow 2Cr^{3+}(aq) + 7H_2O(l)$$

a) Write an overall equation for the reaction of $Sn^{2+}(aq)$ with $Cr_2O_7^{2-}(aq)$.

b) Calculate the concentration of $Sn^{2+}(aq)$ in the solution in mol dm⁻³.

c) Calculate the percentage by mass of tin in the solder. *(6)*

19 Rhubarb leaves contain poisonous ethanedioic acid, $(COOH)_2$. A dose of about 24 g of ethanedioic acid could be fatal if consumed by an adult. On heating, ethanedioate ions react with potassium manganate(VII) solution and so this forms the basis of a titration. The ethanedioate ions are oxidised to carbon dioxide.

$$(COO)_2^{2-}(aq) \rightarrow 2CO_2(g) + 2e^-$$

Four large rhubarb leaves are heated with water to extract the ethanedioic acid. The solution is filtered and then diluted to 250.0 cm³ in a volumetric flask. 25.0 cm³ samples of the solution are acidified with dilute sulfuric acid. 23.90 cm³ of 0.0200 mol dm⁻³ potassium manganate(VII) are required to reach the end point in a titration.

a) Write an overall equation for the reaction between ethanedioate ions and acidified manganate(VII) ions.

b) Use the titration result to calculate the amount (in moles) of ethanedioic acid in the 250.0 cm³ solution.

c) Calculate the mass of ethanedioic acid in the rhubarb leaves.

d) Suggest the number of rhubarb leaves that, when eaten, would be sufficient to kill an adult. *(6)*

20 When excess potassium iodide solution is added to 25.0 cm³ of a solution of bromine in water, the iodine formed reacts exactly with 24.75 cm³ of a 0.800 mol dm⁻³ solution of sodium thiosulfate. Calculate the concentration of the bromine solution in g dm⁻³. *(7)*

21 To be safe, swimming-pool water should contain between 1 g and 2 g of chlorine in every 1000 dm³ of water.
In an investigation, a 500.0 cm³ sample of swimming-pool water is analysed by treating it first with excess potassium iodide solution and then titrating the iodine released against a 1.500×10^{-3} mol dm⁻³ solution of sodium thiosulfate. 10.25 cm³ of sodium thiosulfate is required to reach the end point.
Would it be safe to swim in this swimming pool? Explain your answer. Show all of your working. *(6)*

22 An element, X, is able to form a compound of formula $NaXO_3$. An aqueous solution of $NaXO_3$ of concentration 0.0500 mol dm⁻³ is prepared. When excess potassium iodide solution is added to 25.0 cm³ of this solution, iodine is produced. When this is titrated against a solution of sodium thiosulfate of concentration 13.63 g dm⁻³, 29.00 cm³ of the solution is required to react completely with the iodine. Deduce the change in oxidation state for element X in the reaction. (You may assume that element X is not iodine.) *(7)*

23 Explain how you would use chemical tests to distinguish between solutions of:

a) manganese chloride and manganese nitrate

b) ammonium sulfate and ammonium nitrate

c) copper(II) nitrate and chromium(III) nitrate

d) iron(II) sulfate and iron(III) sulfate

e) manganese(II) chloride and iron(II) chloride. *(12)*

Challenge

24 4.23g of a hydrated Ni^{2+} salt, $NiY_2.xH_2O$, was heated to constant mass and produced 2.31g of anhydrous salt. The anhydrous salt was dissolved in a dilute acid and excess ammoniacal dimethylglyoxime was added. 5.14g of a red precipitate, $Ni(C_4H_7N_2O_2)_2$ was precipitated. Deduce a possible formula for the hydrated Ni^{2+} salt. Show all of your working.　　(11)

25 Ligand exchange can sometimes be used as a means of determining the end point of a complexometric titration. An example of this is a titration to determine the concentration of calcium ions in water. The presence of magnesium ions is essential for the method — if they are not present naturally, some must be added.
This titration could be used to estimate the calcium ion concentration in a soil extract made by shaking a sample of soil with water and filtering. It is unnecessary to add magnesium ions in this case as enough will be present naturally.
The method is as follows.
A solution of the complexing agent EGTA is made up to a specified concentration.

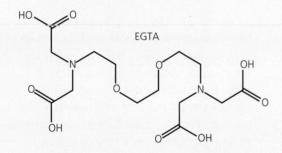

EGTA

A known volume of EGTA solution is then added to a measured volume of the soil extract in a conical flask. The volume of EGTA must be large enough to ensure that all the calcium and magnesium ions in the soil extract are complexed with EGTA and that some free EGTA remains.

A small volume of the complexing agent thymolphthalexone is added and the blue colour of a magnesium thymolphthalexone complex is observed.
A titration is then carried out by running a solution of lead nitrate of known concentration from a burette into the mixture in the conical flask.
During the course of this titration, the blue colour fades. The end point is when the solution becomes completely colourless.
The amount (in moles) of calcium ions in the volume of soil extract is then equal to the amount (in moles) of EGTA added originally minus the amount (in moles) of lead ions added during the titration.
Use the description above to answer the following questions.

a) Place each set of complex ions below in order of decreasing stability:
　i)　calcium–water, calcium–EGTA
　ii)　lead–water, lead–EGTA
　iii)　magnesium–thymolphthalexone, lead–thymolphthalexone
　iv)　magnesium–water, magnesium–EGTA, magnesium–thymolphthalexone
　v)　magnesium–EGTA, calcium–EGTA, lead–EGTA

b) What is the colour of a solution of thymolphthalexone?

c) At the end of the titration, in what type of complex will magnesium ions, calcium ions and lead ions be? If they could be in more than one type of complex, give all the forms.

d) Give a brief explanation of why the amount (in moles) of calcium ions is as explained in the final bullet point.　　(15)

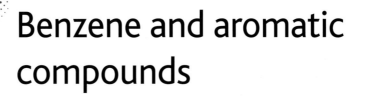

Chapter 7

Benzene and aromatic compounds

Prior knowledge

In this chapter it is assumed that you are familiar with:
- organic chemistry covered in the first year of this course
- mole calculations.

For example, you should be aware that alkenes are unsaturated hydrocarbons, contain a C=C double bond and react readily with electrophiles via an electrophilic addition mechanism. You should be able to state and to apply Markownikoff's rule to the reactions of asymmetric alkenes with reagents that have the general formula HX.

Test yourself on prior knowledge

1 Write equations for the following reactions. State essential reaction conditions, if any, needed for each reaction and name any products formed.
   a) propene and bromine
   b) but-2-ene and hydrogen bromide
   c) cyclobutene and water.
2 Describe the mechanism of the reaction between propene and bromine.
3 **a)** State Markownikoff's rule.
   b) Explain fully why 2-bromobutane is the major product when but-1-ene reacts with HBr.

Arenes

This chapter covers the bonding and structure of benzene and related compounds. Arenes undergo electrophilic substitution reactions, the ease of which depends on the structure of the arene.

The simplest arene is benzene. Benzene has a composition by mass of 92.3% carbon and 7.7% hydrogen. Its relative molecular mass is 78.0. This information shows that the empirical formula of benzene is CH and its molecular formula is C_6H_6.

Benzene was first isolated in 1825 by Michael Faraday. Traditionally chemists have called arenes 'aromatics', ever since the German chemist Friedrich Kekulé was struck by the fragrant smell of oils such as benzene. The name 'arene' is used today; the '**ar**' comes from **ar**omatic and the '**ene**' indicates that, like alk**ene**s, they are unsaturated hydrocarbons.

It is now known that benzene is a carcinogen – exposure to benzene increases the risk of cancer. Benzene levels are carefully monitored and

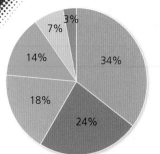

Figure 7.1 Uses of benzene.

Ethylbenzene
Cyclohexane
Cumene
Alkylbenzenes
Chlorobenzenes
Other uses

34%
24%
18%
14%
7%
3%

Figure 7.2 Kekulé claims to have discovered the ring shape of the benzene molecule when dreaming about a snake seizing its own tail. This is a common symbol in some ancient cultures and is known as 'ouroboros'.

you will not use benzene in any experiments you perform in the laboratory. Nevertheless, benzene is an important industrial chemical essential in the manufacture of many day-to-day commodities.

Benzene is used to make alkylbenzenes, for example:

● ethylbenzene – used to make phenylethene, which is the monomer for poly(phenylethene) or polystyrene

● (1-methylethyl)benzene/cumene – used to make phenol and propanone

● dodecylbenzene – used to make detergents.

Benzene is also an important feedstock for the production of cyclohexane in the manufacture of nylon. It is also used in the manufacture of a variety of dyes, medicines and explosives.

Some uses of benzene are shown in Figure 7.1.

Structure of benzene

The cyclic structure of benzene was not established fully until the late nineteenth century and the exact structure of benzene is still open to interpretation today.

Kekulé suggested that benzene was a cyclic molecule (Figure 7.3) with alternating C=C double bonds and C–C single bonds. The C=C double bonds are formed by the sideways overlap of adjacent p-orbitals (see page 208 in the Year 1 book).

The carbons join together to form a hexagonal ring in which each C is bonded, by a σ-bond, to two other C and to 1 H as shown in **a)**.

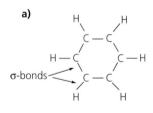

Each C has a electron in a p-orbital which is at a right angle to the hexagonal plane as shown in **b)**.

Kekulé suggested that adjacent p-orbitals overlap to form π-bonds (as they did in alkenes) such that there were alternating double and single bonds around the ring as shown in **c)** and **d)**.

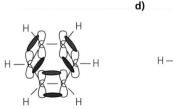

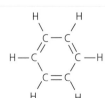

Figure 7.3

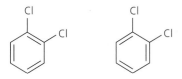

Figure 7.4 Two possible predicted isomers of 1,2-dichlorobenzene.

Kekulé's structure (Figure 7.3) explains many of the properties of benzene but there are some properties that are difficult to explain using his model.

Kekulé suggested that there should be two isomers of 1,2-dichlorobenzene (Figure 7.4), but in practice these have never been separated as there is only one form of 1, 2-dichlorobenzene.

It might be thought that benzene would react in the same way as an alkene, such as cyclohexene, C_6H_{10}, and readily undergo electrophilic addition reactions. Benzene does not do this. For example, compounds that contain a C=C double bond readily decolourise bromine. Benzene only reacts with bromine when boiled and exposed to ultraviolet light, or in the presence of a suitable catalyst. This casts doubt on the existence of C=C double bonds in benzene.

Experimental data from measuring the enthalpy change when unsaturated cyclohexene reacts with hydrogen to produce cyclohexane shows that ΔH is $-120\,kJ\,mol^{-1}$.

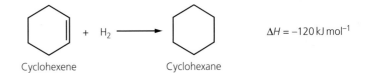

It follows that the enthalpy of hydrogenation of benzene ought to be three times ($-360\,kJ\,mol^{-1}$) that of cyclohexene.

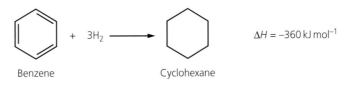

However, when measured experimentally, the enthalpy of hydrogenation of benzene was found to be $-208\,kJ\,mol^{-1}$, which is about $150\,kJ\,mol^{-1}$ less exothermic than predicted from the theoretical alternating double bond–single bond model.

This enthalpy data (Figure 7.5) casts further doubts on the validity of the Kekulé structure.

Using X-ray diffraction techniques, it is possible to measure the bond lengths within a molecule. It is found that the average C–C single bond has a length of 154 pm while the average C=C double bond length is 134 pm (pm = picometre = $1 \times 10^{-12}\,m$). If Kekulé's structure is correct, the bond lengths in benzene should alternate between long (154 pm) and short (134 pm). However, all the bonds in benzene are the same length: 139 pm. This suggests an intermediate bond, somewhere between a double bond and a single bond.

There is sufficient doubt about the Kekulé model that an alternative model has been suggested. The current model of the structure of benzene suggests that each carbon atom contributes one electron to a π-delocalised ring of electrons above and below the plane of atoms. Each carbon has one p-orbital at right angles to the plane of atoms (Figure 7.6a). Each p-orbital overlaps with adjacent p-orbitals in such a way that the delocalisation is extended over all six carbon atoms (Figure 7.6b).

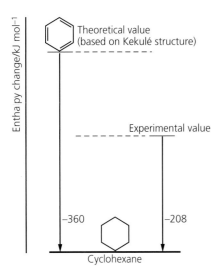

Figure 7.5 Enthalpy of hydrogenation data relating to benzene.

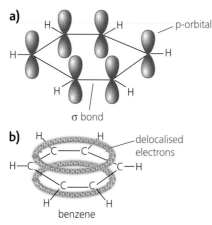

a)

σ bond

b)

benzene — delocalised electrons

Figure 7.6

Figure 7.6a shows benzene with normal covalent sigma bonds (σ bonds) between its carbon and hydrogen atoms. Each carbon atom uses three of its electrons to form three σ bonds with its neighbours, leaving each carbon atom with one electron in an atomic p-orbital.

The six p-orbital electrons do not pair up to form three carbon–carbon double bonds as in the Kekulé structure. Instead, they are shared evenly between all six carbon atoms giving rise to circular clouds of negative charge above and below the ring of carbon atoms (see Figure 7.6b). This is an example of a delocalised π-electron system and the π electrons are free to move anywhere within the system.

Molecules with delocalised electrons, in which the charge is spread over a larger region, are usually more stable than might otherwise be expected. The π-delocalised ring accounts for the increased stability of benzene and explains its low reactivity with bromine. In addition, it explains why all six carbon–carbon bond lengths are identical. Students may represent the structure of benzene as:

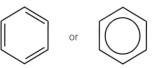

The right-hand structure is used in equations and will be used in this book, but both are acceptable when describing mechanisms

The development of ideas concerning the structure of benzene illustrates the way in which theories develop and get modified as new knowledge becomes available.

Naming arenes

Benzene, C_6H_6, is regarded as the 'parent' arene and other arenes can be derived from benzene by replacing one or more H with other atoms or groups. The names and structures of some common derivatives of benzene are shown in Table 7.1.

Table 7.1 The names used for compounds with a benzene ring can be confusing. C_6H_5- is named with either 'benzene' as a suffix or 'phenyl' as a prefix.

Chlorobenzene	C_6H_5-Cl
Methylbenzene	$C_6H_5-CH_3$
Nitrobenzene	$C_6H_5-NO_2$
Phenol	C_6H_5-OH
Phenylamine	$C_6H_5-NH_2$

When more than one hydrogen atom is substituted, numbers are used to indicate the position of the substituents. The ring is numbered either clockwise or anticlockwise to give the lowest possible numbers. One of the substituents will always occupy position '1' on the ring.

In phenyl compounds like phenol and phenylamine the −OH and the −NH$_2$ are assumed to be in position 1 (Figure 7.7).

methylbenzene
(no number is required
as there is only one hydrogen
substituted and all positions
are equivalent)

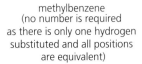

1,3-dichlorobenzene
(count anticlockwise to get
the lowest possible numbers)

1-chloro-3-methylbenzene
(if the two substituents are different
alphabetically chloro gets precedence over methyl)

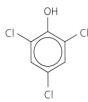

2,4,6 -trichlorophenol
(phenol takes precedence and
does not need a number: it is
assumed the OH is at position 1)

3,5-dimethylphenylamine
(phenylamine takes precedence and
does not need a number: it is
assumed the NH$_2$ is at position 1)

2,5 -dichloro-1-bromobenzene
(alphabetically bromine takes position1
and the Cl are numbered to give the lowest
numbers)

Figure 7.7 A range of aromatic compounds, illustrating naming conventions.

Test yourself

1 Describe, with the aid of suitable labelled diagrams:
  a) the Kekulé model of benzene
  b) the delocalised model of benzene.
2 Draw the following arenes:
  a) 2,6-dibromophenol
  b) 2,4,6-trinitromethylbenzene
  c) 1,3–dimethyl benzene
  d) phenylethene.

Electrophilic substitution

As discussed on page 145, benzene does not readily react with and decolourise bromine. However, it does react with bromine in the presence of a suitable catalyst.

Analysis of the product shows that benzene undergoes substitution reactions rather than addition reactions.

Benzene is stable and retains the π-delocalised ring. The product is formed by replacing one (or more) of the hydrogen atoms in C_6H_6, so that the product is of the form C_6H_5X, where X is an electrophile.

The most common electrophile is H^+ but it is clearly pointless reacting benzene, C_6H_6, with H^+. Catalysts are used to generate different electrophiles such as NO_2^+, Cl^+, Br^+ and $^+CH_3$.

The general equation is:

$$C_6H_6 + X^+ \rightarrow C_6H_5X + H^+$$

where X^+ is the electrophile.

Nitration

Nitroarenes are versatile compounds that are used in the preparation of drugs and dyes. The explosive trinitrotoluene, TNT, is made by the nitration of methylbenzene (toluene is the old name for methylbenzene).

Tip

An electrophile is a species that can accept an electron pair, resulting in the formation of a (dative) covalent bond.

Tip

Metal ions such as Na^+ cannot behave as electrophiles because they do not form covalent bonds.

Nitration of benzene

The nitration of benzene requires the use of a nitrating mixture consisting of concentrated nitric acid and concentrated sulfuric acid.

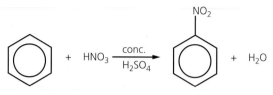

The reaction is heated gently at a temperature of between 50 °C and 55 °C. It is essential to control the temperature to minimise the possibility of forming dinitro- and trinitro-benzenes. (The temperature required varies from one arene to another.)

The mechanism for the reaction has **three** distinct steps.

Step 1 Formation of the electrophile
The electrophile in this reaction is the nitronium ion, NO_2^+, which is generated by the reaction between sulfuric acid and nitric acid. This complex reaction can be summarised as:

$$H_2SO_4 + HNO_3 \rightarrow HSO_4^- + H_2NO_3^+$$

$$H_2NO_3^+ \rightarrow H_2O + NO_2^+$$

Sulfuric acid also donates a proton to water forming H_3O^+ and HSO_4^-. The net reaction can be written as:

$$2H_2SO_4 + HNO_3 \rightarrow 2HSO_4^- + H_3O^+ + NO_2^+$$

Step 2 Electrophilic attack at the benzene ring
The mechanism can be described using the delocalised model of benzene:

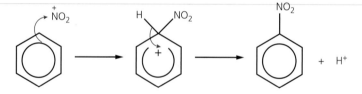

The mechanism can also be described using the Kekulé model of benzene.

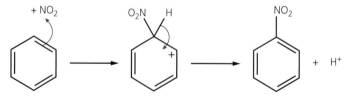

Step 3 Regeneration of the catalyst
The H^+ formed in the second stage of the mechanism then reacts with the hydrogen sulfate ion, HSO_4^-, to reform sulfuric acid, H_2SO_4. Therefore, the H_2SO_4 is a catalyst in the reaction.

$$H^+ + HSO_4^- \rightarrow H_2SO_4$$

Friedel–Crafts reactions

Friedel–Crafts reactions were developed by Charles Friedel and James Crafts in the late nineteenth century. There are three main types: halogenation, alkylation and acylation. They are all electrophilic substitution reactions.

The general reaction can be written as:

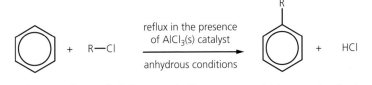

The mechanism for each follows a similar pattern which can also be broken down into three steps:

Step 1 Formation of the electrophile

The reagent R–Cl is polarised by the catalyst, $AlCl_3$, and generates an electrophile:

$$R-Cl + AlCl_3 \rightarrow \overset{\delta+}{R}\!-\!\overset{\delta-}{Cl} --- AlCl_3$$

which can be written as

$$R-Cl + AlCl_3 \rightarrow R^+ + AlCl_4^-$$

$R^{\delta+}$ or R^+ behave as electrophiles and react with benzene in step 2.

Step 2 Substitution of a hydrogen by the electrophile

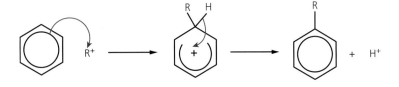

Step 3 Reform the halogen carrier (catalyst)

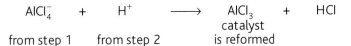

$$\underset{\text{from step 1}}{AlCl_4^-} \quad + \quad \underset{\text{from step 2}}{H^+} \quad \longrightarrow \quad \underset{\substack{\text{catalyst} \\ \text{is reformed}}}{AlCl_3} \quad + \quad HCl$$

Table 7.2 summarises the three possible types of Friedel–Crafts reactions.

Table 7.2

Type of Friedel–Crafts reaction	Reagent, R–Cl	Organic product
halogenation	Cl — Cl	Cl (on benzene ring)
alkylation	CH₃ — Cl	CH₃ (on benzene ring)
acylation	O=C(–CH₃)–Cl	O=C–CH₃ (on benzene ring)

Chlorination of benzene

Benzene undergoes an electrophilic substitution reaction with both chlorine and bromine. If chlorine is bubbled through benzene at room temperature in the presence of a halogen carrier, chlorobenzene is formed.

The mechanism for the reaction has three distinct steps.

Step 1 Formation of the electrophile

The halogen carrier is either iron, iron(III) chloride, $FeCl_3$, or aluminium chloride, $AlCl_3$. If iron is used it reacts initially with the chlorine to form $FeCl_3$. Halogen carriers work by polarising the Cl–Cl bond and promoting heterolytic fission of the halogen bond:

$$Cl{-}Cl \qquad Fe^{3+}(Cl^-)_3 \qquad \longrightarrow \qquad \overset{\delta+}{Cl}{-}\overset{\delta-}{Cl} {-}{-}{-} Fe^{3+}(Cl^-)_3$$

Cl_2 is non-polar

Fe^{3+} has high charge density and can polarise the Cl_2

The Cl–Cl bond is now polarised and the $Cl^{\delta+}$ can behave as an electrophile. $FeCl_3$ (and $AlCl_3$) have vacant orbitals that can accept an electron pair and form a dative covalent bond. The halogen bond is broken heterolytically and the net reaction can be written as:

$$Cl_2 + FeCl_3 \rightarrow Cl^+ + FeCl_4^-$$

Compounds like $FeCl_3$ and $AlCl_3$ appear to carry the halogen and are known as 'halogen carriers'. Both $FeCl_3$ and $AlCl_3$ react with water and it is essential that the reactions are carried out in anhydrous conditions. Iron can also function as a halogen carrier due to iron initially reacting with chlorine to form $FeCl_3$, which in turn then generates the Cl^+ electrophile by reacting with Cl_2.

Step 2 Electrophilic attack at the benzene ring

As with the nitration of benzene, the mechanism is usually written using the delocalised model of benzene:

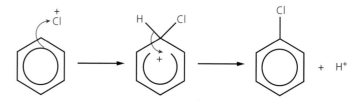

The mechanism can also be described using the Kekulé model of benzene.

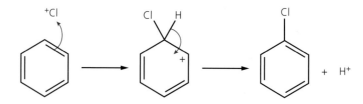

Step 3 Regeneration of the catalyst

The H^+ formed in the second stage of the mechanism then reacts with the $FeCl_4^-$ ion to reform iron (III) chloride, $FeCl_3$. Therefore, $FeCl_3$ is a catalyst in the reaction:

$$H^+ + FeCl_4^- \rightarrow FeCl_3 + HCl$$

Tip

$AlCl_3$, $AlBr_3$, $FeCl_3$, $FeBr_3$ and Fe can all behave as halogen carriers.

Activity

Studying the reaction of benzene with chlorine

Figure 7.8 shows the apparatus which might once have been used to prepare chlorobenzene by heating benzene with chlorine gas in the presence of iron filings. This preparation is now banned in teaching laboratories in schools and colleges.

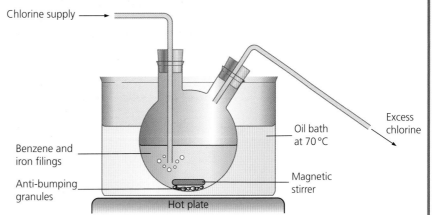

Figure 7.8 Preparing chlorobenzene.

1 Why is this reaction banned in teaching laboratories in schools?

2 a) Why should the reaction be carried out in a fume cupboard?

   b) Why is a hot plate used?

   c) Why is the oil bath at 70 °C?

3 During the reaction, iron reacts with chlorine to form iron(III) chloride, which then acts as an electron pair acceptor, polarising the Cl_2 molecules as $\overset{\delta+}{Cl}-\overset{\delta-}{Cl}$. Write equations to show:

   a) the formation of iron(III) chloride

   b) the polarisation of Cl_2 as $\overset{\delta+}{Cl}-\overset{\delta-}{Cl}$ by iron(III) chloride.

4 Write a mechanism for the reaction of polarised $\overset{\delta+}{Cl}-\overset{\delta-}{Cl}$ molecules with benzene to form an intermediate cation plus Cl^- in the first step, and then chlorobenzene plus HCl in the second step.

5 The reaction shown in Figure 7.8 is just as effective if aluminium chloride or iron(III) chloride is used in place of iron. These three substances (Fe, $FeBr_3$ and $AlCl_3$) are often described as 'catalysts' and 'halogen carriers'.

   a) Why are these substances described as halogen carriers for the reaction?

   b) Why is it correct to describe aluminium chloride and iron(III) chloride as catalysts?

   c) Why is it incorrect to describe iron as a catalyst for the reaction?

6 Aluminium chloride acts as a catalyst for the chlorination of benzene by polarising Cl_2 molecules in the same way as iron(III) chloride.

   a) Do you think aluminium chloride will be more effective or less effective than iron(III) chloride?

   b) Explain your answer to part **a)**.

7 Aluminium chloride can also be used with haloalkanes, such as chloromethane, CH_3Cl, to add alkyl groups to benzene and other arenes. The aluminium chloride increases the polarity of the $\overset{\delta+}{C}-\overset{\delta-}{Cl}$ bond of chloromethane and promotes the action of the methyl group, $\overset{\delta+}{CH_3}$, as an electrophile.

Write a possible mechanism for the electrophilic action of $\overset{\delta+}{CH_3}-\overset{\delta-}{Cl}$ on benzene to form an intermediate cation, which then breaks up to form methylbenzene and hydrogen chloride.

Alkylation of benzene

Benzene undergoes an electrophilic substitution reaction with haloalkanes (also known as alkyl chlorides) like chloromethane, CH_3Cl. The mechanism follows the same pattern as the chlorination of benzene and can, likewise, be broken down into three steps.

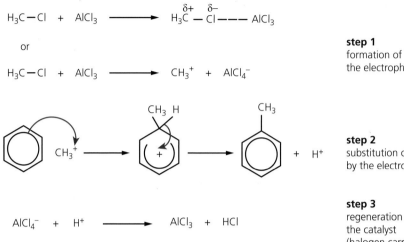

step 1
formation of the electrophile

step 2
substitution of H by the electrophile

step 3
regeneration of the catalyst (halogen carrier)

Acylation of benzene

Benzene undergoes an electrophilic substitution reaction with acyl chlorides (also known as acid chlorides) like ethanoyl chloride, CH_3COCl. The mechanism follows the same pattern as the chlorination of benzene and can likewise be broken down into three steps.

Tip

Alkylation and acylation both increase the number of carbon atoms in the molecule.

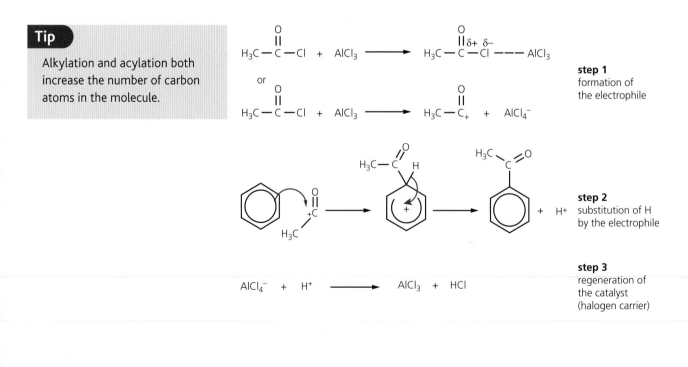

step 1
formation of the electrophile

step 2
substitution of H by the electrophile

step 3
regeneration of the catalyst (halogen carrier)

Test yourself

3 a) Describe the mechanism for the reaction between bromine and benzene in the presence of $FeBr_3(s)$.

b) Explain why $FeBr_3(s)$ can be regarded as a catalyst.

c) $FeBr_3(s)$ can be replaced by $Fe(s)$.

d) Explain why $Fe(s)$ is not regarded as a catalyst.

4 Under certain conditions benzene can be nitrated to form a mixture of isomers each with molecular formula $C_6H_4N_2O_4$. Draw and name the isomers.

5 Write equations and state the conditions for each of the following conversions.

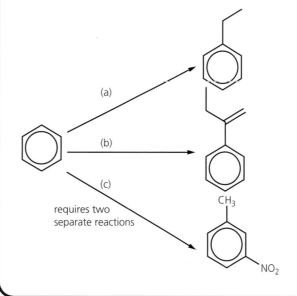

(a)

(b)

(c)

requires two
separate reactions

CH_3

NO_2

Bromination of alkenes and arenes

Alkenes, such as cyclohexene, react readily with bromine, in the absence of sunlight, and undergo electrophilic addition reactions. The reaction is rapid and is initiated by the induced dipole in bromine. Alkenes like cyclohexene are able to induce a dipole in bromine because of the high electron density in the carbon-to-carbon double bond, C=C, which is sufficient to induce a dipole in the Br–Br bond:

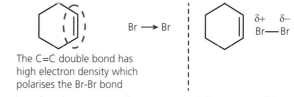

The C=C double bond has
high electron density which
polarises the Br-Br bond

The mechanism for the reaction between cyclohexene and bromine is shown below:

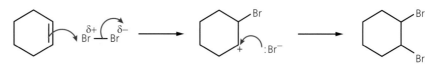

Benzene also reacts with bromine but it is more resistant and reacts much less readily. The electron density in benzene is insufficient to induce a dipole in the Br–Br bond because the p-electrons are delocalised over the six carbon atoms in the ring. It follows that polarisation of the Br–Br bond requires the presence of a halogen carrier to generate the electrophile. The resultant reaction is electrophilic substitution, *not* electrophilic addition. This is explained by the stability of the π-delocalised ring of electrons that is retained in most reactions of arenes.

> **Test yourself**
>
> 6 Phenylethene, $C_6H_5CHCH_2$, reacts with Br_2. The products depend on the reaction conditions. Suggest what would be formed when:
>   a) phenylethene reacts with Br_2 in the presence of UV light, and name the mechanism involved in this reaction
>   b) phenylethene reacts with Br_2 in the presence of $AlBr_3$, and name the mechanism(s) involved.

Phenols

Phenols occur widely in nature and are examples of compounds with a functional group directly attached to a benzene ring. In phenols, the functional group is the hydroxy group, –OH, which is also found in alcohols. You might, therefore, expect phenols to react in the same way as alcohols. Phenols and alcohols do have some reactions in common, but phenols behave differently to alcohols in many reactions.

Experiments show that the –OH group affects the behaviour of the benzene ring, while the benzene ring also modifies the properties of the –OH group. As a result of this, phenols have some distinctive and useful properties.

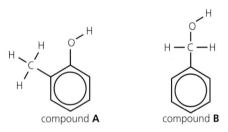

compound **A**   compound **B**

The formula of both compounds A and B is C_7H_8O. They are isomers. Compound A is a phenol; compound B is an alcohol.

Derivatives of phenol are named in a similar way to those of benzene – by numbering the carbon atoms in the benzene ring, starting from the –OH group. The numbering runs clockwise or anticlockwise to give the lowest possible numbers for the substituted groups.

> **Tip**
>
> The OH group in all phenols occupies position 1 on the ring.

  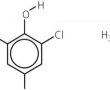

phenol   2-methylphenol   2,4,6-trichlorophenol   2,4-dimethylphenol

Reactions of phenols

Phenols are weak acids, unlike alcohols which are not acidic.

In phenol, the $-C_6H_5$ ring is electron deficient and one of the lone pairs of electrons on the oxygen in the $-OH$ group is incorporated and delocalised into the ring.

The overlap of the p-orbitals leads to a delocalisation that extends from the ring out over the oxygen atom. As a result, the negative charge is no longer entirely localised on the oxygen, but is spread out around the whole ion.

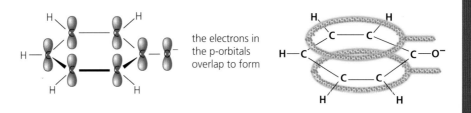

the electrons in the p-orbitals overlap to form

This has three effects:

- it weakens the O−H bond such that phenol is able to donate a H^+
- it stabilises the phenoxide ion, $C_6H_5O^-$, that is formed
- the increased electron density in the π-delocalised system means that phenol undergoes electrophilic substitution reactions much more readily than benzene does.

Activity

Manufacturing phenol

Phenol is manufactured from benzene, propene and oxygen in two stages. The procedure is called the cumene process (Figure 7.9).

- The first stage of the process involves the acid-catalysed electrophilic substitution of benzene with propene to form cumene.
- The second stage involves the air oxidation of cumene.

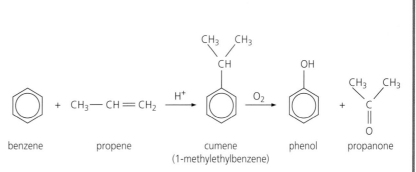

benzene   propene   cumene (1-methylethylbenzene)   phenol   propanone

Figure 7.9 The manufacture of phenol by the cumene process.

This produces equimolar amounts of phenol and propanone, a valuable co-product. About 100 000 tonnes of phenol are manufactured each year in the UK using the cumene process.

1 Benzene and propene are obtained from crude oil for use in the cumene process. What processes, starting with crude oil, are used to produce:
   a) benzene
   b) propene?

2 In the first stage of the cumene process, H+ ions react with propene to produce electrophiles.
   a) Write the formulae of two possible electrophiles produced when H+ ions react with propene.
   b) Explain why one of these electrophiles is more stable than the other.
   c) Name and draw the structure of a second possible product of this first stage, besides cumene.

3 a) Write an equation for the reaction of the stable electrophile, identified in **2b)**, with benzene to produce cumene.
   b) Why is this reaction described as 'acid catalysed'?

4 Write an equation for the second stage of the process, in which cumene is oxidised to phenol and propanone.

5 The overall yield in the cumene process is 85%. Calculate the mass of benzene required to manufacture 1 tonne of phenol, and the mass of propanone formed at the same time.

6 In the USA, some phenol is manufactured by refluxing chlorobenzene with concentrated sodium hydroxide solution at 400 °C and a pressure of 150 atmospheres (Figure 7.10).
   a) What kind of reagent is the OH− ions in this reaction?
   b) Why do you think the reaction requires such harsh conditions to produce phenol?

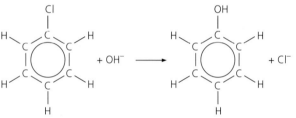

Figure 7.10 The reaction of chlorobenzene with hydroxide ions.

Reaction with sodium hydroxide and with sodium

As phenol is weakly acidic,

$$C_6H_5OH \rightleftharpoons C_6H_5O^- + H^+$$

it forms salts when it reacts with both sodium hydroxide and with sodium, but phenols are not sufficiently acidic to react with carbonates.

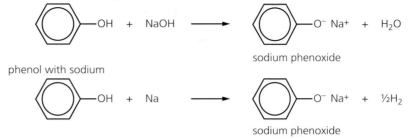

Ethanol, C_2H_5OH, also reacts with sodium to produce a salt (sodium ethoxide) and hydrogen.

$$C_2H_5OH + Na \rightarrow C_2H_5O^-Na^+ + \tfrac{1}{2}H_2$$

Ethanol is not acidic so it does not react with NaOH.

It is worth noting that phenols and alcohols both contain the OH group but phenols react differently to alcohols. Phenols do not:

- react with carboxylic acids to form esters

- undergo elimination reactions

- oxidise in the same way as alcohols. They can be oxidised but they tend to form complex polymeric materials.

Electrophilic substitution reactions of phenol

Phenol also reacts with electrophiles and readily undergoes nitration, halogenation, alkylation and acylation. Phenol reacts more readily than benzene and requires less stringent conditions. This is best illustrated in its reaction with bromine.

Reaction with bromine

Phenol reacts readily with bromine. When bromine is dripped into an aqueous solution of phenol, the bromine is immediately decolourised and white crystals of 2,4,6-tribromophenol are instantly formed (see Figure 7.11 below).

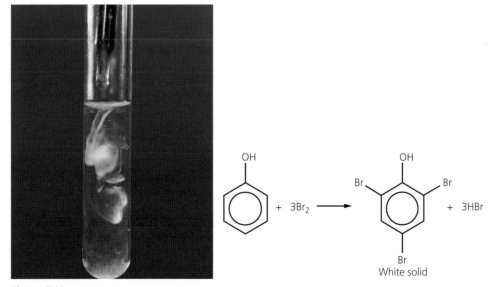

Figure 7.11

Unlike benzene, phenol does not require a halogen carrier. This is because the ring is activated by one of the lone pairs of electrons on the oxygen atom in the −OH group. This increases the electron density in the ring, so that it induces a dipole in the Br−Br bond, thereby generating an electrophile, $Br^{\delta+}$. The increased electron density of the ring attracts the electrophile, leading to a rapid reaction at room temperature.

Substitution occurs at the 2, 4 and 6 positions on the ring.

By contrast, the bromination of benzene takes place only in the presence of a suitable halogen carrier.

The increased rate of reaction between phenol and bromine is also true for any other electrophilic substitution reaction. This is most evident in the conditions required for nitration.

● Benzene can be nitrated using a mixture of concentrated nitric acid and concentrated sulfuric acid at a temperature of approximately 55 °C.

● Phenol can be nitrated using dilute nitric acid. A mixture of 2-nitrophenol and 4-nitrophenol is produced. If concentrated acids are used, it is possible to form 2,4,6-trinitrophenol (also known as picric acid).

> **Test yourself**
>
> 7 Phenol and ethanol both contain the −OH functional group.
>   **a)** Identify one reaction in which phenol and ethanol react in a similar way.
>   **b)** Identify one reaction in which phenol and ethanol react in different ways.
> 8 Phenol reacts rapidly with dilute nitric acid to produce a mixture of 2-nitrophenol and 4-nitrophenol.
>   Draw the structures of the products.

Directing effect of substituted aromatic compounds

Benzene, C_6H_6, undergoes electrophilic substitution reactions. Substituted aromatic compounds, C_6H_5X, can also be nitrated, chlorinated, alkylated or acylated in the same way as benzene. However, the side chain, X, influences the rate of reaction and the position of substitution within the C_6H_5X ring. Common side chains, X, include:

● CH_3- (or any alkyl group R, C_nH_{2n+1}), which is electron releasing

● $HO-$, which is electron releasing (a lone pair of electrons from the O can be delocalised into the ring)

● H_2N-, which is electron-releasing (a lone pair of electrons from the N can be delocalised into the ring)

● O_2N-, which is electron withdrawing (due to the high electronegativity of the O compared to the N)

● $-C=O$, which is electron withdrawing (due to the high electronegativity of the O compared to the C)

● $-COOH$, which is electron withdrawing.

Electron-releasing side chains increase the electron density of the ring and, therefore, increase the attraction for the electrophile, such that the rate of reaction increases. The side chains in compounds like $C_6H_5CH_3$, C_6H_5OH and $C_6H_5NH_2$ are described as 'activating' and all react faster than benzene.

Electron-withdrawing side chains are 'deactivating' – they decrease the electron density of the ring and, therefore, the attraction for the electrophile is also decreased, such that the rate of reaction in compounds like $C_6H_5NO_2$, is slower than for benzene.

The side chains also influence the position of substitution on the ring. There are three possible isomers of a di-substituted benzene compound (Figure 7.12).

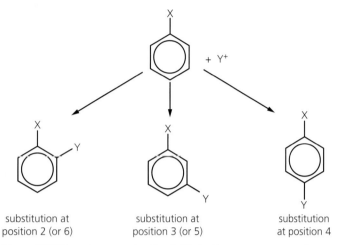

substitution at position 2 (or 6)

substitution at position 3 (or 5)

substitution at position 4

Figure 7.12 Three isomers of a di-substituted benzene compound.

If the side chain, X, is electron releasing, substitution is directed to the 2 and 4 positions and the side chain is 2,4-directing. (Position 6 is equivalent to position 2, so substitution can occur at all of 2, 4 and 6 positions on the ring.)

If methylbenzene is nitrated, the major products are 2-nitromethylbenzene and 4-nitromethylbenzene and very little 3-nitromethylbenzene is formed (Figure 7.13).

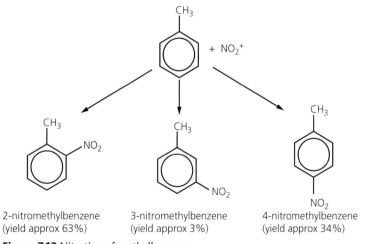

2-nitromethylbenzene (yield approx 63%)

3-nitromethylbenzene (yield approx 3%)

4-nitromethylbenzene (yield approx 34%)

Figure 7.13 Nitration of methylbenzene.

If, however, nitrobenzene is nitrated, very little, if any, 1,2-dinitrobenzene and 1,4-dinitrobenzene are formed; the major product is 1,3-dinitrobenzene. (Figure 7.14).

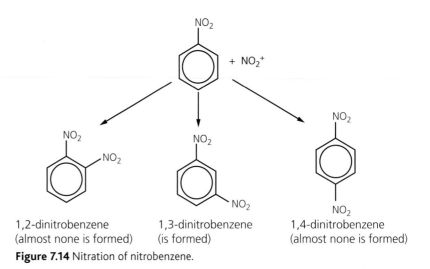

1,2-dinitrobenzene 1,3-dinitrobenzene 1,4-dinitrobenzene
(almost none is formed) (is formed) (almost none is formed)

Figure 7.14 Nitration of nitrobenzene.

Table 7.3 summarises the effect on the rate of reaction and the directing effect of some common side chains.

Table 7.3 The effect on the rate of reaction and the directing effect of some common side chains.

Side chain	Electron		Activating	Deactivating	Directing to positions				
	releasing	withdrawing			2	3	4	5	6
CH_3	✓		✓		✓		✓		✓
C_2H_5	✓		✓		✓		✓		✓
OH*	✓		✓		✓		✓		✓
NH_2*	✓		✓		✓		✓		✓
NO_2*		✓		✓		✓		✓	
C=O		✓		✓		✓		✓	
COOH		✓		✓		✓		✓	

* Students are only expected to know the groups with an asterisk. If other groups are used, relevant data will be supplied in the examination.

Test yourself

9 Draw and name the organic products formed when
  a) phenol reacts with Cl_2 – assume only **one** H has been substituted
  b) phenol reacts with Cl_2 – assume **two** Hs have been substituted
  c) phenol reacts with Cl_2 – assume **three** Hs have been substituted.
10 Benzene and methylbenzene can both be nitrated using a mixture of concentrated nitric acid and concentrated sulfuric acid. Explain why the temperature to nitrate benzene is approximately 55 °C, whilst the temperature for nitrating methylbenzene is only about 25 °C.
11 Phenol reacts rapidly with dilute nitric acid to produce a mixture of 2-nitrophenol and 4-nitrophenol.
  a) Draw the structures of the two products.
  b) Explain why 3-nitrophenol is not formed as a product.

Practice questions

Multiple choice questions 1–10

1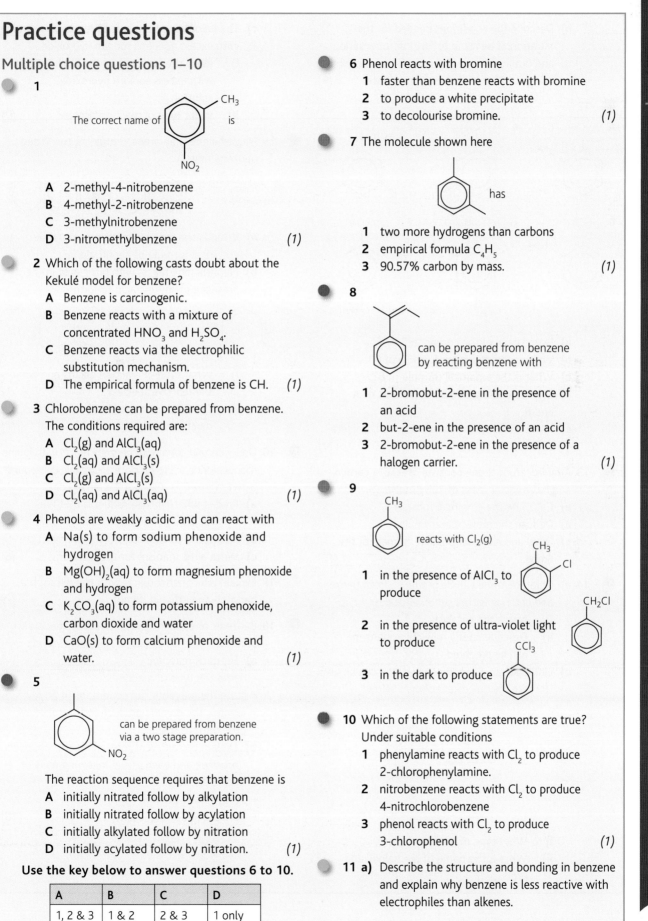

The correct name of [structure with CH₃ and NO₂] is

  A 2-methyl-4-nitrobenzene
  B 4-methyl-2-nitrobenzene
  C 3-methylnitrobenzene
  D 3-nitromethylbenzene   *(1)*

2 Which of the following casts doubt about the Kekulé model for benzene?
  A Benzene is carcinogenic.
  B Benzene reacts with a mixture of concentrated HNO_3 and H_2SO_4.
  C Benzene reacts via the electrophilic substitution mechanism.
  D The empirical formula of benzene is CH.   *(1)*

3 Chlorobenzene can be prepared from benzene. The conditions required are:
  A $Cl_2(g)$ and $AlCl_3(aq)$
  B $Cl_2(aq)$ and $AlCl_3(s)$
  C $Cl_2(g)$ and $AlCl_3(s)$
  D $Cl_2(aq)$ and $AlCl_3(aq)$ *(1)*

4 Phenols are weakly acidic and can react with
  A Na(s) to form sodium phenoxide and hydrogen
  B $Mg(OH)_2(aq)$ to form magnesium phenoxide and hydrogen
  C $K_2CO_3(aq)$ to form potassium phenoxide, carbon dioxide and water
  D CaO(s) to form calcium phenoxide and water.   *(1)*

5 [structure with NO₂] can be prepared from benzene via a two stage preparation.

The reaction sequence requires that benzene is
  A initially nitrated follow by alkylation
  B initially nitrated follow by acylation
  C initially alkylated follow by nitration
  D initially acylated follow by nitration.   *(1)*

Use the key below to answer questions 6 to 10.

A	B	C	D
1, 2 & 3 correct	1 & 2 correct	2 & 3 correct	1 only correct

6 Phenol reacts with bromine
  1 faster than benzene reacts with bromine
  2 to produce a white precipitate
  3 to decolourise bromine.   *(1)*

7 The molecule shown here [structure] has
  1 two more hydrogens than carbons
  2 empirical formula C_4H_5
  3 90.57% carbon by mass.   *(1)*

8 [structure] can be prepared from benzene by reacting benzene with
  1 2-bromobut-2-ene in the presence of an acid
  2 but-2-ene in the presence of an acid
  3 2-bromobut-2-ene in the presence of a halogen carrier.   *(1)*

9 [structure with CH₃] reacts with $Cl_2(g)$
  1 in the presence of $AlCl_3$ to produce [structure with CH₃ and Cl]
  2 in the presence of ultra-violet light to produce [structure with CH₂Cl]
  3 in the dark to produce [structure with CCl₃]

10 Which of the following statements are true? Under suitable conditions
  1 phenylamine reacts with Cl_2 to produce 2-chlorophenylamine.
  2 nitrobenzene reacts with Cl_2 to produce 4-nitrochlorobenzene
  3 phenol reacts with Cl_2 to produce 3-chlorophenol   *(1)*

11 a) Describe the structure and bonding in benzene and explain why benzene is less reactive with electrophiles than alkenes.

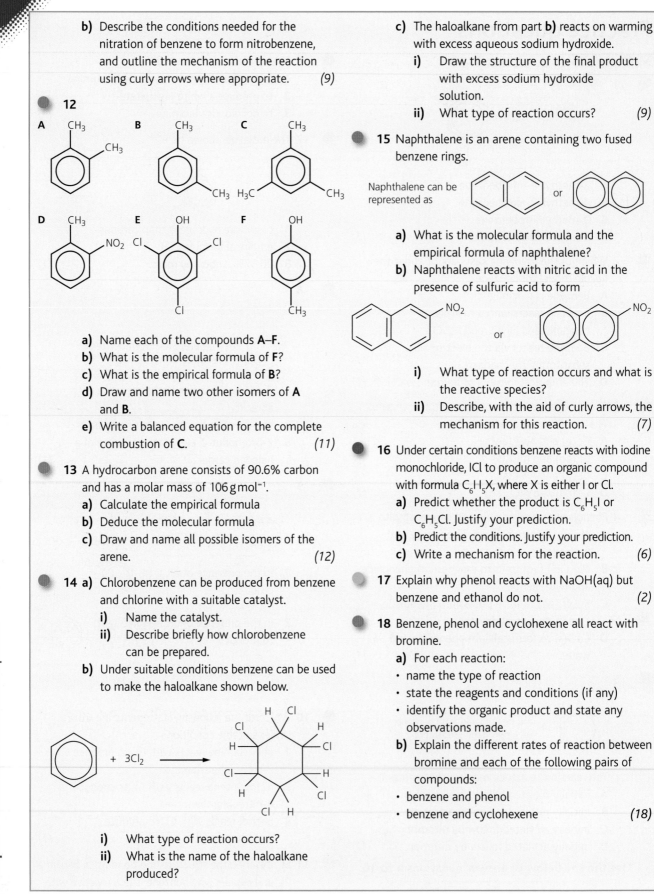

b) Describe the conditions needed for the nitration of benzene to form nitrobenzene, and outline the mechanism of the reaction using curly arrows where appropriate. *(9)*

12

A CH₃ ... CH₃

B CH₃ ... CH₃

C CH₃ H₃C ... CH₃

D CH₃ ... NO₂

E OH Cl ... Cl Cl

F OH ... CH₃

a) Name each of the compounds **A–F**.
b) What is the molecular formula of **F**?
c) What is the empirical formula of **B**?
d) Draw and name two other isomers of **A** and **B**.
e) Write a balanced equation for the complete combustion of **C**. *(11)*

13 A hydrocarbon arene consists of 90.6% carbon and has a molar mass of $106\,\text{g mol}^{-1}$.
a) Calculate the empirical formula
b) Deduce the molecular formula
c) Draw and name all possible isomers of the arene. *(12)*

14 a) Chlorobenzene can be produced from benzene and chlorine with a suitable catalyst.
   i) Name the catalyst.
   ii) Describe briefly how chlorobenzene can be prepared.
b) Under suitable conditions benzene can be used to make the haloalkane shown below.

[benzene] + 3Cl₂ ⟶ [cyclohexane structure with H and Cl substituents]

   i) What type of reaction occurs?
   ii) What is the name of the haloalkane produced?

c) The haloalkane from part **b)** reacts on warming with excess aqueous sodium hydroxide.
   i) Draw the structure of the final product with excess sodium hydroxide solution.
   ii) What type of reaction occurs? *(9)*

15 Naphthalene is an arene containing two fused benzene rings.

Naphthalene can be represented as [naphthalene structure] or [naphthalene structure]

a) What is the molecular formula and the empirical formula of naphthalene?
b) Naphthalene reacts with nitric acid in the presence of sulfuric acid to form

[naphthalene with NO₂] or [naphthalene with NO₂]

   i) What type of reaction occurs and what is the reactive species?
   ii) Describe, with the aid of curly arrows, the mechanism for this reaction. *(7)*

16 Under certain conditions benzene reacts with iodine monochloride, ICl to produce an organic compound with formula C_6H_5X, where X is either I or Cl.
a) Predict whether the product is C_6H_5I or C_6H_5Cl. Justify your prediction.
b) Predict the conditions. Justify your prediction.
c) Write a mechanism for the reaction. *(6)*

17 Explain why phenol reacts with NaOH(aq) but benzene and ethanol do not. *(2)*

18 Benzene, phenol and cyclohexene all react with bromine.
a) For each reaction:
• name the type of reaction
• state the reagents and conditions (if any)
• identify the organic product and state any observations made.
b) Explain the different rates of reaction between bromine and each of the following pairs of compounds:
• benzene and phenol
• benzene and cyclohexene *(18)*

19 Under suitable conditions phenol can be converted into nitrophenols.

a) State the reaction conditions for converting phenol into nitrophenol.

When phenol is nitrated, three isomers are formed but the amount of each isomer formed differs:

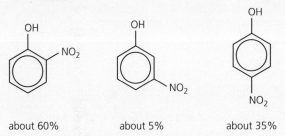

about 60%   about 5%   about 35%

b) Explain why only about 5% of 3-nitrophenol is formed

c) Suggest why more 2-nitrophenol is formed compared to 4-nitrophenol   (5)

20 2,4,6-trinitromethylbenzene (TNT) can be used as an explosive. When detonated, it forms CO_2(g), H_2O(l) and N_2(g). Calculate the volume of gas produced when 9.08 g of TNT is exploded. Assume all measurements were taken at RTP (1 mol of gas occupies $24.0 \, dm^3$).   (6)

21 TCP, 2,4,6-trichlorophenol, can be used as an antiseptic. A $100 \, cm^3$ solution of TCP was titrated with a $0.400 \, mol \, dm^{-3}$ NaOH(aq) solution. Exactly $25.0 \, cm^3$ NaOH(aq) was required.

a) Calculate the mass of TCP dissolved in the $100 \, cm^3$ solution of TCP.

b) Calculate the concentration ($mol \, dm^{-3}$) of the TCP.   (6)

22 4-nitromethylbenzene can be prepared from benzene via a two-step process. Write equations for each step, state the conditions and describe, with the aid of curly arrows, the mechanism for each step. Justify the sequence of the steps you have used.   (11)

Challenge

23 Benzene reacts with alkenes in the presence of an acid catalyst to form:

$$C_6H_6 + C_nH_{2n} \rightarrow C_6H_5C_nH_{2n+1}$$

The initial step in the reaction is the formation of a carbonium ion (carbocation) when the H^+ catalyst reacts with the alkene.

Benzene reacts with propene in the presence of an acid catalyst to form (1-methylethyl)benzene, $C_6H_5CH(CH_3)_2$:

$$C_6H_6 + C_3H_6 \xrightarrow{H^+} C_6H_5CH(CH_3)_2$$

a) Draw the displayed formula of (1-methylethyl) benzene, $C_6H_5CH(CH_3)_2$, and predict all the bond angles.

b) Outline, with the aid of curly arrows, the reaction between benzene, propene and H^+.

c) In the initial reaction between propene and H^+ it is possible to form two different carbonium ions, one primary and the other secondary. Explain which of these is more stable and which is, therefore, more likely to react.   (10)

24 Picric acid is a nitrophenol compound which is used extensively in explosives. When it detonates it produces carbon dioxide, steam and nitrogen. The volume of carbon dioxide produced is four times greater than the steam produced. Picric acid contains 18.34% by mass of nitrogen. Assume the molecular formula of picric acid is $C_wH_xN_yO_z$.

a) Deduce the numerical value of w, x, y and z and hence work out the molecular formula of picric acid. Show your working.

b) Suggest a structural formula for picric acid. Justify your answer.

c) Write an equation for the detonation of picric acid.   (13)

Chapter 8

Carbonyl compounds

Prior knowledge

In this chapter it is essential that you are familiar with:

- classification of alcohols
- hydrogen bonding
- oxidation of alcohol
- infrared spectra of alcohols, aldehydes, ketones and carboxylic acids.

You should be aware that alcohols can be classified as either primary, secondary or tertiary. Alcohols have higher than expected boiling points because of the intermolecular hydrogen bonds formed between alcohol molecules. You should know that primary, secondary and tertiary alcohols react differently with an oxidising mixture such as acidified dichromate. Infrared spectroscopy can be used to detect certain functional groups in organic compounds.

Test yourself on prior knowledge

1 Draw, name and classify the alcohols with formula $C_5H_{11}OH$.
2 Explain, with the aid of a labelled diagram, how hydrogen bonds are formed between two molecules of:
   a) water
   b) ethanol, CH_3CH_2OH.
3 Draw and name all possible oxidation products, if any, for each of the following:
   a) butan-1-ol
   b) butan-2-ol
   c) 2-methylpropan-2-ol
   d) 2-methylpropan-1-ol.
4 The three infrared spectra below are for propan-1-ol, propanal and propanoic acid. Identify which is which and explain your answer.

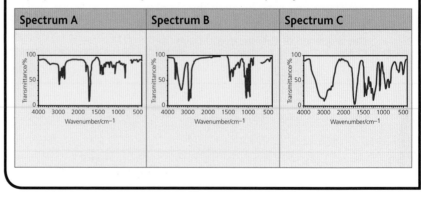

| Spectrum A | Spectrum B | Spectrum C |

Alcohols, aldehydes and ketones

You have already studied some of the reactions of alcohols, one of which is that primary and secondary alcohols can be oxidised to form aldehydes and ketones. It is essential that you revise the chemistry of alcohols as you will be expected to recall the preparation of carbonyl compounds from alcohols.

Aldehyde and ketone groups occur widely in sugars such as glucose and fructose, and they also contribute to the distinctive odours of many plants and foods.

The carbonyl group is C=O and, like the C=C double bond in alkenes, is formed by the overlap of adjacent p-orbitals (Figure 8.1). Like the alkenes, carbonyl compounds are unsaturated molecules and undergo addition reactions. However, unlike alkenes, carbonyls are polar because of the difference in electronegativity between the carbon and oxygen atoms.

The position of the C=O on the carbon chain determines whether or not the compound is classified as an aldehyde or a ketone. An aldehyde has the carbonyl on the end of the carbon chain; in a ketone, the carbonyl group can be anywhere, other than on the terminal carbon atoms.

An aldehyde contains the functional group:

Ketones look similar, but the functional group is:

where R and R' = alkyl group, for example $-CH_3$ or $-CH_2CH_3$.

Figure 8.1 The p-orbitals overlap to form a π-bond.

Formation of carbonyl compounds and carboxylic acids

These are formed by the oxidation of alcohols. The ease and extent of oxidation depends on the type of alcohol:

● Tertiary alcohols are resistant to oxidation.

● Primary and secondary alcohols undergo oxidation when warmed with an oxidising mixture, such as acidified dichromate, $Cr_2O_7^{2-}/H^+$ (e.g. $K_2Cr_2O_7/H_2SO_4$).

● A secondary alcohol is oxidised to a ketone.

● A primary alcohol can be oxidised to form either an aldehyde or a carboxylic acid.

When oxidising a primary alcohol, such as propan-1-ol, the choice of apparatus is important. The alcohol and the oxidising mixture have to be heated. This can be achieved by either distillation (Figure 8.2, page 167) or reflux.

If propan-1-ol is oxidised it is possible to produce either propanal or propanoic acid. The boiling points are shown in Table 8.1.

Table 8.1 Boiling points of propan-1-ol, propanal and propanoic acid.

Name	Formula	Boiling point/°C
Propan-1-ol	$CH_3CH_2CH_2OH$	97
Propanal	CH_3CH_2CHO	49
Propanoic acid	CH_3CH_2COOH	141

In general, alcohols and carboxylic acids have considerably higher boiling points than aldehydes. Alcohols and carboxylic acids both contain $-O-H$ groups and, therefore, both form intermolecular hydrogen bonds.

In order to isolate the aldehyde, a distillation is usually carried out. Distillation involves evaporation followed by condensation and allows the most volatile component to be separated.

Oxidation of a primary alcohol to an aldehyde

Methanol is oxidised to methanal:

$$CH_3OH + [O] \rightarrow HCHO + H_2O$$

Methanol   Methanal

Ethanol is oxidised to ethanal:

$$CH_3CH_2OH + [O] \rightarrow CH_3CHO + H_2O$$

Ethanol   Ethanal

Oxidation of a secondary alcohol to a ketone

Propan-2-ol is oxidised to propanone (propan-2-one):

$$CH_3CH(OH)CH_3 + [O] \rightarrow CH_3COCH_3 + H_2O$$

Propan-2-ol   Propanone

Butan-2-ol is oxidised to butanone (butan-2-one):

$$CH_3CH_2CH(OH)CH_3 + [O] \rightarrow CH_3CH_2COCH_3 + H_2O$$

Butan-2-ol   Butanone

Oxidation of a primary alcohol to a carboxylic acid

The primary alcohol is oxidised in two steps:

● The first step creates an aldehyde.

● In the second step, the second C–H bond is converted into a C–O–H.

Overall, the carboxylic acid functional group is generated:

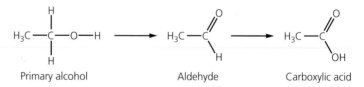

Primary alcohol   Aldehyde   Carboxylic acid

All balanced equations for the oxidation of a primary alcohol to a carboxylic acid are similar. Water is always formed and the reaction can be represented as:

primary alcohol + 2[O] → carboxylic acid + water

Methanol is oxidised to methanoic acid:

$$CH_3OH + 2[O] \rightarrow HCOOH + H_2O$$

Methanol   Methanoic acid

Ethanol is oxidised to ethanoic acid:

$$CH_3CH_2OH + 2[O] \rightarrow CH_3COOH + H_2O$$

Ethanol   Ethanoic acid

Activity

Converting an alcohol to an aldehyde

A 3 g sample of sodium dichromate(VI) is added to $10\,cm^3$ of dilute sulfuric acid in a pear-shaped flask. Then $1.5\,cm^3$ of propan-1-ol is added, a few drops at a time. The flask is gently shaken to mix the contents until all the solid has dissolved. The flask is fitted with a condenser as shown in Figure 8.2. Then 2–$3\,cm^3$ of liquid is distilled into a small flask from the reaction mixture.

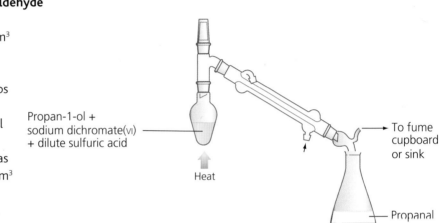

Propan-1-ol + sodium dichromate(VI) + dilute sulfuric acid

Heat

To fume cupboard or sink

Propanal

Figure 8.2 Apparatus used to oxidise an alcohol.

1 What type of alcohol is propan-1-ol?
2 What type of reagent is an acidic solution of sodium dichromate(VI)?
3 Write an equation for the reaction.
4 Suggest a reason for adding the propan-1-ol a few drops at a time.
5 Explain why the reaction mixture is not heated in a flask fitted with a reflux condenser before distilling off the product.
6 Describe two tests that can be used to show that the product of this reaction is an aldehyde and not a carboxylic acid.

Chemicals in perfumes

The perfume 'Chanel No. 5' was innovative when produced for the first time in 1921. As well as natural extracts from flowers, the scent includes a high proportion of synthetic aldehydes such as undecanal (Figure 8.3). This produces a highly original perfume.

The people who devise new perfumes think of the mixture as a sequence of 'notes'. You first smell the 'top notes', but the main effect depends on the 'middle notes', while the more lasting elements of the perfume are the 'end notes'. The overall balance of the three is critical. This means that the volatility of perfume chemicals is of great importance to the perfumer (Table 8.2).

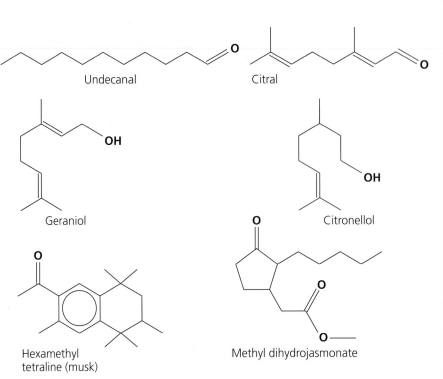

Undecanal

Citral

Geraniol

Citronellol

Hexamethyl tetraline (musk)

Methyl dihydrojasmonate

Figure 8.3 Skeletal formulae of some perfume chemicals.

Table 8.2 Natural and synthetic chemicals used to make perfumes.

Note	Natural chemicals	Synthetic chemicals	Boiling point or melting point
Top	Citrus oils Lavender	Octanal (citrus) Undecanal (green)	b.p. 168 °C b.p. 117 °C
Middle	Rose Violet	Geraniol (floral) Citronellol (rosy)	b.p. 146 °C b.p 224 °C
End	Balsam Musk	Indane (musk) Hexamethyl tetraline (musk)	m.p. 53 °C m.p. 55 °C

1 Draw the skeletal formula of octanal.
2 Suggest two advantages for the perfumer of using synthetic chemicals instead of chemicals extracted from living things.
3 Identify the carbonyl compounds among the compounds shown in Figure 8.3. In each case state whether the compound includes the functional group of an aldehyde or of a ketone.
4 Like many perfume chemicals, geraniol is a terpene. Terpene molecules are built from units derived from 2-methylbuta-1,3-diene.
   a) Draw the structure of 2-methylbuta-1,3-diene.
   b) How many 2-methylbuta-1,3-diene units are needed to make up the hydrocarbon skeleton of geraniol?
5 Use your knowledge of intermolecular forces to explain why:
   a) aldehydes are useful as top notes, while alcohols are more commonly used as middle notes
   b) the musks used as 'end notes' also help to 'fix', or retain, the more volatile components of a perfume
   c) geraniol is more soluble in water than undecanal.

Test yourself

1 Write equations for any possible oxidation reactions for each of the following. Use [O] to represent the oxidising agent.
  a) 4,4–dimethylpentan-2-ol
  b) 4,4-dimethylpentan-1-ol
  c) 2-methylpentan-3-ol

2 Menthone, see below, is a naturally occurring ketone which can be extracted from leaves of the peppermint plant and it is used in aromatherapy. It can also be prepared in the laboratory by the oxidation of menthol.

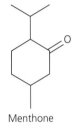

Menthone

  a) What is the molecular formula of menthone?
  b) Draw the skeletal formula of menthol.
  c) Write an equation for the oxidation of menthol.

Reactions of carbonyl compounds

Aldehydes and ketones both contain the C=O group and, therefore, have some reactions in common. However, aldehydes can be oxidised easily whereas, ketones cannot, so this type of reaction can be used to distinguish between them.

Reactions common to both aldehydes and ketones

Reduction

Aldehydes or ketones can be reduced to their respective alcohols using an aqueous solution of sodium tetrahydridoborate(III), $NaBH_4$, as the reducing agent.

All balanced equations for the reduction of an aldehyde or a ketone are similar. The reaction can be represented as:

aldehyde/ketone + 2[H] → alcohol

An aldehyde is reduced to a primary alcohol, for example:

$CH_3CH_2CHO + 2[H] \rightarrow CH_3CH_2CH_2OH$

Propanal   Propan-1-ol

A ketone is reduced to a secondary alcohol:

$CH_3COCH_3 + 2[H] \rightarrow CH_3CH(OH)CH_3$

Propanone   Propan-2-ol

Tip

[H] is used to represent the reducing agent in equations.

Nucleophilic addition reactions

$\overset{\delta+}{C}=\overset{\delta-}{O}$ is a polar bond because the carbon and oxygen atoms have different electronegativities. The carbon in the carbonyl group is $\delta+$ and can be attacked by a nucleophile such as the hydride ion, $:H^-$, or the cyanide ion, $:CN^-$. The carbonyl group, C=O, is unsaturated and, hence, undergoes addition reactions. The mechanism is, therefore, nucleophilic addition.

Nucleophilic addition with the hydride ion, :H⁻

This reduction reaction is complex but can be regarded as a nucleophilic addition reaction with the reducing agent, $NaBH_4$, providing the hydride ion, $:H^-$, which behaves as a nucleophile.

The reaction can be broken down into two steps:

- nucleophilic attack to produce an intermediate anion
- reaction with a proton, H^+, from either water or from an acid to give the product.

An example is the reduction of ethanal by aqueous $NaBH_4$:

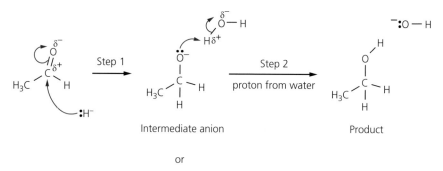

or

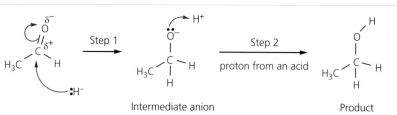

Ketones behave similarly — for example, consider the reduction of propanone by aqueous $NaBH_4$:

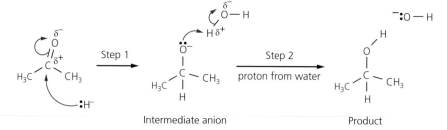

or

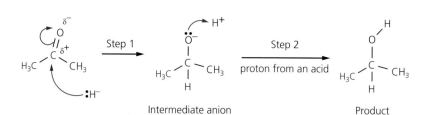

Nucleophilic addition with the cyanide ion, :CN⁻

Hydrogen cyanide can be used as the source of the cyanide nucleophile, **:CN⁻**, but the rate of reaction is very slow. It is more common to use potassium cyanide, KCN, and an acid.

The mechanism can again be broken down into two steps:

- nucleophilic attack to produce an intermediate anion

- reaction with a proton, H⁺, from either water or from an acid to give the product.

The product contains two functional groups; an alcohol (–OH) and the cyanide group which is known as a nitrile (–C≡N). The organic product is always a 2-hydroxynitrile.

Aldehydes react with KCN in the presence of water or an acid to form a 2-hydroxynitrile.

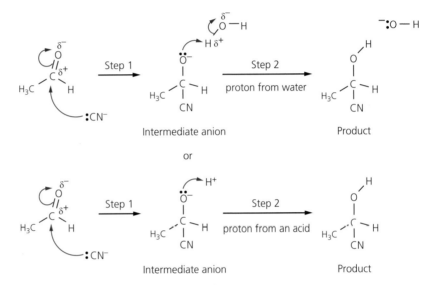

In this reaction ethanal is converted into 2-hydroxypropanenitrile. You will note that the product contains one more carbon atom than the reagent. This reaction can be used to extend the length of the carbon chain.

Ketones behave similarly — they react with KCN in the presence of an acid to form a 2-hydroxynitrile.

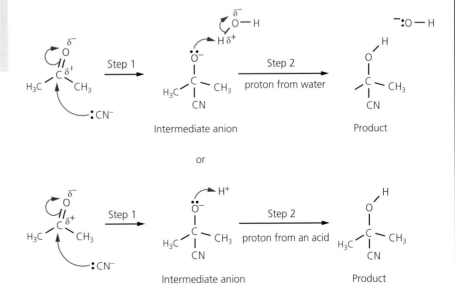

> **Tip**
>
> When naming a nitrile, the nitrile group always takes position 1 on the chain and always appears at the end of the name. Alcohols usually end with 'ol', but 'hydroxy' at the start of the name can also be used.

In this reaction propanone is converted into 2-hydroxy-2-methylpropanenitrile. You will note that the product again contains one more carbon atom than the reagent.

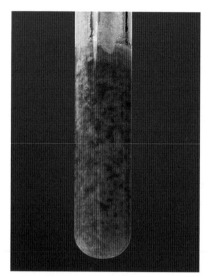

Figure 8.4 A precipitate of 2,4-dinitrophenylhydrazine.

Characteristic tests for carbonyl compounds

Reaction with 2,4-dinitrophenylhydrazine

Aldehydes and ketones both react with 2,4-dinitrophenylhydrazine.

Carbonyl compounds react with an excess of 2,4-dinitrophenylhydrazine to produce a bright red, orange or yellow precipitate. The precipitates formed are derivatives of 2,4-dinitrophenylhydrazine and are known as 2,4-dinitrophenylhydrazones.

The reactions of carbonyl compounds with 2,4-dinitrophenylhydrazine (2,4-DNPH) are important for several reasons:

● As the organic product is a bright red, orange or yellow precipitate it can be used to identify the presence of a carbonyl group (aldehyde and ketone).

● The organic product (the 2,4-DNPH derivative) is relatively easy to purify by recrystallisation. The melting point of the brightly coloured precipitate can then be determined.

● Each derivative has a different melting point, which may be used to identify a specific carbonyl compound. Table 8.3 shows the melting points of the derivatives of a few common carbonyl compounds.

The formula of 2,4-dinitrophenylhydrazine is used here to illustrate that, as in very many organic reactions, water is often formed as a product.

> **Tip**
>
> You are not expected to recall the formula of 2,4-dinitrophenylhydrazine. The abbreviation 2,4-DNPH is acceptable.

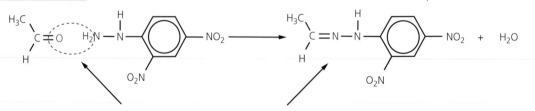

Remove water and join the two organic molecules together to produce the product

If in doubt over the product of an organic reaction, make water and then deduce the organic product.

Table 8.3 Melting points of some carbonyl compounds.

Carbonyl compound	Boiling point/°C	Melting point of the 2,4-DNPH derivative/°C
Ethanal	20	148
Propanal	49	156
Butanal	75	123
Methylpropanal	64	182
Propanone	56	128
Butanone	80	115
Pentan-2-one	102	142
Pentan-3-one	102	153
3-Methylbutan-2-one	94	124

The preparation and purification of a 2,4-dinitrophenylhydrazine derivative can be used to identify a specific aldehyde or ketone. The process involves the initial preparation of the derivative, which is filtered and then recrystallised and dried. The melting point of the derivative is then determined and checked against data values to identify the original carbonyl compound. This is particularly useful when trying to distinguish between isomers such as pentan-2-one and pentan-3-one, which have the same boiling point.

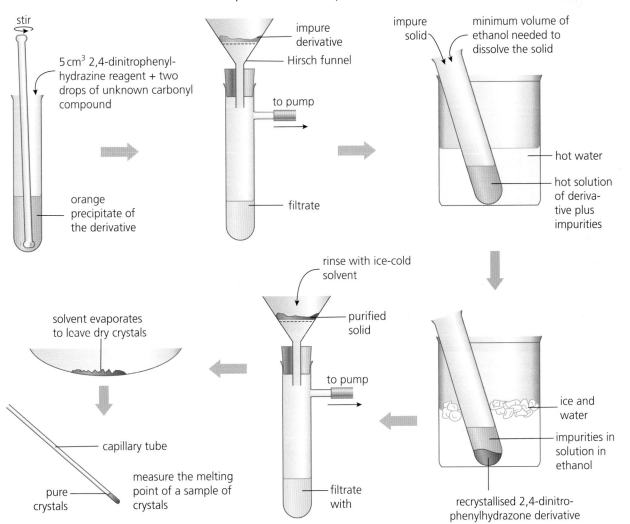

Figure 8.5 Making a pure crystalline derivative of a carbonyl compound.

Activity

Identifying an unknown carbonyl compound

Figure 8.5 shows stages in making, purifying and identifying a carbonyl compound.

1 Why is it necessary to purify the derivative before measuring its melting point?
2 Explain how the procedure illustrated in Figure 8.5 removes impurities from the derivative.
3 In this instance, ethanol is the solvent used for recrystallising the derivative. What determines the choice of solvent?
4 When measuring the melting point, what are the signs that the derivative is pure?
5 Identify the carbonyl compound which forms a 2,4-dinitrophenylhydrazone that melts at 115 °C. It boils at 80 °C. The compound does not form a silver mirror with Tollens' reagent.

Reactions of aldehydes only

Aldehydes and ketones can be distinguished by a series of redox reactions. Aldehydes are oxidised readily to carboxylic acids; ketones are not oxidised easily.

There are two common oxidising mixtures that can be used:

1 **Oxidising mixture**: acidified dichromate ($H^+/Cr_2O_7^{2-}$)

   Conditions: warm

   Observation: when reacted with an aldehyde there is a colour change from orange to green.

2 **Oxidising mixture**: an aqueous solution of Ag^+ ions in excess ammonia, $Ag(NH_3)_2^+$ (Tollens' reagent)

   Conditions: warm gently in a water bath at about 60 °C

   Observation: silver metal is precipitated forming a 'silver mirror'.

Tollens' reagent is prepared by adding a few drops of dilute sodium hydroxide solution to a solution of silver nitrate. A brown precipitate of silver oxide, $Ag_2O(s)$, is produced immediately. Dilute ammonia is then added until the $Ag_2O(s)$ just redissolves. This creates an ammoniacal solution of silver nitrate, $Ag(NH_3)_2^+$, which acts as the oxidising agent.

In this reaction, the Ag^+ ions are reduced to silver metal (Figure 8.6) and the aldehyde is oxidised to a carboxylic acid. The reaction with ethanal is given below.

Figure 8.6 Tollens' reagent oxidises an aldehyde to form a carboxylic acid, also producing the silver mirror effect as $Ag^+(aq)$ is reduced to $Ag(s)$.

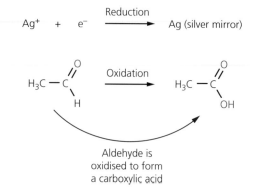

Aldehyde is oxidised to form a carboxylic acid

Tip

Infrared spectroscopy was first introduced in the Year 1 book (see Chapter 15 page 253).

The oxidation of an aldehyde to a carboxylic acid can be identified using infrared spectroscopy.

The relevant absorptions to pick out are those of the C=O group, which is present in both aldehydes and ketones and which has a strong absorption in the range 1640–1750 cm^{-1} (Figure 8.7). Carboxylic acids also have a C=O absorption but, in addition, have a characteristic broad absorption within the region 2500–3300 cm^{-1}, which is due to the O–H group in the carboxylic acid (Figure 8.8).

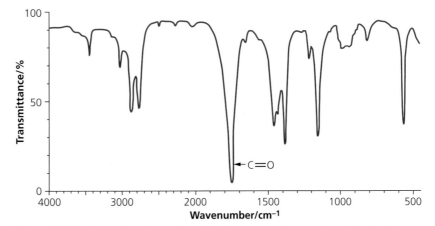

Figure 8.7 The absorption due to the carbonyl group can be seen clearly in the IR spectrum of ethanal.

The spectrum below shows the absorption for both the C=O and the O–H groups, confirming that the ethanal has been oxidised to ethanoic acid.

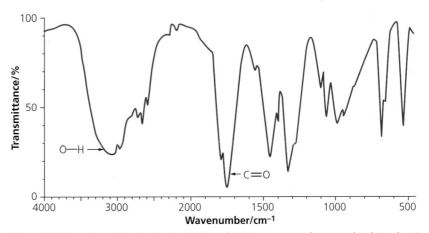

Figure 8.8 The absorption due to the O-H and C=O groups can be seen clearly in the IR spectrum of ethanoic acid.

Practice questions

Multiple choice questions 1–10

1 2,2-dimethylpropanal can be made by the oxidation of:
- **A** 2,2-dimethylpropan-2-ol under reflux
- **B** 2,2-dimethylpropan-2-ol under distillation
- **C** 2,2-dimethylpropan-1-ol under reflux
- **D** 2,2-dimethylpropan-1-ol under distillation *(1)*

2 When propanone reacts with HCN in the presence of an acid the organic product is
- **A** 2-hydroxypropanenitrile
- **B** 2-hydroxybutanenitrile
- **C** 2-hydroxy-2-methylpropanenitrile
- **D** 2-hydroxy-2-methylbutanenitrile *(1)*

3 The mechanism for the reaction between propanone and $NaBH_4$ is shown below. It contains numerous errors.

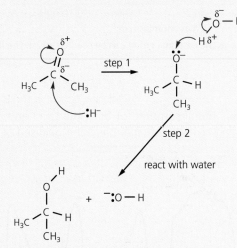

The number of errors in the mechanism is
- **A** 1
- **B** 2
- **C** 3
- **D** 4 *(1)*

4 If cooking oil is heated to its smoke point, a nasty smelling compound, acrolein, is produced. The molar mass of acrolein is $56\,g\,mol^{-1}$. Acrolein produces a silver mirror when reacted with Tollens' reagent and also decolourises bromine. The formula of acrolein is

a)

b) ![structure] OH

c) $H_2C = CH — CHO$

d) $H_3C — C \equiv C — OH$ *(1)*

5 Muscone, shown below, is obtained from musk, a glandular secretion of the musk deer and it has been used in perfumery and medicine for very many years.

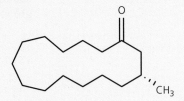

Which of the statements about muscone is false?
- **A** It has 16 C atoms.
- **B** It has 30 H atoms.
- **C** It reacts with 2,4-dinitrophenylhydrazine to give a orange–yellow precipitate.
- **D** It reacts with $NaBH_4$ to produce a tertiary alcohol. *(1)*

6 2-hydroxy-2-ethylbutanenitrile can be made by reacting acidified sodium cyanide with
- **A** pentan-3-one
- **B** pentan-2-one
- **C** 3-ethylpentan-2-one
- **D** 3-methylbutan-2-one. *(1)*

Use the key below to answer questions 7–10.

A	B	C	D
1, 2 & 3 correct	1 & 2 correct	2 & 3 correct	1 only correct

7 Propane-1,2-diol can be oxidised to form a number of possible products. Which of the following could be formed?
1. $CH_3CH(OH)CHO$
2. CH_3COCH_2OH
3. $CH_3COOHCH_2OH$ *(1)*

8 Citronellal, shown below, is obtained from the oil of kaffir lime leaves.

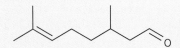

Which of the following statements are true?
1. Citronellal has a stereoisomer.
2. Citronellal reacts with Tollens' reagent.
3. Citronellal reacts with steam to produce a tertiary alcohol. *(1)*

9 Which of the following statements about the compound shown below are correct?

O≈∕∖∕≈O

1 The empirical formula and the molecular formula of the compound are the same.
2 The compound can be oxidised to produce a diol.
3 The compound can be oxidised to form butane-1,4-dioic acid. *(1)*

10 2-methylpropane-1,2-diol is refluxed with **excess** acidified dichromate. Which of the following are true?
1 A carboxylic acid is formed.
2 An aldehyde is formed.
3 A ketone is formed. *(1)*

11 Butan-2-ol can be prepared by the reduction of a carbonyl compound. Identify the compound, state the reagents and conditions required, write a balanced equation and describe, with the aid of curly arrows, the mechanism. *(7)*

12

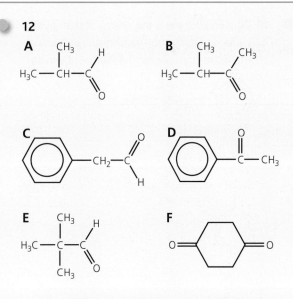

a) Name each of the compounds **A–F**.
b) Classify each of **A** to **F** as either an aldehyde or a ketone.
c) What is the molecular formula of compound **F**?
d) Identify the alcohols from which **C** and **D** could be prepared.
e) Compounds **B** and **E** are isomers of each other. Draw the structures of three other carbonyl compounds that are isomers of **B** and **E**. *(18)*

13 Copy and complete the table to show three different reactions of propanal.

Reactant		Reagent	Conditions (if any)	Organic product	
Name	Structure			Name	Structure
propanal		Tollens' reagent	water bath at 60 °C		
	CH₃CH₂CH₂CHO			butanoic acid	
	CH₃CH₂COCH₃	NaBH₄			
		KCN	H⁺		(CH₃)₂C(OH)CN

(14)

14 The molar mass of a hydrocarbon W is 56 g mol⁻¹ and it contains 85.7% carbon. W reacts with hydrogen bromide to form X. Heating X under reflux with aqueous sodium hydroxide produces Y. Heating Y with an acidified solution of potassium dichromate(VI) converts it to Z. Z gives a yellow precipitate with 2,4-dinitrophenylhydrazine but it does not give a silver mirror with Tollens' reagent.
a) Identify W, X, Y and Z and give your reasoning.
b) Explain why it is not possible to unambiguously identify compound W.
c) Draw a flow chart connecting together W, X, Y and Z. *(13)*

15 Describe the mechanisms for the reactions of:
a) propene with bromine
b) propanone with NaBH₄.
Identify similarities and differences between the two mechanisms with reference to the nature of the bonds and bond breaking, the reagents involved, the formation of intermediates and the overall effects of the change. *(14)*

16 Cinnamaldehyde is the chemical that gives cinnamon its flavour. Vanillin is the chemical mainly responsible for the flavour of vanilla.

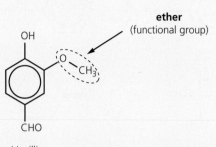

Cinnamaldehyde

Vanillin

a) i) What is the molecular formula of cinnamaldehyde?

ii) Name the two functional groups (other than the ether) in vanillin.

b) Copy and complete the table below.

	Reagent	Organic product
Reagent that reacts with both cinnamaldehyde and vanillin		
Reagent that reacts with cinnamaldehyde only		
Reagent that reacts with vanillin only		

c) i) Draw the structure of the product after treating vanillin with $NaBH_4$.

ii) Draw the structure of the compound formed when this product reacts with excess aqueous sodium hydroxide solution.

d) Cinnamaldehyde, C_9H_8O, is a yellow liquid with a powerful odour.

i) Copy and complete the reaction grid below for the reactions of cinnamaldehyde.

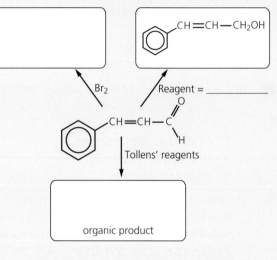

organic product

ii) Write an equation for the oxidation of cinnamaldehyde by Tollens' reagent.

iii) There are two possible isomers of cinnamaldehyde. Draw their structures and state the type of isomerism. *(21)*

17 Butanone can be converted into 2-hydroxy-2-methylbutanenitrile.

a) Draw butanone and 2-hydroxy-2-methylbutanenitrile.

b) Identify the reagents and conditions used to convert butanone to 2-hydroxy-2-methylbutanenitrile.

c) Construct an equation for this reaction.

d) Write the mechanism for this reaction, including curly arrows and any relevant dipoles and lone pairs of electrons. *(9)*

Challenge

18 In the presence of NaOH, benzaldehyde, C_6H_5CHO, is both oxidised and reduced. The products are sodium benzoate and phenylmethanol:

$$2C_6H_5CHO + NaOH \rightarrow C_6H_5COO^-Na^+ + C_6H_5CH_2OH$$

sodium benzoate   phenylmethanol

The mechanism for this reaction is shown below with the curly arrows missing. Copy the mechanism and add the curly arrows to track the movement of the electron pairs in the reaction. *(12)*

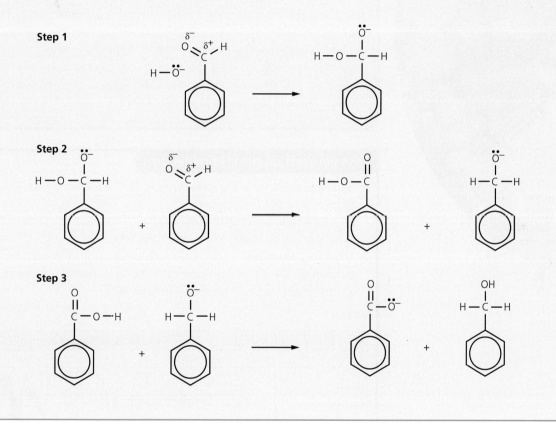

Step 1

Step 2

Step 3

Carboxylic acids and derivatives

Test yourself on prior knowledge

1 Write equations, including state symbols, for the following reactions
  a) nitric acid and calcium oxide
  b) sulfuric acid and lithium carbonate
  c) aluminium hydroxide and hydrochloric acid.
2 An alcohol was heated under reflux with excess acidified potassium dichromate. The infrared spectrum and the mass spectrum of the alcohol and the organic product are shown in Table 9.1.

Table 9.1

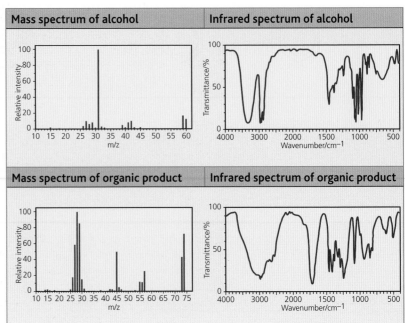

a) Deduce the molecular formula of the alcohol, draw and name all possible isomers of the alcohol.
b) Use the spectra to identify the organic product. Explain your reasoning.
c) Deduce the identity of the alcohol.
d) Show how the fragmentation pattern in the mass spectrum of the original alcohol confirms the identity of the alcohol.

Carboxylic acids

Carboxylic acids can be prepared via the oxidation of a primary alcohol, by refluxing the alcohol with acidified dichromate.

All carboxylic acids contain the functional group –COOH. Carboxylic acids are present in many foods. Ethanoic acid and citric acid are responsible for the sharp tastes of vinegar and lemons, respectively. Benzoic acid is used as a preservative and as a flavouring in fizzy drinks. Ethanedioic acid (oxalic acid) is found in the leaves of many plants including rhubarb. The leaves of rhubarb are rich in oxalic acid and can cause the formation of kidney stones (calcium oxalate); the leaves of rhubarb should not be eaten (Figure 9.1).

Figure 9.1 The leaves of rhubarb are rich in oxalic acid.

Naming carboxylic acids

Chemists name carboxylic acids by changing the ending of the corresponding alkan**e** to '...**oic acid**', such that methan**e** becomes methan**oic acid**. This applies to the systematic names only. Many carboxylic acids retain their old trivial names, as illustrated below.

The formulae of some carboxylic acids are given below.

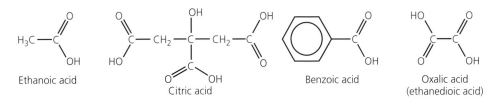

At first glance, the carboxylic acid might be regarded as a carbonyl group, C=O, attached to an alcohol group, –OH. If this were so, carboxylic acids would react with 2,4-dinitrophenylhydrazine to give a brightly coloured precipitate, just like aldehydes and ketones. In practice, carboxylic acids do not react in the same way as aldehydes and ketones and it is essential to regard the carboxylic acid as a single functional group with distinctive properties.

The carboxylic acid group, −COOH is polar

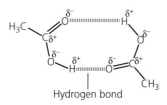

and gives rise to hydrogen bonding.

Oxygen is more electronegative than either carbon or hydrogen. This results in both the C=O and the O−H, in the −COOH group, being polar. Carboxylic acids form hydrogen bonds with other carboxylic acids and with water.

In the absence of water, carboxylic acids form dimers. This reduces their volatility and, hence, increases their boiling points.

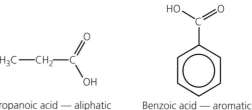

Hydrogen bond

The ability to form hydrogen bonds and to undergo dipole−dipole interaction also explains why methanoic acid and ethanoic acid are soluble in water. Solubility decreases with increasing molar mass − benzoic acid, C_6H_5COOH, is soluble in hot water, but insoluble in cold water.

Carboxylic acids are acidic and can, therefore, donate protons. However, they are weak acids and dissociate only partially into their ions.

$$CH_3COOH(aq) \rightleftharpoons CH_3COO^-(aq) + H^+(aq)$$

The carboxylic acid group can be attached to either a chain (aliphatic) or to a ring (aromatic), for example:

Propanoic acid — aliphatic   Benzoic acid — aromatic

Test yourself

1 The odour of goats contains a blend of three unbranched carboxylic acids with 6, 8 and 10 carbon atoms. The trivial names are based on the Latin word 'caper' meaning goat. The three acids are caproic acid (6C), caprylic acid (8C) and capric acid (10C). For each of the three acids:
   a) state the molecular formula,
   b) draw the skeletal formula,
   c) give the systematic name.
2 Draw a diagram to show the hydrogen bonding between:
   a) ethanoic acid and water molecules,
   b) two molecules of propanoic acid.
3 Suggest why ethanoic acid is a liquid whilst ethanedioic acid (oxalic acid) (HOOC-COOH) is a solid.

Reactions of carboxylic acids

Aqueous solutions of carboxylic acids display typical reactions of an **acid** and can form **salts** (carboxylates).

Salt formation can occur by any of the following reactions, illustrated by the formation of ethanoate compounds of Group 1, $CH_3COO^-Na^+$, or Group 2, $(CH_3COO^-)_2Ca^{2+}$, from ethanoic acid CH_3COOH:

$$acid + base \rightarrow salt + water$$

$$CH_3COOH(aq) + NaOH(aq) \rightarrow CH_3COO^-Na^+(aq) + H_2O(l)$$

$$2CH_3COOH(aq) + Ca(OH)_2(aq) \rightarrow (CH_3COO^-)_2Ca^{2+}(aq) + 2H_2O(l)$$

$$acid + (reactive)\ metal \rightarrow salt + hydrogen$$

$$CH_3COOH(aq) + Na(s) \rightarrow CH_3COO^-Na^+(aq) + \tfrac{1}{2}H_2(g)$$

$$2CH_3COOH(aq) + Ca(s) \rightarrow (CH_3COO^-)_2Ca^{2+}(aq) + H_2(g)$$

$$acid + carbonate \rightarrow salt + water + carbon\ dioxide$$
(Figure 9.2)

$$2CH_3COOH(aq) + Na_2CO_3(aq) \rightarrow 2CH_3COO^-Na^+(aq) + H_2O(l) + CO_2(g)$$

$$2CH_3COOH(aq) + CaCO_3(s) \rightarrow (CH_3COO^-)_2Ca^{2+}(aq) + H_2O(l) + CO_2(g)$$

The reaction with a carbonate can be used as a test for a carboxylic acid. When an acid is added to a solution of a carbonate, bubbles (effervescence) of carbon dioxide are seen.

Figure 9.2 Bubbles of carbon dioxide are produced when ethanoic acid is added to sodium carbonate.

Reaction with alcohols to form esters

A carboxylic acid can react with an alcohol to form an ester. This type of reaction is known as **esterification**. It is a reversible reaction that is usually carried out in the presence of a concentrated acid catalyst, such as sulfuric acid. The general reaction can be summarised as follows:

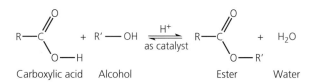

Carboxylic acid   Alcohol   Ester   Water

The name of an ester is derived from the names of the carboxylic acid and the alcohol from which it is formed. The first part of the name relates to the alcohol and the second part of the name relates to the acid – for example:

● methyl ethanoate

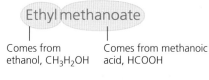

Comes from methanol, CH_3OH Comes from ethanoic acid, CH_3COOH

● ethyl methanoate

Ethyl methanoate

Comes from ethanol, CH_3H_2OH Comes from methanoic acid, $HCOOH$

In an organic reaction, it is the functional groups that react. In the reaction between a carboxylic acid and an alcohol it is helpful to write the formulae, so that the functional groups 'face each other'. The alcohol group reacts with the carboxylic acid group to produce water and the ester can be deduced by simply joining the two organic parts together:

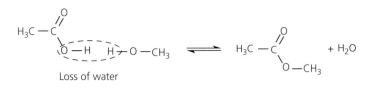

Loss of water

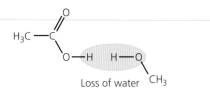

Loss of water

There are two ways in which the ester could be formed, because there are two different bonds in the alcohol that might break. The bond between the hydrogen and the oxygen in the alcohol might break:

$$H_3C - C\begin{smallmatrix}O\\\\O-H \quad H-O\\\quad\quad\quad CH_3\end{smallmatrix}$$

Loss of water

or the bond between the carbon and the oxygen in the alcohol could also break:

$$H_3C - C\begin{smallmatrix}O\\\\O-H \quad H-O\\\quad\quad\quad CH_3\end{smallmatrix}$$

Loss of water

To decide which is correct, the reaction has been studied using alcohols containing the ^{18}O isotope. The two possible routes are:

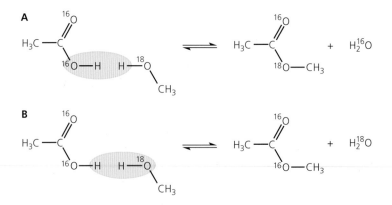

When the ester is analysed using mass spectrometry, the ^{18}O isotope is always found in the ester and not in the water.

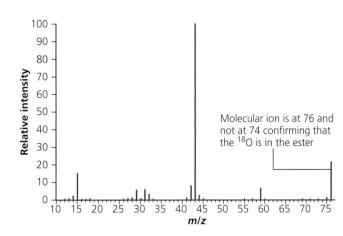

Molecular ion is at 76 and not at 74 confirming that the ^{18}O is in the ester

This indicates that the reaction proceeds as shown in route **A**, confirming that the O–H bond in the alcohol and the C–O bond in the carboxylic acid break to produce the water.

Reaction with halogenating agents to form acid halides (acyl halides)

Carboxylic acids react with sulfur dichloride oxide (thionyl chloride) to form acid halides (acyl halides).

$$CH_3COOH(l) + SOCl_2(l) \rightarrow CH_3COCl(l) + SO_2(g) + HCl(g)$$

It is essential that the reaction between $SOCl_2$ and the carboxylic acid is carried out in anhydrous conditions, as both $SOCl_2$ and the acid chloride react with water.

Test yourself

4 Ethanoic acid, CH_3COOH, lactic acid, $CH_3CH(OH)COOH$, and ethanedioic acid, HOOC-COOH, are all naturally occurring carboxylic acids. Deduce the formula of the following salts:
   a) potassium ethanoate
   b) calcium lactate
   c) disodium ethanedioate
   d) magnesium ethanoate.
5 Name the following esters:

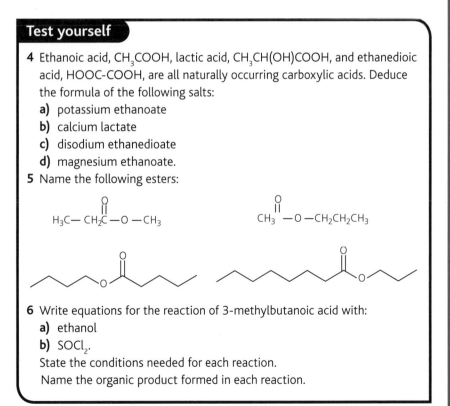

6 Write equations for the reaction of 3-methylbutanoic acid with:
   a) ethanol
   b) $SOCl_2$.
   State the conditions needed for each reaction.
   Name the organic product formed in each reaction.

Derivatives of carboxylic acids

Tip

An **acyl** group consists of all the parts of a carboxylic acid except the −OH group. For example the acyl group from ethanoic acid, CH_3COOH, is the ethanoyl group, CH_3CO-.

The hydroxyl group (−OH) in carboxylic acids can be replaced by other functional groups forming a series of compounds that contain the group, RCO−:

Carboxylic acid → Carboxylic acid derivative

Table 9.2 below illustrates the versatility of carboxylic acids, which can form a wide range of derivatives.

Table 9.2 Derivatives of carboxylic acids

X $R-\overset{\overset{O}{\|}}{C}-?$	Functional group	Acid derivative	Example
−OH	$R-\overset{\overset{O}{\|}}{C}-OH$		$H_3C-\overset{\overset{O}{\|}}{C}-OH$ ethanoic acid
−OR	$R-\overset{\overset{O}{\|}}{C}-O-R$	ester	$H_3C-\overset{\overset{O}{\|}}{C}-O-CH_3$ methyl ethanoate
−Cl	$R-\overset{\overset{O}{\|}}{C}-Cl$	acyl chloride (can also be fluoride, bromide or iodide)	$H_3C-\overset{\overset{O}{\|}}{C}-Cl$ ethanoyl chloride
−OCOR	$R-\overset{\overset{O}{\|}}{C}-O-\overset{\overset{O}{\|}}{C}-R$	acid anhydride	$H_3C-\overset{\overset{O}{\|}}{C}-O-\overset{\overset{O}{\|}}{C}-CH_3$ ethanoic anhydride
−NH$_2$	$R-\overset{\overset{O}{\|}}{C}-NH_2$	amide	$H_3C-\overset{\overset{O}{\|}}{C}-NH_2$ ethanamide
−NHR	$R-\overset{\overset{O}{\|}}{C}-\underset{\underset{H}{\|}}{N}-R$	secondary amide	$H_3C-\overset{\overset{O}{\|}}{C}-\underset{\underset{H}{\|}}{N}-CH_3$ methylethanamide

Tip

Each derivative is named in a slightly different way. It is important that you are familiar with each naming system.

Test yourself

7 Name each of the following:

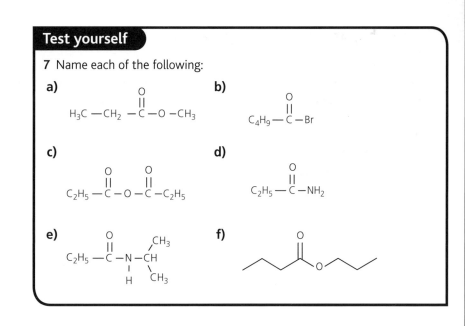

a)

$$H_3C\!-\!CH_2\!-\!\overset{\overset{\displaystyle O}{\|}}{C}\!-\!O\!-\!CH_3$$

b)

$$C_4H_9\!-\!\overset{\overset{\displaystyle O}{\|}}{C}\!-\!Br$$

c)

$$C_2H_5\!-\!\overset{\overset{\displaystyle O}{\|}}{C}\!-\!O\!-\!\overset{\overset{\displaystyle O}{\|}}{C}\!-\!C_2H_5$$

d)

$$C_2H_5\!-\!\overset{\overset{\displaystyle O}{\|}}{C}\!-\!NH_2$$

e)

$$C_2H_5\!-\!\overset{\overset{\displaystyle O}{\|}}{C}\!-\!\underset{\underset{\displaystyle H}{|}}{N}\!-\!\overset{\nearrow CH_3}{\underset{\searrow CH_3}{CH}}$$

f)

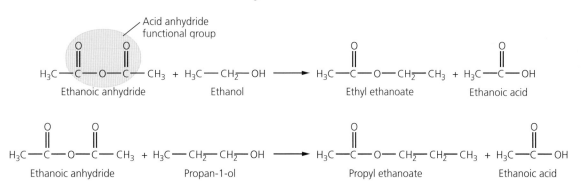

Esters

Preparation of esters

A range of esters can be formed by changing either the alcohol or the carboxylic acid, or both — for example:

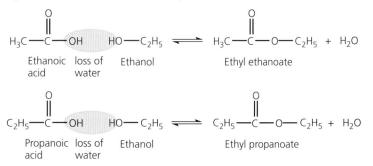

An ester can also be prepared by the reaction of an acid anhydride with an alcohol. Acid anhydrides contain the $-$OCOCO$-$ functional group. Ethanoic anhydride is the most common. Equations for the preparation of two esters from ethanoic anhydride are shown below.

Acid anhydride functional group

$$H_3C\!-\!\overset{\overset{\displaystyle O}{\|}}{C}\!-\!O\!-\!\overset{\overset{\displaystyle O}{\|}}{C}\!-\!CH_3 + H_3C\!-\!CH_2\!-\!OH \longrightarrow H_3C\!-\!\overset{\overset{\displaystyle O}{\|}}{C}\!-\!O\!-\!CH_2\!-\!CH_3 + H_3C\!-\!\overset{\overset{\displaystyle O}{\|}}{C}\!-\!OH$$

Ethanoic anhydride   Ethanol   Ethyl ethanoate   Ethanoic acid

$$H_3C\!-\!\overset{\overset{\displaystyle O}{\|}}{C}\!-\!O\!-\!\overset{\overset{\displaystyle O}{\|}}{C}\!-\!CH_3 + H_3C\!-\!CH_2\!-\!CH_2\!-\!OH \longrightarrow H_3C\!-\!\overset{\overset{\displaystyle O}{\|}}{C}\!-\!O\!-\!CH_2\!-\!CH_2\!-\!CH_3 + H_3C\!-\!\overset{\overset{\displaystyle O}{\|}}{C}\!-\!OH$$

Ethanoic anhydride   Propan-1-ol   Propyl ethanoate   Ethanoic acid

The advantages of preparing an ester from an acid anhydride are that the reaction is not reversible and it occurs readily.

Preparation of an ester

This sequence of diagrams shows the procedure for preparing a small sample of an ester (Figure 9.3).

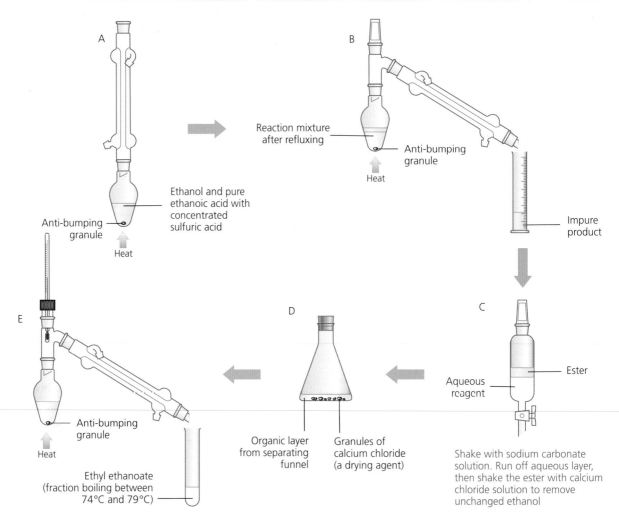

Figure 9.3 Stages in the preparation of an ester.

1 Identify what is happening in each of the stages A, B, C, D and E.
2 Write an equation for the reaction which forms the ester and name the product.
3 What is the purpose of the concentrated sulfuric acid?
4 What are the visible signs of reaction during stage C, and what practical precautions are necessary during this stage?

The calcium chloride in step C removes any unreacted ethanol by forming a complex ion with the ethanol.

5 A volatile by-product distils off in the boiling range 35–40 °C before the ester in stage E. Suggest a structure for this by-product, which has the molecular formula $C_4H_{10}O$.
6 Calculate the percentage yield of the ester if the actual yield is 50 g from 40 g ethanol and 52 g ethanoic acid.

Physical properties

Esters have a characteristic 'fruity' smell. They are used in both perfumes and artificial flavourings. Naturally occurring fruits contain a complex mix of chemicals, many of which contribute to the overall smell or scent. By mixing together different chemicals, including esters, it is possible to manufacture artificial flavours. The esters shown below are used to generate the flavours pineapple, pear and apple.

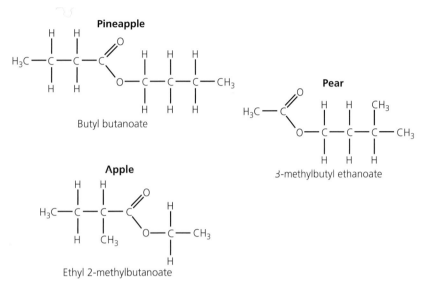

Pineapple

Butyl butanoate

Pear

3-methylbutyl ethanoate

Apple

Ethyl 2-methylbutanoate

Hydrolysis of esters

In the presence of a catalyst, esters react with water (hydrolysis), breaking down to form two new products.

When a chemical is **hydrated** it reacts with water to form a single new product. You discovered this in Year 1 when you studied the chemistry of alkenes, when ethene reacted with steam to produce ethanol:

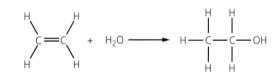

The hydrolysis of an ester is a slow reaction that requires heating and is carried out in the presence of either an acid, H⁺(aq), or a base, OH⁻(aq). Acid-catalysed hydrolysis leads to the formation of the carboxylic acid and the alcohol – for example, using dilute sulfuric acid as the catalyst:

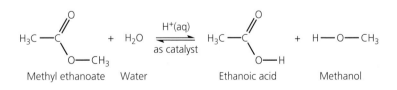

Methyl ethanoate   Water   Ethanoic acid   Methanol

Hydrolysis in the presence of a base leads to the formation of the salt of the carboxylic acid (the carboxylate) and the alcohol – for example, using an aqueous solution of NaOH(aq) as the base:

Tip

Do not confuse *hydrolysis* with *hydration*.

Tip

Think back to the section on equilibrium and to Le Chatelier's principle (page 26). Why is a dilute acid, rather than a concentrated acid, used as the catalyst?

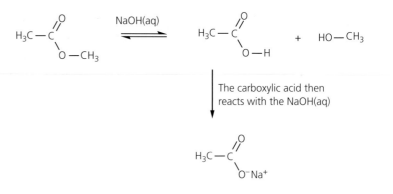

The carboxylic acid then reacts with the NaOH(aq)

The net reaction of the base hydrolysis of methyl ethanoate can be written as:

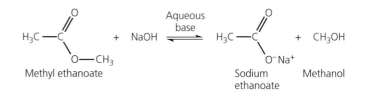

Methyl ethanoate   Sodium   Methanol
                                        ethanoate

Test yourself

8 Write equations for each of the following reactions. Name the organic product in each reaction (except in part **d**).
  a) ethanoic acid + butan-1-ol
  b) butane-1,4-dioic acid + ethanol
  c) 2-hydroxypropanoic acid + ethanol
  d) 2-hydroxypropanoic acid + ethanoic acid.
9 Esters can be hydrolysed using either a dilute acid or a dilute alkali. Identify the organic products when each of the following esters is hydrolysed. Give the products for **i)** acid hydrolysis and **ii)** alkali hydrolysis, using sodium hydroxide.
  a) ethyl methanoate
  b) methyl ethanoate
  c) phenyl propanoate
  d) Explain why, in acid hydrolysis, the acid is a catalyst.
  e) Explain why, in alkali hydrolysis, the alkali cannot be regarded as a catalyst.

Acyl halides (acid halides)

Preparation of acyl chlorides

Carboxylic acids, like alcohols, can be chlorinated by reaction with sulfur dichloride oxide (thionyl chloride), $SOCl_2$.

$$CH_3COOH + SOCl_2 \rightarrow CH_3COCl + SO_2 + HCl$$

Physical properties of acyl halides

The lower acyl halides are colourless liquids with pungent odours. They fume in moist air and are readily hydrolysed to the corresponding carboxylic acid and halogen acid (Figure 9.4).

Reactions of acyl chlorides

Acyl halides are very reactive and they are hard to control, but they can be used to make a variety of other chemicals. The overall reaction of acyl chlorides, RCOCl, can be summarised as:

$$RCO{-}Cl + H{-}Y \rightarrow RCO{-}Y + H{-}Cl$$

The reagent, H–Y, can be any of those given in Table 9.3.

Table 9.3

H–Y	Example reagent
H–OH	H_2O (water)
H–OR	CH_3OH (alcohol)*
H–NH$_2$	NH_3 (ammonia)
H–NH-R	CH_3NH_2 (amine)

*Phenols also react with acyl halides.

The reactions of acyl halides are summarised in Figure 9.5. Note that HCl is formed in each reaction.

Figure 9.4

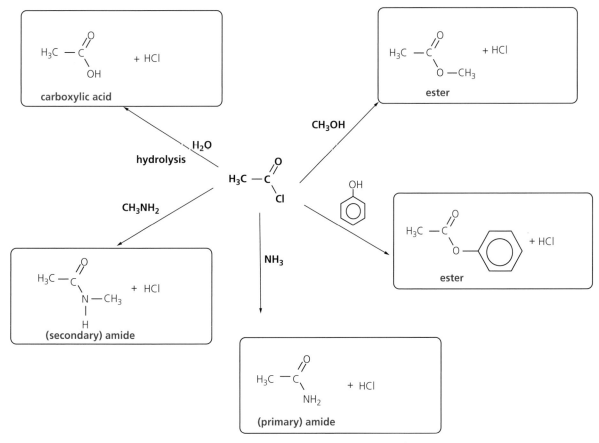

Figure 9.5

Acyl halides are very versatile and can be used to produce a wide range of other functional groups. You will recall that esters can be made by reacting alcohols with carboxylic acids, but the reaction is reversible and the yield of the ester is, therefore, low. Alcohols react readily with acyl chlorides to produce esters. The yield is much higher as the reaction is not reversible.

Test yourself

10 Write equations for the following reactions:
  a) propanoic acid + sulfur dichloride oxide, $SOCl_2$
  b) benzoic acid, C_6H_5COOH, + sulfur dichloride oxide, $SOCl_2$
  c) ethanoyl chloride + H_2O
  d) propanoyl chloride + phenol
  e) 2,2-dimethylpropanoyl chloride + NH_3
  f) butanoyl chloride + propanamide.

Practice questions

Multiple choice questions 1–10

1 Hydrolysis of ethyl propanoate with NaOH(aq) produces:
  A ethanol and propanoic acid
  B propanol and ethanoic acid
  C ethanol and sodium propanoate
  D propanol and sodium ethanoate. *(1)*

2 Ethanoyl chloride can be prepared by reacting ethanoic acid with $SOCl_2$.

$$CH_3COOH + SOCl_2 \rightarrow CH_3COCl + SO_2 + HCl$$

The atom economy for the reactions is
  A 48.33%
  B 56.17%
  C 43.83%
  D 51.67%. *(1)*

3

could be made by reacting together:

  A 2-methylpentanoic acid + phenol
  B 4-methylpentanoic acid + phenol
  C 2-methylpentanoyl chloride + phenol
  D 4-methylpentanoyl chloride + phenol. *(1)*

4 $CH_3COO(CH_2)_2CH(CH_3)_2$ contributes to the flavour and smell of pears. The systematic name of $CH_3COO(CH_2)_2CH(CH_3)_2$ is:
  A dimethylpropylethanoate
  B 3-methylbutylethanoate
  C 2-methylbutylethanoate
  D ethyl-3-methylbutanoate. *(1)*

Questions 5–7 refer to the reaction scheme shown below.

5 Compound W is

(1)

6 Reagents X and Z are:
  A H_2O and C_6H_5OH respectively
  B H_2O and $C_6H_5CO_2H$ respectively
  C NaOH(aq) and C_6H_5OH respectively
  D NaOH(aq) and $C_6H_5CO_2H$ respectively. *(1)*

7 Reagent Y is
  A ammonia
  B 3-methylpropylamine
  C 2-methylpropylamine
  D methyl propanamide. *(1)*

Use the key below to answer questions 8, 9 and 10.

A	B	C	D
1, 2 & 3 correct	1 & 2 correct	2 & 3 correct	1 only correct

8 Phenyl ethanoate can be prepared by reacting
  1 phenol with ethanoic acid
  2 phenol with ethanoic anhydride
  3 phenol with ethanoyl chloride. *(1)*

9

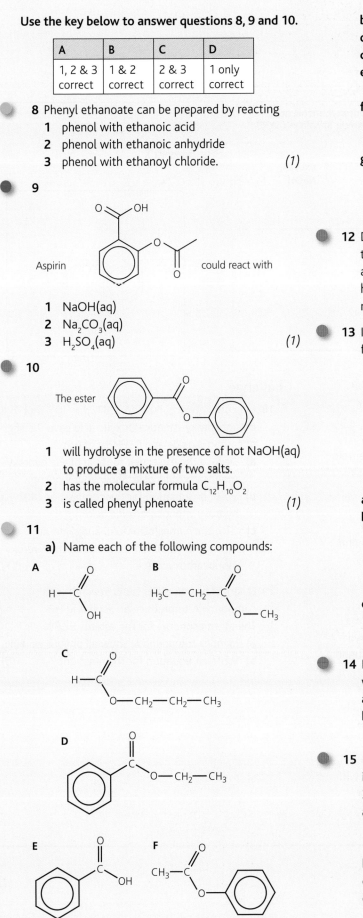

  1 NaOH(aq)
  2 Na$_2$CO$_3$(aq)
  3 H$_2$SO$_4$(aq) *(1)*

10

  1 will hydrolyse in the presence of hot NaOH(aq) to produce a mixture of two salts.
  2 has the molecular formula C$_{12}$H$_{10}$O$_2$
  3 is called phenyl phenoate *(1)*

11
  a) Name each of the following compounds:

b) What is the molecular formula of E?
c) What is the empirical formula of F?
d) Calculate the percentage of carbon in D.
e) Explain, with the aid of a diagram, why compound A is soluble in water.
f) An ester can be prepared by reacting an alcohol with a carboxylic acid. Write an equation for the formation of C.
g) Write a balanced equation for the reaction of compound A with:
  i) Na(s)
  ii) NaHCO$_3$(aq). *(14)*

12 Describe two examples of test-tube reactions that you could use to show the similarities and differences between ethanoic acid and hydrochloric acid. Write equations for the reactions that occur. *(6)*

13 Ibuprofen is a painkiller with this skeletal formula:

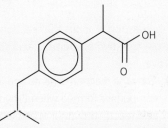

  a) What is the molecular formula of ibuprofen?
  b) Draw the formula of the organic products when ibuprofen reacts with:
    i) dilute sodium hydroxide solution
    ii) ethanol and a little concentrated sulfuric acid on warming.
  c) Suggest whether or not ibuprofen is soluble or insoluble in water. Explain your answer in terms of intermolecular forces. *(5)*

14 Propyl ethanoate can be hydrolysed by refluxing with NaOH(aq).
  a) Explain what is meant by 'refluxing'.
  b) Write a balanced equation for the reaction. Name the products. *(5)*

15 12.0 g of ethanoic acid was reacted with methanol in the presence of concentrated sulfuric acid. 3.70 g of the ester, methyl ethanoate, was isolated.
  a) Write a balanced equation for this reaction and explain the role of the concentrated sulfuric acid.
  b) Calculate the percentage yield.
  c) Calculate the atom economy.
  d) Suggest why the percentage yield is so low. *(9)*

193

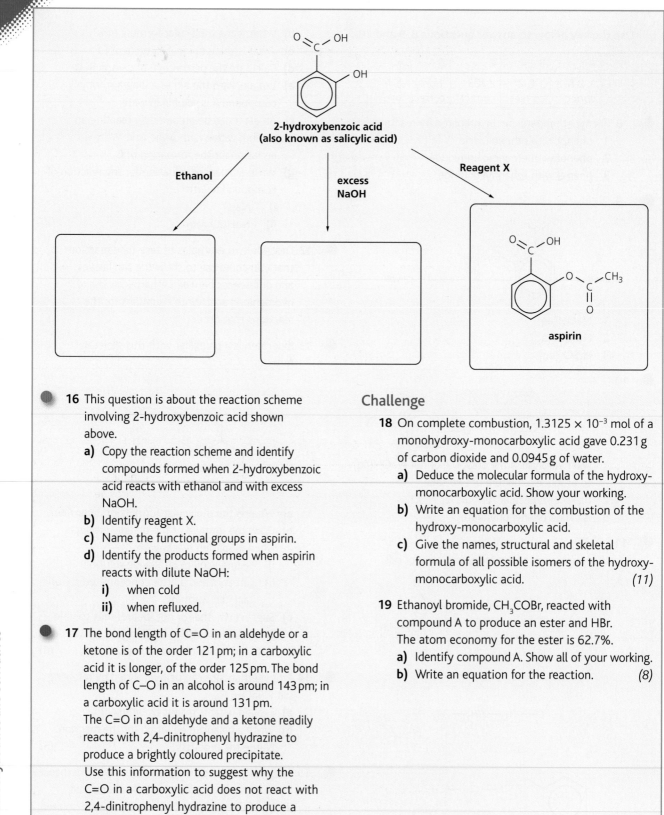

**2-hydroxybenzoic acid
(also known as salicylic acid)**

Ethanol

excess
NaOH

Reagent X

aspirin

16 This question is about the reaction scheme involving 2-hydroxybenzoic acid shown above.
   a) Copy the reaction scheme and identify compounds formed when 2-hydroxybenzoic acid reacts with ethanol and with excess NaOH.
   b) Identify reagent X.
   c) Name the functional groups in aspirin.
   d) Identify the products formed when aspirin reacts with dilute NaOH:
      i) when cold
      ii) when refluxed.

17 The bond length of C=O in an aldehyde or a ketone is of the order 121 pm; in a carboxylic acid it is longer, of the order 125 pm. The bond length of C–O in an alcohol is around 143 pm; in a carboxylic acid it is around 131 pm.
The C=O in an aldehyde and a ketone readily reacts with 2,4-dinitrophenyl hydrazine to produce a brightly coloured precipitate.
Use this information to suggest why the C=O in a carboxylic acid does not react with 2,4-dinitrophenyl hydrazine to produce a brightly coloured precipitate. *(5)*

Challenge

18 On complete combustion, 1.3125×10^{-3} mol of a monohydroxy-monocarboxylic acid gave 0.231 g of carbon dioxide and 0.0945 g of water.
   a) Deduce the molecular formula of the hydroxy-monocarboxylic acid. Show your working.
   b) Write an equation for the combustion of the hydroxy-monocarboxylic acid.
   c) Give the names, structural and skeletal formula of all possible isomers of the hydroxy-monocarboxylic acid. *(11)*

19 Ethanoyl bromide, CH_3COBr, reacted with compound A to produce an ester and HBr. The atom economy for the ester is 62.7%.
   a) Identify compound A. Show all of your working.
   b) Write an equation for the reaction. *(8)*

Chapter 10

Nitrogen compounds

> ## Prior knowledge
>
> *In this chapter it is essential that you are familiar with:*
> - acids, bases and salts
> - nucleophilic substitution reactions
> - reactions of halogenoalkanes
> - reactions of carboxylic acids
> - infrared spectroscopy to identify functional groups.
>
> You should know the definitions of an acid, a base and a salt and be able to write appropriate equations. Haloalkanes undergo nucleophilic substitution reactions with a suitable nucleophile including OH^-, H_2O, $^-C \equiv N$, NH_3 and RNH_2. Infrared spectroscopy can be used to detect the presence of various functional groups, including the amine group.
>
> Carboxylic acids are weak acids and can react with bases, reactive metals or carbonates to produce salts. Amines are bases and react with acids to produce salts.
>
> You should revisit and check your understanding of structural and *E/Z* isomerism.

Test yourself on prior knowledge

1 a) Describe, with the aid of curly arrows, the mechanism for the hydrolysis of chloromethane with aqueous sodium hydroxide.
   b) Ethanol can also be produced by the hydrolysis of chloroethane using water but the reaction is very slow.
      The mechanism for this reaction is shown below.

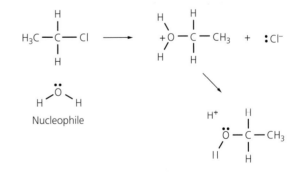

   Copy the mechanism and add curly arrows to track the movement of electron pairs in the reaction.

2 Write equations for the following reactions:
   a) ethanoic acid and sodium
   b) ethanoic acid and lithium hydroxide
   c) ethanoic acid and calcium carbonate.

3 a) Identify the ions in each of the following substances. State which is the anion and which is the cation.
      i) $CaCO_3(aq)$
      ii) $HNO_3(aq)$
      iii) $NaHCO_3(aq)$

b) A solution of $H_2SO_4(aq)$ contains one cation and two possible anions. Identify each ion.

c) A solution of $H_3PO_4(aq)$ contains one cation and three possible anions. Identify each ion.

4 The three infrared spectra below are for an amine, an alcohol and an aldehyde. Identify which is which. Explain your answer.

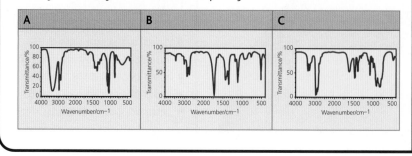

Amines

Amines can be very smelly. Ethylamine, for example, has a fishy smell. However, the importance of the amine functional group is not its smell, but the role it plays in biochemistry and medicine. The active component of an asthma inhaler is salbutamol, which contains the amine group. The amine group is also present in amino acids, the monomers for proteins. As a result of this, the amine group plays an important part in metabolism and it appears in the structures of many medical drugs. In the chemical industry, aromatic amines have commercial value because they are the basis of the manufacture of a wide range of colourful dyes.

You should recall from Year 1 chemistry (see page 101 in the Year 1 book) that ammonia, NH_3, is pyramidal and that the bond angle is approximately 107°. Amines contain the functional group $-NH_2$. The nitrogen atom has a lone pair of electrons and the shape around it is also pyramidal. The bond angle in amines is similar, but varies depending on the group attached to the $-NH_2$.

The amine group can be attached to either a chain or a ring. If it is attached to a chain, the amine is **aliphatic**; if it is attached to a ring, the amine is **aromatic**.

Naming amines

Simple aliphatic and aromatic amines end with the suffix −**amine** and include:

- CH_3NH_2 methyl**amine**

- $CH_3CH_2NH_2$ ethyl**amine**

Aromatic amines include phenyl**amine**, $C_6H_5NH_2$.

Propylamine is ambiguous and could relate to $CH_3CH_2CH_2NH_2$ or to $CH_3CH(NH_2)CH_3$. It is easiest to distinguish between these two isomers by using the prefix **amino–**.

- $CH_3CH_2CH_2NH_2$ 1-**amino**propane
- $CH_3CH(NH_2)CH_3$ 2-**amino**propane

Amines can also be classified as primary, secondary and tertiary.

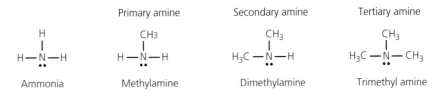

Test yourself

1 Draw the structures of:
   a) dimethylamine
   b) ethylmethylamine
   c) dimethylethylamine
   d) ethylmethylpropylamine.
2 a) The following three amines all have medical applications. Classify each as either primary, secondary or tertiary.

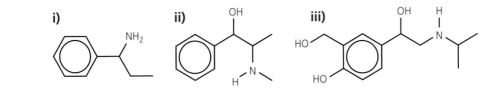

   b) Deduce the molecular formula of each of the amines above.

Properties and reactions of amines

The physical and chemical properties of amines are similar to those of ammonia (Figure 10.1).

Hydrogen bonds in ammonia

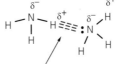

The H-bond is formed between the lone pair on the N and the δ^+ on a H in an adjacent molecule

Hydrogen bonds in methylamine

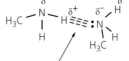

Like ammonia the H-bond is formed between the lone pair on the N and the δ^+ on a H in an adjacent molecule

Figure 10.1 Hydrogen bonding in ammonia and an amine.

Alkyl amines with short hydrocarbon chains are soluble in water and, like ammonia, the aqueous solutions of amines are alkaline.

Methylamine and other simple amines dissolve in water because they react with it.

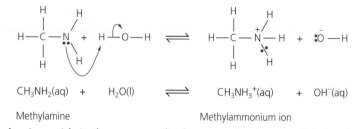

CH$_3$NH$_2$(aq)   +   H$_2$O(l) $\rightleftharpoons$ CH$_3$NH$_3^+$(aq)   +   OH$^-$(aq)

Methylamine   Methylammonium ion

Phenylamine, with its large non-polar benzene ring is only slightly soluble in water.

Test yourself

3 Write equations to show each of the following reacting with water:
   a) ammonia
   b) methylamine
   c) dimethylamine
   d) trimethylamine.

Preparation of aliphatic amines

Haloalkanes with ammonia or with amines

Aliphatic amines such as aminoethane, CH$_3$CH$_2$NH$_2$, can be prepared by heating a haloalkane with ammonia dissolved in ethanol. The reaction is carried out in a sealed flask. It is essential that the solvent is ethanol; if any water is present the NH$_3$ is protonated and the ammonium ion, NH$_4^+$ is formed.

The reaction can be written as:

CH$_3$CH$_2$Cl + NH$_3$(alc) $\rightarrow$ CH$_3$CH$_2$NH$_2$ +HCl

but the HCl produced reacts with either:

● any excess NH$_3$ to produce the ammonium salt, NH$_4$Cl

CH$_3$CH$_2$NH$_2$ + HCl + excess NH$_3$(alc) $\rightarrow$ CH$_3$CH$_2$NH$_2$ + NH$_4$Cl

● or the amine to produce the amine salt, CH$_3$CH$_2$NH$_3^+$Cl$^-$

CH$_3$CH$_2$NH$_2$ + HCl $\rightarrow$ CH$_3$CH$_2$NH$_3^+$Cl$^-$

The addition of aqueous sodium hydroxide produces the amine by reacting with any amine salt produced.

The overall reaction is shown below.

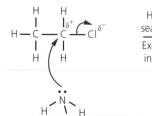
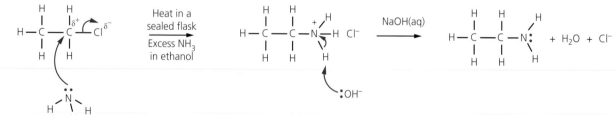

The nitrogen in the amine RNH$_2$ has a lone pair of electrons that allows the RNH$_2$ to behave as a nucleophile, so the amine can react further. Depending on the conditions, further reactions can result in the formation of secondary and tertiary amines or a quaternary amine salt – (CH$_3$CH$_2$)$_2$NH, (CH$_3$CH$_2$)$_3$N and (CH$_3$CH$_2$)$_4$N$^+$, respectively.

The secondary amine, $(CH_3CH_2)_2NH$, is produced via a two-stage mechanism in which primary amine is produced initially. The nitrogen in the primary amine has a lone pair of electrons and can behave as a nucleophile, reacting with a second molecule of chloroethane. The two stages are shown below.

Initial reaction between chloroethane and ammonia

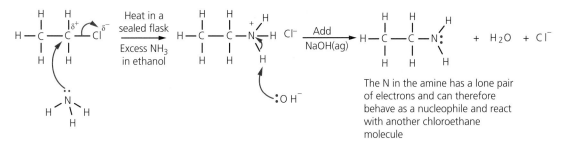

The N in the amine has a lone pair of electrons and can therefore behave as a nucleophile and react with another chloroethane molecule

Additional reaction between another chloroethane and the amine formed in the initial step

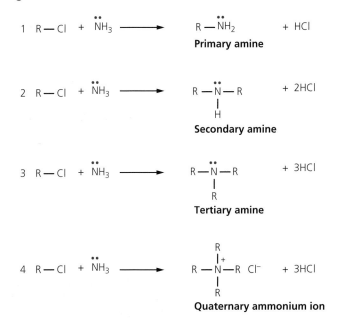

The preparation of an amine can result in a number of different products depending on the conditions.

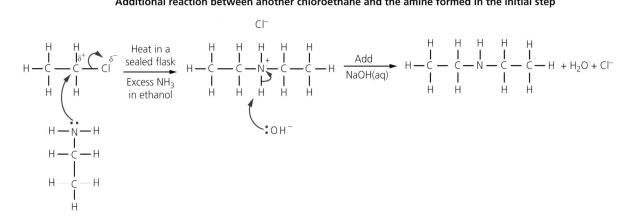

1 $R-Cl + \overset{..}{N}H_3 \longrightarrow R-\overset{..}{N}H_2 + HCl$

Primary amine

2 $R-Cl + \overset{..}{N}H_3 \longrightarrow R-\underset{H}{\overset{..}{N}}-R + 2HCl$

Secondary amine

3 $R-Cl + \overset{..}{N}H_3 \longrightarrow R-\underset{R}{\overset{..}{N}}-R + 3HCl$

Tertiary amine

4 $R-Cl + \overset{..}{N}H_3 \longrightarrow R-\underset{R}{\overset{R}{\overset{|+}{N}}}-R \; Cl^- + 3HCl$

Quaternary ammonium ion

- If excess ammonia is used, the major product will be the primary amine.

- If excess chloroethane is used, the major product will be the quaternary salt, $(CH_3CH_2)_4N^+Cl^-$.

Reduction of nitriles

Primary amines, $R\text{-}CH_2NH_2$, can also be formed by the reduction of a nitrile, $R\text{-}C{\equiv}N$. The reducing agent is either sodium and alcohol, lithium tetrahydridoaluminate, $LiAlH_4$ or hydrogen and a nickel catalyst. The reducing agent can be represented as [H] in an equation:

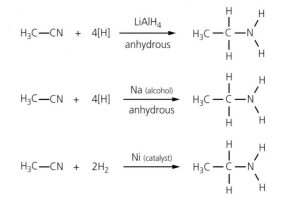

Preparation of aromatic amines

Aromatic amines such as phenylamine, $C_6H_5NH_2$, can be prepared by heating nitrobenzene, under reflux, with tin and concentrated hydrochloric acid (Figure 10.2).

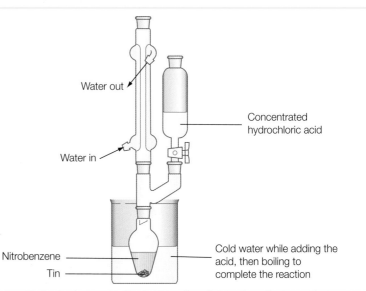

Water out

Concentrated hydrochloric acid

Water in

Nitrobenzene

Tin

Cold water while adding the acid, then boiling to complete the reaction

Figure 10.2 Reducing nitrobenzene to phenylamine by refluxing with tin and hot, concentrated hydrochloric acid.

This is a reduction reaction in which the reducing agent is formed from the reaction between tin and concentrated hydrochloric acid.

Using [H] to represent the reducing agent, the overall equation for this reaction is:

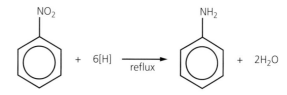

Nitrobenzene is reduced to form phenylamine which immediately reacts with the HCl acid to form a salt. Phenylamine is reformed by the addition of NaOH. The reduction of nitrobenzene to form phenylamine can be regarded as a two-step process.

Two step preparation of aromatic amines

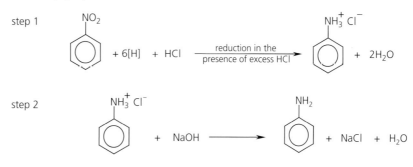

Typical reactions of amines

Reaction with acids

Amines are weak bases, since the lone pair of electrons on the nitrogen can accept a proton. Consequently, amines react with acids to form salts:

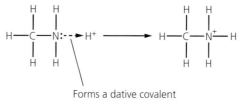

Forms a dative covalent bond with the proton

Two examples of the formation of salts from amines are:

$C_2H_5NH_2 + HCl \rightarrow C_2H_5NH_3{}^+Cl^-$

ethylamine

$C_6H_5NH_2 + HCl \rightarrow C_6H_5NH_3{}^+Cl^-$

phenylamine

Phenylamine is only slightly soluble in water, but dissolves very easily in solutions of hydrochloric acid.

Amino acids

Amino acids contain two functional groups, the amine ($-NH_2$) and the carboxylic acid ($-COOH$). The general formula of all amino acids is $H_2NCH(R)COOH$, such that the two functional groups are bonded to the same central carbon atom, the amino acid is described as an α-amino acid.

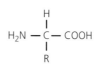

Four common amino acids are shown below:

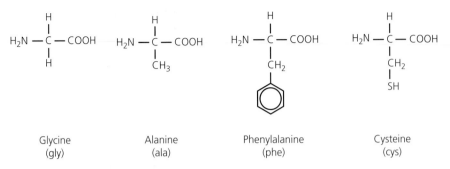

Glycine	Alanine	Phenylalanine	Cysteine
(gly)	(ala)	(phe)	(cys)

When systematically naming amino acids, the carbon in the COOH acid group is always assigned as carbon atom 1, such that the amino group is always attached to carbon atom 2.

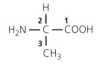

The systematic name is
2-aminopropanoic acid

All naturally occurring amino acids are 2-amino carboxylic acids. Amino acids are usually known by their trivial names and the 2-amino carboxylic acid is usually referred to as an α-amino acid.

There are about 20 naturally occurring α-amino acids but you are not expected to recall the R group side chain of any of the amino acids.

Properties of amino acids

An amino acid contains two functional groups – the carboxylic acid group and the amine group. They are described as 'bi-functional'; an amino acid can behave as both an acid and a base.

Reactions as an acid

The carboxylic acid group can react with a base to form a salt, and with an alcohol to form an ester:

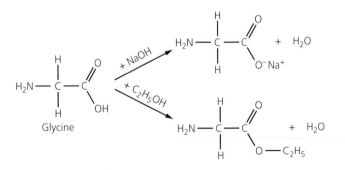

As well as reacting with alkalis, amino acids also react with metals, carbonates and metal oxides:

$$2H_2NCH(R)COOH + 2Na \rightarrow 2H_2NCH(R)COO^- Na^+ + H_2$$

$$2H_2NCH(R)COOH + K_2CO_3 \rightarrow 2H_2NCH(R)COO^-K^+ + H_2O + CO_2$$

$$2H_2NCH(R)COOH + MgO \rightarrow (H_2NCH(R)COO^-)_2Mg^{2+} + H_2O$$

Reactions as a base

The amine group is a base and can react with an acid, such as HCl, to produce a salt.

Glycine

The amine group can also behave as a nucleophile and can react with compounds that contain the $C^{\delta+}–X^{\delta-}$ group.

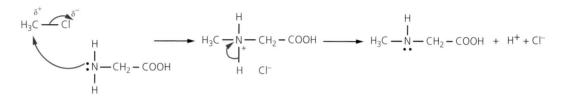

There are numerous other reactions that depend on both functional groups, the most important being the formation of peptide links.

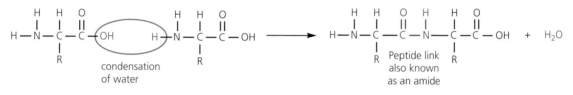

For chemists the peptide bond is simply an example of an amide bond (see page 215).

A dipeptide contains one peptide link (–CONH–) and two side chains (R groups). Dipeptides can react further with additional amino acids, thus extending the chain length. This leads to the formation of tripeptides, polypeptides and proteins:

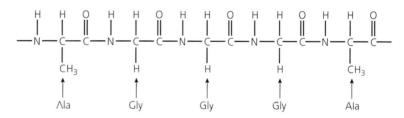

The reactions mentioned above are important in a biological context, but are not covered within the specification. You will not be expected to remember the formula of any particular amino acid, but you may be tested on your understanding of how the two functional groups react.

Amides

Primary amides have the general formula R-$CONH_2$ and include compounds such as:

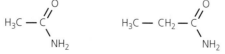

Ethanamide   Propanamide   Phenylethanamide

It is possible to form secondary amides such as:

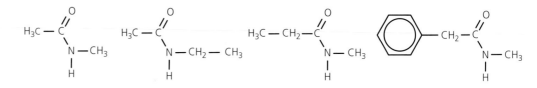

Amides are polar groups and, like esters, can be hydrolysed by reaction with either an acid or a base.

Hydrolysis by heating with a mineral acid, usually HCl(aq)

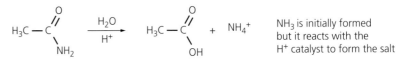

NH_3 is initially formed but it reacts with the H^+ catalyst to form the salt

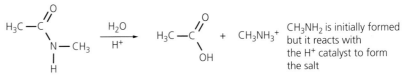

CH_3NH_2 is initially formed but it reacts with the H^+ catalyst to form the salt

Hydrolysis by heating with an alkali such as NaOH(aq)

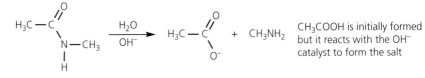

CH_3COOH is initially formed but it reacts with the OH^- catalyst to form the salt

CH_3COOH is initially formed but it reacts with the OH^- catalyst to form the salt

Test yourself

6 Write equations for the reaction of the amino acids below. For each amino acid the R group is given.
  a) alanine with Na   (R group is $-CH_3$)
  b) glutamic acid with NaOH   (R group is $-CH_2CH_2COOH$)
  c) alanine with HCl   (R group is $-CH_3$)
  d) phenylalanine with Na_2CO_3(aq)   (R group is $-CH_2C_6H_5$)
  e) aspartic acid with excess
     ethanol and H_2SO_4 catalyst   (R group is $-CH_2COOH$)
  f) lysine with excess HCl   (R group is $-(CH_2)_4NH_2$)
7 Draw the products formed when $CH_3CH_2CONHCH_2CH_3$ is hydrolysed using
  a) HCl(aq)
  b) NaOH(aq)

8 Draw the products formed when the tripeptide shown below is hydrolysed

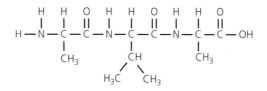

a) by acid hydrolysis using HCl(aq)
b) by base hydrolysis using NaOH(aq).

Chirality and optical isomerism

Key term

Chiral molecules are asymmetric molecules that are non-superimposable mirror images of each other. Asymmetric molecules are molecules with no centre, axis or plane of symmetry.

Every molecule has a mirror image. Generally, the mirror image of a molecule can be turned around to show that it is identical to the original molecule. Sometimes, however, it turns out that a molecule and its mirror image are not quite the same – the molecule and its mirror image cannot be superimposed. A molecule is chiral if, like one of your hands, it cannot be superimposed on its mirror image. The word 'chiral' comes from the Greek for 'hand'.

Chiral molecules are **asymmetric**. This means that they have mirror image forms that are not identical.

Test yourself

9 Identify the objects in the figure below that are not superimposable.

The commonest chiral compounds are organic molecules in which there is a carbon atom attached to four different atoms or groups. Look closely at the two molecules of alanine (Figure 10.3).

The two molecules in Figure 10.3 each have the same four different atoms or groups attached to their central carbon atom – a CH_3 group, an NH_2 group, a COOH group and an H atom. It is impossible to superimpose the mirror images of alanine. You can use a molecular modelling set to construct two models of alanine (Figure 10.4), so that each of the four atoms or groups is represented by a different coloured sphere. For example, white = H, blue = NH_2, red = COOH and maroon = CH_3.

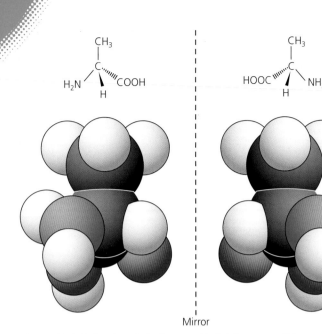

Mirror

Figure 10.3 Molecules of alanine are chiral. It is not possible to superimpose the two mirror image molecules.

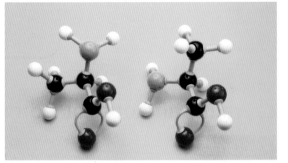

Figure 10.4 No matter how you turn the molecules around, you cannot get the two to be identical, with groups and atoms in the same position in space.

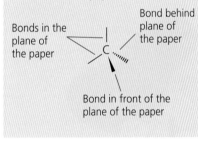
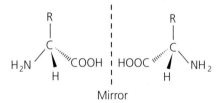

Mirror

Chirality in amino acids

The two forms of alanine behave identically in all their chemical reactions and all their physical properties – except for their effect on polarised light. Light is made up of electromagnetic radiation that has oscillations in all directions. Plane-polarised light is created by restricting the oscillations to one plane only. When plane polarised light is passed through one isomer, the plane is rotated to the left for example, while the other isomer rotates the plane to the right. This optical property is the only way of telling the two forms of alanine apart. So, chemists call them optical isomers. The term 'enantiomers' is also used to describe mirror image molecules which are optical isomers. The word 'enantiomer' comes from a Greek word meaning 'opposite'.

All the natural amino acids which occur in proteins, except glycine, have a central carbon atom attached to four different groups. So, except for glycine, all these amino acids have chiral molecules which can exist as mirror images. Chemists represent amino acids by drawing three dimensional representations as shown on the left.

Some amino acids have more than one chiral centre (carbon) and the number of optical isomers (enantiomers) varies depending on the number of chiral centres, as shown in Table 10.1.

Table 10.1 If there are 'n' chiral centres in a molecule, there will be 2^n optical isomers.

Number of chiral centres	Number of optical isomers
1	2 (2^1)
2	4 (2^2)
3	8 (2^3)

Test yourself

10 Which of the following alcohols are chiral: butan-1-ol, butan-2-ol, pentan-1-ol, pentan-2-ol, pentan-3-ol?

11 Copy the three amino acids shown below and label each chiral carbon with an asterisk*.

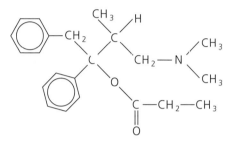

Threonine   Isoleucine   Cysteine

12 The chemists who synthesise new drugs must pay close attention to chirality. Dextropropoxyphene, for example, is a painkiller.

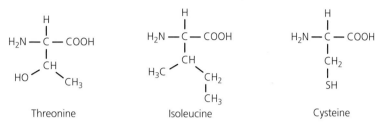

The molecule has two asymmetric carbon atoms. Its mirror image is useless for treating pain, but it is a useful ingredient in cough mixtures. Redraw the structure of dextropropoxyphene and indicate the chiral centres with asterisks.

Practice questions

Multiple choice questions 1–10

1 If chloroethane, CH_3CH_2Cl, is reacted with a large excess of ammonia, the main product is likely to be:

A $CH_3CH_2NH_2$

B $(CH_3CH_2)_2NH$

C $(CH_3CH_2)_3N$

D $(CH_3CH_2)_4N^+Cl^-$ *(1)*

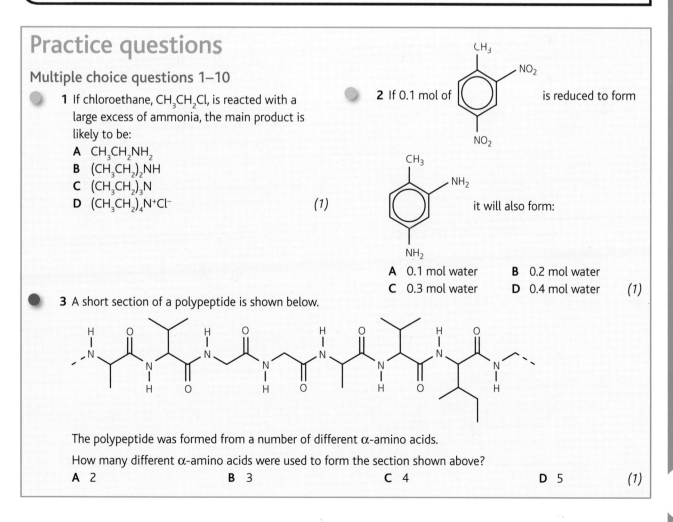

2 If 0.1 mol of [structure with CH_3, NO_2, NO_2] is reduced to form [structure with CH_3, NH_2, NH_2] it will also form:

A 0.1 mol water   B 0.2 mol water

C 0.3 mol water   D 0.4 mol water   *(1)*

3 A short section of a polypeptide is shown below.

The polypeptide was formed from a number of different α-amino acids.

How many different α-amino acids were used to form the section shown above?

A 2   B 3   C 4   D 5   *(1)*

4 Ethylamine can behave as a base.

$$CH_3CH_2NH_2 + HCl \rightarrow CH_3CH_2NH_3^+ Cl^-$$

When this reaction takes place the C-N-H bond angle changes by approximately:

A 2°   B 5°

C 8°   D 10°   *(1)*

5 When the alkaline hydrolysis of

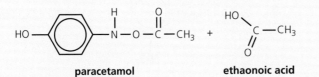

takes place, the organic products contain:

A 2 cations and 1 anion

B 1 cation and 2 anions

C 2 cations and 1 molecule

D 2 anions and 1 molecule   *(1)*

6 Paracetamol can be prepared by shaking an aqueous solution of 4-aminophenol with excess ethanoic anhydride.

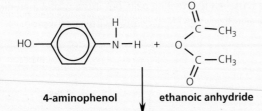

4-aminophenol   ethanoic anhydride

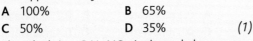

paracetamol   **ethaonoic acid**

A student carried out an experiment and reacted 10.9 g of 4-aminophenol with excess ethanoic anhydride. The student produced 10.9 g of paracetamol. The student's percentage yield was approximately:

A 100%   B 65%

C 50%   D 35%   *(1)*

7 Phenylephrine, $C_9H_{13}NO_2$, is shown below.

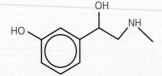

Which of the following is false?

A It reacts with acidified dichromate.

B It reacts with sodium hydroxide.

C It reacts with sodium carbonate.

D It reacts with hydrochloric acid.   *(1)*

Use the key below to answer questions 8, 9 and 10.

A	B	C	D
1, 2 & 3 correct	1 & 2 correct	2 & 3 correct	1 only correct

8 Isoleucine is an α-amino acid. Its structure is shown below.

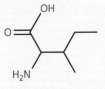

Which of the following statements are true?

1 It could react with sulfuric acid.

2 It could react with a solution of sodium carbonate.

3 It has one chiral centre.   *(1)*

9 Procaine, shown below, is a local anaesthetic and can be used in dentistry to reduce pain.

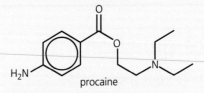

procaine

Which of the following statements are true?

1 It contains an amide group.

2 Its molecular formula is $C_{13}H_{20}N_2O_2$.

3 It can be hydrolysed.   *(1)*

10 Salbutamol, shown below, is used in inhalers and in nebulizers to relieve bronchospasms.

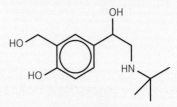

Which of the following statements are true?

1 It contains a chiral carbon.

2 It reacts with an acid.

3 It reacts with a base.   *(1)*

11 a) Name each of the following compounds:

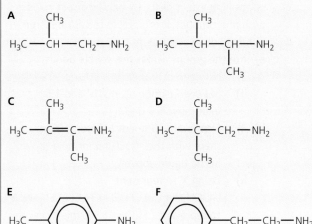

A

$H_3C - CH - CH_2 - NH_2$
     |
     CH_3

B

$H_3C - CH - CH - NH_2$
     | |
     CH_3 CH_3

C

$H_3C - C = C - NH_2$
     | |
     CH_3 CH_3

D

$H_3C - C - CH_2 - NH_2$
     |
     CH_3
(with CH_3 above the central C)

E

$H_3C - \langle benzene \rangle - NH_2$

F

$\langle benzene \rangle - CH_2 - CH_2 - NH_2$

b) Draw the skeletal formulae of **A**, **B** and **C**.

c) What is the molecular formula of compound **E**?

d) Starting from 1-chloro-2-methylpropane, show, with the aid of an equation, how compound **A** could be prepared.

e) Starting from benzene, explain how compound **E** could be prepared. Give reagents, conditions and equations for each step. *(21)*

12 Describe the main reactions of amines, pointing out the similarities and differences between aliphatic amines, such as 1-aminobutane, and aromatic amines, such as phenylamine. Write balanced equations where appropriate. *(9)*

13 A series of tests was carried out on three organic compounds A, B and C. The results of the tests are described below. State the deductions you can make from the tests on each of A, B and C. You are not expected to identify compounds A, B and C.

- A is a colourless liquid which does not mix with water.
- After warming a few drops of A with aqueous sodium hydroxide, the resulting solution was acidified with nitric acid. Silver nitrate solution was then added and a cream-coloured precipitate formed.
- A solution of B turns universal indicator red.
- A solution of B reacts with aqueous sodium carbonate to produce a colourless gas which turns limewater milky.
- When a little of B is warmed with ethanol and one drop of concentrated sulfuric acid, a sweet-smelling product can be detected on pouring the reaction mixture into cold water.
- C is a liquid which burns with a very smoky, yellow flame.
- C does not react with sodium carbonate solution.
- C fizzes with sodium and gives off a gas which produces a 'pop' with a burning splint. *(9)*

14 The diagram below shows a series of reactions beginning with the amine, cadaverine. Cadaverine is formed when proteins decompose.

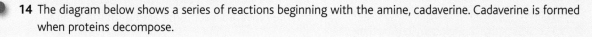

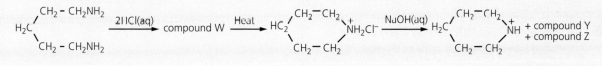

Cadaverine   + compound X   Piperidine

a) i) What characteristic physical property of cadaverine would you expect to notice if you were provided with a sample of it?

ii) What is the systematic name of cadaverine?

iii) Draw the structural formula of compound W.

iv) Write the name and formula of compound X.

v) Write the formulae of compounds Y and Z.

b) Amines are classed as primary, secondary and tertiary.

i) Describe the differences in structure between the three types of amine.

ii) Which category/categories do cadaverine and piperidine belong to? *(11)*

209

15 The artificial sweetener aspartame is shown below.

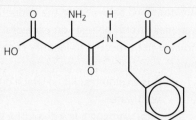

a) Deduce the molecular formula of aspartame.
b) Name the functional groups present in aspartame, other than the arene.
c) Why do you think that soft drinks sweetened by aspartame carry a warning for people with a genetic disorder which means they must not consume phenylalanine?
d) Suggest why aspartame cannot be used for food that will be cooked.
e) Aspartame can be hydrolysed. Name the functional groups in aspartame that can be hydrolysed.
f) Draw the products formed when aspartame is hydrolysed in:
  i) acidic conditions
  ii) alkaline conditions. *(15)*

16 Explain the term stereoisomerism and describe the different types of stereoisomerism, illustrating your answer with suitable examples. *(4)*

17 Draw three-dimensional diagrams of the optical isomers of each of the following:
a) valine, $H_2NCH(CH(CH_3)_2)COOH$
b) threonine, $H_2NCH(CH(OH)CH_3)COOH$
c) 2,3-dimethyl-2,3-diphenylbutanedioic acid. *(9)*

18 The female silk moth secretes a pheromone called bombycol which attracts the male silk moth strongly.

$$HO\overset{1}{C}H_2\ \overset{2}{C}H_2\ \overset{3}{C}H_2\ CH_2\ CH_2\ CH_2\ CH_2\ CH_2\ \overset{9}{C}H_2$$

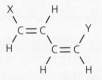

Chemists are interested in pheromones because they offer an alternative to pesticides for controlling insect pests. By baiting insect traps with pheromones, it is possible to capture large numbers of insects before they mate.

a) Are the groups across the double bonds
  i) between carbon atoms 10 and 11
  ii) between carbon atoms 12 and 13
  in bombycol *E* or *Z*?
b) i) How many different *E/Z* isomers are there with the structural formula shown in the figure above?
  ii) Using **X** to represent $HOCH_2CH_2CH_2CH_2CH_2CH_2CH_2CH_2$ and **Y** to represent $CH_2CH_2CH_3$ in this molecule

$$\underset{H}{\overset{X}{\diagdown}}C=C\underset{\underset{H}{\diagup}C=C\underset{H}{\overset{H}{\diagdown}}}{\overset{H}{\diagup}}\overset{Y}{\diagup}$$

  and draw the *E/Z* stereoisomers.
c) Write the systematic name of bombycol, assuming that the straight-chain alkane with 16 carbon atoms is called hexadecane.
d) Why are there no optical isomers for bombycol? *(9)*

By the time you have completed the organic chemistry section of this course, you should be able to describe steps 1, 2, 3 and 5 in the sequence for the conversion of benzene into benzocaine (see below). Step 4 is an oxidation reaction using $KMnO_4$ as the oxidising agent, which is not on the syllabus.

As you cover more of the syllabus, this question will be repeated, each time asking for more detail.

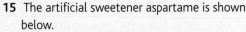

19 Benzocaine is a local anaesthetic with a wide range of applications. As the name suggests it is a derivative of benzene and could be made in the laboratory by the multi-stage process shown below.

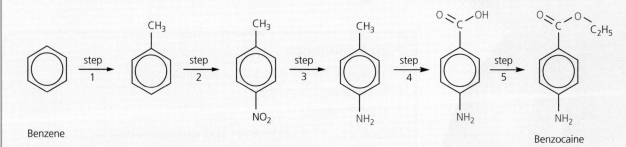

Describe the reagents, the conditions, the equation and:
a) the mechanism for step 1
b) the mechanism for step 2
c) step 3   d) step 5.   (14)

Challenge

20 Aliphatic amines can be prepared by the reaction between chloroalkanes and ammonia. The ammonia must be dissolved in ethanol.
a) Explain why the ammonia must **not** be dissolved in water.
b) Suggest why ammonia dissolved in a saturated aqueous solution of NH_4Cl might be a suitable reagent.
c) Chloromethane reacts with ammonia to produce methylamine. Complete the following mechanism by adding curly arrows, relevant dipoles and lone pairs of electrons.

Step 1:

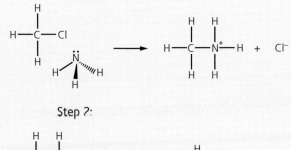

Step 2:

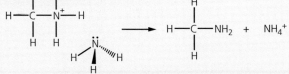

d) In the reaction between chloromethane and ammonia, various different organic products are formed including:

- $CH_3CH_2NH_2$
- $(CH_3CH_2)_2NH$
- $(CH_3CH_2)_3N$
- $(CH_3CH_2)_4N^+$

Suggest how you might manipulate the reaction conditions to ensure that you obtain mainly:
i) $CH_3CH_2NH_2$
ii) $(CH_3CH_2)_4N^+$ (11)

21 Polymers such as poly(propene) can exist in different forms:

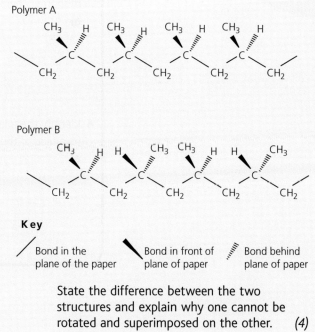

Key

/ Bond in the plane of the paper   ◣ Bond in front of plane of paper   ⫰ Bond behind plane of paper

State the difference between the two structures and explain why one cannot be rotated and superimposed on the other.   (4)

Polymers

Prior knowledge

In this chapter it is assumed that you are familiar with:

- how alkenes form polymers
- why most polymers are not biodegradable.

For example, you should be aware that alkenes are unsaturated hydrocarbons and contain a C=C double bond. Alkenes can add together to form long chain polymers. Most addition polymers are non-biodegradable and disposal of addition polymers is an on-going problem.

Test yourself on prior knowledge

1 The covalent bond in bromine, Br_2, can be broken by either homolytic fission or by heterolytic fission. Write equations to illustrate:
  a) homolytic fission
  b) heterolytic fission.
  c) Use the products from a) and b) to identify:
    i) a radical
    ii) an electrophile
    iii) a nucleophile.
2 State three ways in which we dispose of addition polymers.
3 Draw two repeat units of the polymer chain formed from each of the monomers below and write the systematic name of the polymer.
  a)   b) propenamide

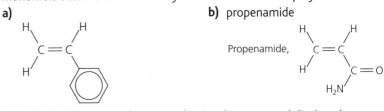

4 Identify the following polymers and write the name and displayed formulae of their monomers.
  a)   b)

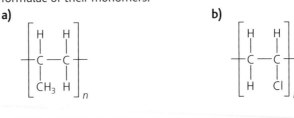

Introduction to polymers

Synthetic polymers are chain-like molecules made from small individual units called monomers. Many polymers are plastics that play an important part in everyday life. The amount of plastic we use is enormous and growing. Plastics are easy to make, easy to colour and can be chemically tailored to have specific properties. They are often cheaper to produce than the materials that they replace. There are many instances where plastic is used instead of wood, metal, stone, glass, leather or a natural fabric, such as cotton or silk. However, there are problems with their disposal and there is growing concern over the environmental consequences.

In this course, two categories of polymer are important:

- addition polymers
- condensation polymers.

Tip

Common monomers and their addition polymers were discussed the Year 1 book (page 218).

Addition polymers

Addition polymers are formed when alkenes undergo a reaction in which one alkene molecule joins to another and a long molecular chain is built up. The individual alkene molecule is referred to as a monomer; the long chain molecule is known as the polymer.

Condensation polymers

Condensation polymers are formed when monomers link together to form a long chain in which each link is accompanied by the loss, or condensation, of a small molecule such as H_2O or HCl. The two main types of condensation polymer are polyesters and polyamides.

Polyesters

Terylene® (or PET, **p**ol**y**ethylene **t**erephthalate) is a common polyester used in synthetic fibres and in containers for food and beverages. It is one of the most important raw materials of synthetic fibres. It is made by reacting the monomers benzene-1,4-dicarboxylic acid and ethane-1,2-diol:

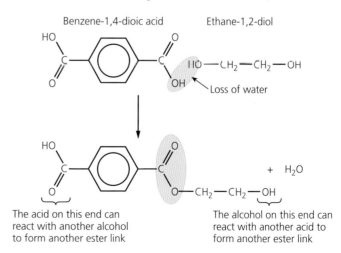

Benzene-1,4-dioic acid   Ethane-1,2-diol

Loss of water

The acid on this end can react with another alcohol to form another ester link

The alcohol on this end can react with another acid to form another ester link

Tip

It is important to be able to identify the simplest repeat unit of a polymer.

Both monomers react at each end, building up a long-chain molecule held together by a large number of ester linkages – hence, producing a polyester.

213

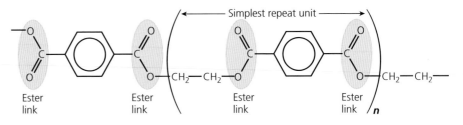

The traditional names for the monomers benzene-1,4-dicarboxylic acid and ethane-1,2-diol are **ter**ephthalic acid and eth**ylene** glycol hence the commercial name **Terylene®**.

The resulting polymer is a regular, approximately linear, structure. The polymer chains pack closely together, so there are strong intermolecular forces.

Polyesters can be formed by the reaction between any dioic acid and any diol, but perhaps the most important development in polymer chemistry in recent years is the polymerisation of hydroxycarboxylic acids which contain both the alcohol and the carboxylic acid functional groups. Hydroxycarboxylic acids, such as 2-hydroxypropanoic acid (lactic acid), $CH_3CH(OH)COOH$, are probably the most versatile of the new biodegradable plastics.

A molecule of lactic acid has a chiral centre, $CH_3{}^*CH(OH)COOH$, and contains an alcohol group (–OH) and a carboxylic acid group (–COOH).

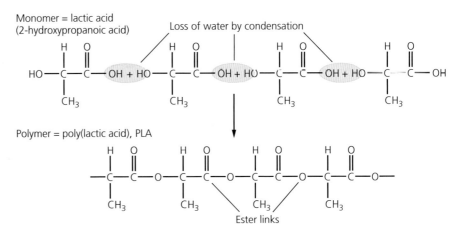

Chemists have a role in minimising our impact on the environment through the use of renewable resources and the development of degradable polymers, such as polylactic acid (PLA). Polylactic acid is particularly attractive as a sustainable alternative to products derived from petrochemicals, since the monomer can be produced by the bacterial fermentation of agricultural by-products, such as corn starch or sugar cane, which are renewable feedstocks.

PLA is more expensive than many petroleum-derived commodity plastics, but its price has been falling as more production comes online. The degree to which the price will fall and the degree to which PLA will be able to compete with non-sustainable petroleum-derived polymers is uncertain.

Table 11.1

Monomer	Repeat unit of polymer	Properties
lactic acid $$H-O-\underset{\underset{CH_3}{\mid}}{\overset{\overset{\displaystyle H}{\mid}}{C}}-\overset{\overset{\displaystyle O}{\|}}{C}-O-H$$	poly(lactic acid) $$\left[\begin{array}{c}\overset{\overset{\displaystyle H}{\mid}}{\underset{\underset{CH_3}{\mid}}{C}}-\overset{\overset{\displaystyle O}{\|}}{C}-O\end{array}\right]$$	Compostable/decomposes to form CO_2 and H_2O. Polymers that contain groups such as the carbonyl group, $-C=O$, may also be photodegradable because the C=O bond absorbs radiation.
Uses		
Biomedical applications — internal stitches and drug-release systems 'Bioplastic' used in food packaging, female hygiene products and disposable nappies Fibres and non-woven textiles Possible material for tissue engineering		

Polyamides

Polyamides can be prepared from two monomers, one with an amine group at each end and the other with a carboxylic acid group at each end. The monomers are linked by an amide bond:

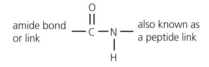

Nylon

Nylon is the general name for a family of condensation polymers first produced by Wallace Carothers in 1935. Today, nylon is one of the most common polymers used as a fibre. Nylons are formed by reacting equal parts of a diamine and a dicarboxylic acid, so that an amide (or peptide) bond is formed at both ends of each monomer. Diacyl dichlorides, $ClOC(CH_2)_nCOCl$, can be used in place of the dioic acids.

Different nylons have a different numerical suffix, which specifies the numbers of carbons in each monomer; the diamine is first and the diacid second. The most common variant is nylon-6,6, the name of which indicates that the diamine (1,6-diaminohexane) and the dicarboxylic acid (hexanedioic acid) both donate six carbons to each repeat unit of the polymer chain.

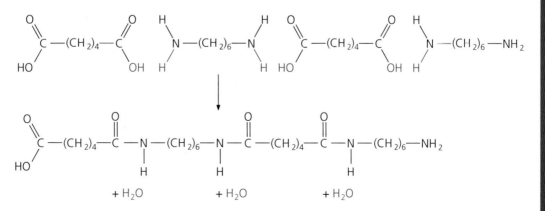

The overall equation for nylon-6,6 can be represented by:

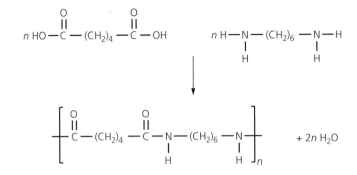

Other nylons include nylon-6,10, which is made from the monomers 1,6-diaminohexane and decane-dicarboxylic acid.

Nylons can be produced more readily in the laboratory by using the more reactive acid chlorides in place of the carboxylic acids. For instance, decanedioyl dichloride reacts with 1,6-diaminohexane at room temperature to produce nylon-6,10 (Figure 11.1).

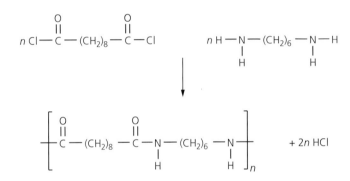

Nylon forms a strong flexible fibre when it is melt-spun. Nylon has a range of uses including clothing, ropes and fishing lines, toothbrush bristles, parachutes, flexible tubing and tights (Figure 11.2).

Figure 11.1 In the nylon rope experiment, a strand of nylon-6,10 can be pulled from the interface of 1,6-diaminohexane and decanedioyl chloride.

Figure 11.2 Nylon products.

Nylon is a collective name for polyamides that contain aliphatic hydrocarbon sections. These aliphatic sections are non-polar and free to rotate and twist. As the hydrocarbon sections become longer, the nylon becomes more flexible. Nylon-6,6 is less flexible than nylon-6,10. Chemists

Tip

You will not be expected to recall the names or formulae of the monomers for making different nylons.

have developed these features of nylon and have replaced the hydrocarbon sections with benzene rings, which reduces the flexibility and increases the strength.

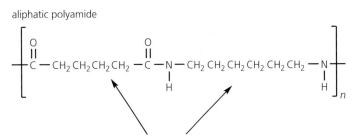

the hydrocarbon sections are long and flexible

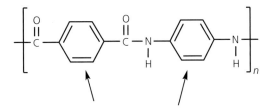

the hydrocarbon sections are shorter and rigid

Kevlar®

Kevlar® is a polyamide. It is stronger than steel and is fire-resistant. It is used for making bulletproof vests, crash helmets and the protective clothing used by firefighters. It is also used for sports equipment and loudspeaker cones.

Kevlar® is made from the monomers benzene-1,4-diamine and benzene-1,4-dicarboxylic acid (Figure 11.3).

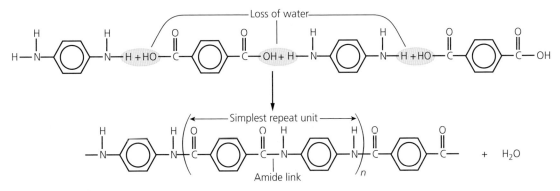

Figure 11.3 Kevlar.

Fibres of Kevlar® consist of long molecular chains and there are many inter-chain forces, which makes the material extremely strong. Kevlar® derives its high strength partly from intermolecular hydrogen bonds formed between the oxygen in the carbonyl groups and protons on neighbouring polymer chains and partly from the partial π-stacking of the aromatic rings. An aromatic, or π–π, interaction is a non-covalent interaction caused by the intermolecular overlapping of p-orbitals in adjacent π-delocalised systems. These interactions become stronger as the number of π-electrons increases.

The extensive hydrogen bonding is shown in Figure 11.4.

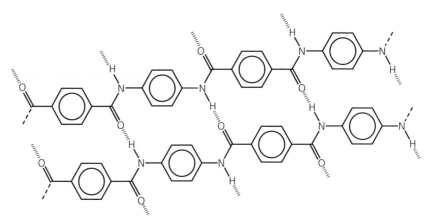

Figure 11.4 Hydrogen bonding between Kevlar polymer chains.

Activity

Modelling and synthesising polyamides

Experiments show that nylons with longer hydrocarbon sections to their chains are more flexible than those with shorter sections. Kevlar® is similar to nylon-6,6, but with benzene rings rather than aliphatic chains linked by the amide group. The repetition of benzene rings in its structure makes Kevlar® exceptionally strong and very inflexible, compared with nylon-6,6. Because of this, it is used extensively in tyres, brakes and clutch fittings, in ropes and cables and in protective clothing (Figure 11.5).

1 Look closely at the structure of one chain of Kevlar® in Figure 11.4.

   a) Explain how Kevlar® is a condensation polymer of benzene-1,4-dioic acid and benzene-1,4-diamine.

   b) Weight for weight, Kevlar® is five times stronger than steel. This exceptional strength is due to hydrogen bonding between the separate chains. Use Figure 11.4 to explain why inter-chain hydrogen bonding is so strong in Kevlar®.

   c) Suggest a reason why Kevlar® is made from monomers with functional groups in the 1 and 4 positions, and not from isomers with functional groups in the 1 and 2, or 1 and 3 positions.

2 A condensation polymer can be prepared by mixing equal amounts of the monomers in Figure 11.6 at room temperature.

   a) Draw the structure of one repeat unit of the polymer formed from the two monomers.

   b) The polymer forms even more rapidly if the reaction mixture contains sodium carbonate. Why is this?

   c) The polymer molecules obtained at room temperature can be linked to one another (cross-linked) by a second reaction. Explain how this cross-linking can be achieved and state the conditions needed for it to happen.

   d) Explain how the choice of reaction conditions can control the extent of polymerisation and the extent of cross-linking.

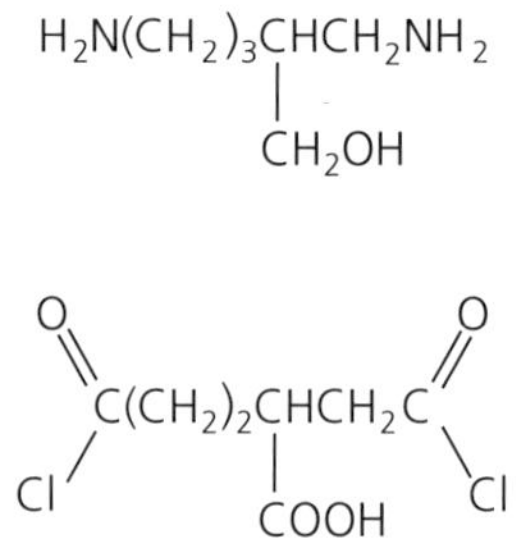

Figure 11.5 This policeman is wearing a bulletproof jacket made from Kevlar®.

Figure 11.6

Test yourself

1 Nylon is a generic name for aliphatic polyamides.
   a) What is meant by aliphatic?
   b) What is an amide?
   c) Draw two repeat sections of nylon-6,10.
   d) Name the nylon below:

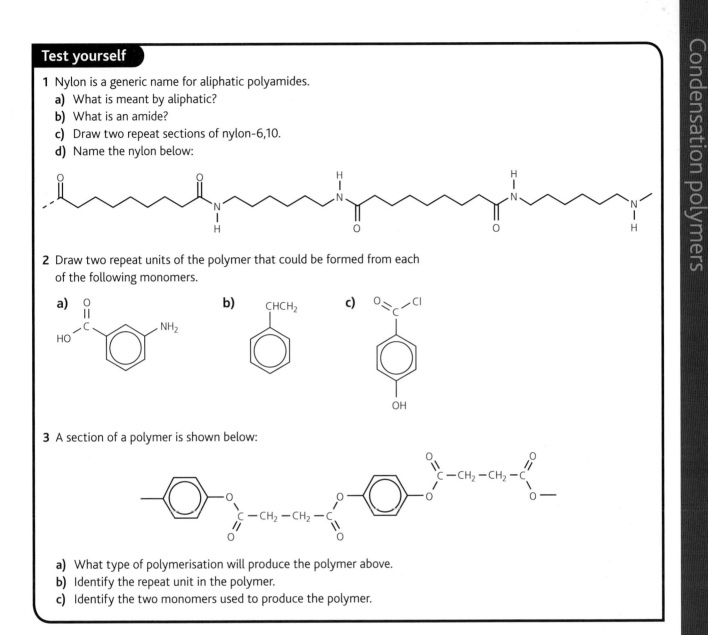

2 Draw two repeat units of the polymer that could be formed from each of the following monomers.

   a)

   b)

   c)

3 A section of a polymer is shown below:

   a) What type of polymerisation will produce the polymer above.
   b) Identify the repeat unit in the polymer.
   c) Identify the two monomers used to produce the polymer.

Hydrolysis of condensation polymers

The ester link in a polyester and the amide link in a polyamide are both polar and are subject to acid-catalysed and base-catalysed hydrolysis.

Acid hydrolysis of a polyester results in the formation of a diol and a dioic acid:

Polyester + H_2O $\xrightarrow{\text{H}^+\text{(aq) catalyst}}$ Diol + Dioic acid

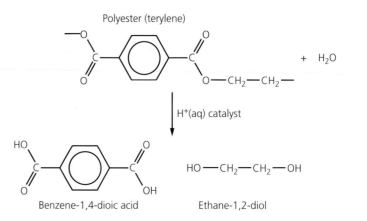

Polyester (terylene)

$+ \quad H_2O$

$H^+(aq)$ catalyst

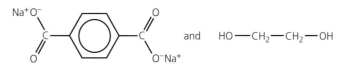

Benzene-1,4-dioic acid   Ethane-1,2-diol

Base hydrolysis of a polyester also forms a diol, but the dioic acid formed then reacts with the base catalyst to form the dioate salt. The products of refluxing Terylene® with an aqueous solution of NaOH(aq) are:

and $HO-CH_2-CH_2-OH$

Polyamides can also be hydrolysed:

● Acid-catalysed hydrolysis results in the formation of the dioic acid and the di-salt of the diamine.

● Base-catalysed hydrolysis results in the formation of the diamine and the di-salt of the dioic acid.

Hydrolysis of polyamides is summarised in the reaction scheme below.

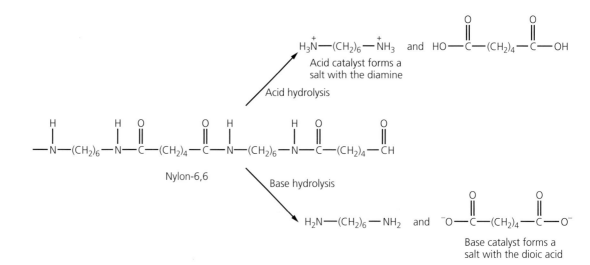

Polyamides, such as nylon, and polyesters containing aromatic groups are fairly resistant to hydrolysis (Figure 11.7), unless the reaction is catalysed by acid or base. But polyesters made from purely aliphatic monomers hydrolyse slowly at pH 7 without acid or base catalysts. This has led to the use of poly(glycolic) acid for stitching internal wounds. Once the stitches (sutures) have been inserted and the surgical incision closed, the stitches dissolve slowly in the patient's tissue fluid.

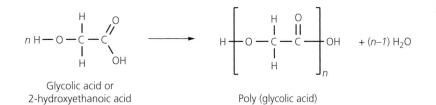

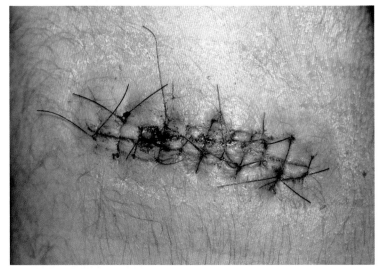

Glycolic acid or
2-hydroxyethanoic acid

Poly (glycolic acid)

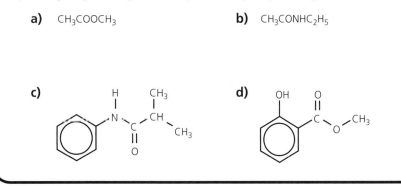

Figure 11.7 Polyamides such as nylon and aromatic polyesters are used for stitches (sutures) on external wounds. These are not degradable and have to be removed manually when the wound has healed.

Test yourself

4 Identify the products formed when each of the following is hydrolysed by
   i) acid hydrolysis using HCl and **ii)** alkaline hydrolysis using NaOH.

   a) CH_3COOCH_3

   b) $CH_3CONHC_2H_5$

   c)

   d)

Comparing addition and condensation polymerisation

Although both addition and condensation polymerisation result in the formation of long-chain organic molecules, known as polymers, from relatively small organic molecules, known as monomers, there are some clear differences between the two processes.

Table 11.2 A comparison of addition and condensation polymerisation.

Differences	Addition polymers	Condensation polymers
The type of reaction	Addition reaction only	Addition reaction followed by elimination reaction, resulting in the condensation of a small molecule such as H_2O or HCl
Type of links along the central chain	C-C single bonds only, which are non-polar	Short aliphatic or aryl sections linked by either ester groups or amide groups, which are polar
The type of monomer involved	All monomers have C=C double bond	The monomers have molecules with at least two functional groups, which may be the same or different
Hydrolysis	Resistant to hydrolysis	Undergo hydrolysis
Conditions for preparation	Require an initiator, together with high temperature and high pressure, unless a catalyst is involved	Do not require initiators and usually occur at much lower temperatures and atmospheric pressure
Polymer properties	Non-polar and resistant to attack	Polar and subject to hydrolysis

As polymer science has grown, chemists and material scientists have learnt how to develop new materials with particular properties. Some of the ways of modifying the properties of polymers include:

- altering the average length of polymer chains

- changing the structure of the monomer to one with different side groups and different intermolecular forces, thereby varying the extent of cross-linking between chains

- selecting a monomer which produces a polymer that is biodegradable or photodegradable

- adding fillers and pigments.

The development of degradable polymers

As it becomes more and more expensive to dump waste in landfill sites, plastics are seen as an increasing problem. The major problem with most plastic waste is that it is non-biodegradable. This means that the only choices for dealing with plastic waste are recycling and energy recovery.

Although some progress has been made, the separation, sorting and recycling of different plastics is difficult to mechanise and automate.

Modern incinerators, which burn plastic waste in order to recover the energy from its combustion, have to meet tough environmental standards. Despite these higher standards, many people remain suspicious of the emissions from incinerators and worry that they are a health risk.

These concerns over incineration and the difficulties in recycling have led chemists to look for other ways of minimising the waste from plastics. The most promising approach involves the development of biodegradable polymers, such as poly(lactic acid), see page 215.

Another important approach is the development and use of condensation polymers that are either:

- photodegradable, in which the C=O bonds in condensation polymers absorb radiation that has sufficient energy to facilitate the decomposition of the polymer

- readily degraded by hydrolysis, such as poly(glycolic acid), see page 223.

Activity

Developing and using poly(lactic acid)

Poly(lactic acid) is an aliphatic, thermoplastic, biodegradable polyester (Figure 11.8). Its name is sometimes abbreviated to PLA.

Polylactic acid can be produced from renewable resources, such as the sugar from canes and the starch from wheat and sweetcorn. As such, it is a sustainable alternative to oil-based plastics. PLA is more expensive to produce than its oil-based alternatives, but its cost has fallen with increased production.

The manufacture of PLA from starch or sugar requires two stages.

- In the first stage, starch or sugar is converted to lactic acid by bacterial fermentation.
- In the second stage, the lactic acid undergoes condensation polymerisation to produce PLA by heating with a catalyst of tin(ıı) octanoate.

Figure 11.8 The skeletal formula of poly(lactic acid).

Like most thermoplastics, PLA can be processed into fibres or film for a variety of uses. Poly(lactic acid) is already used for waste sacks, compost bags, disposable plastic plates, packaging and a number of biomedical applications (Figure 11.9). These include stitches, dialysis bags and capsules containing various medicinal drugs. Packaging made from PLA will degrade to lactic acid in less than two months, but this relatively rapid breakdown is only possible in the ideal conditions of a commercial composting plant.

Figure 11.9 Bags made from poly(lactic acid) – a biodegradable polymer.

1 Explain what is meant by each of the following adjectives when applied to poly(lactic acid):
   a) aliphatic
   b) thermoplastic
   c) biodegradable.
2 Why is poly(lactic acid) described as a sustainable alternative to petroleum-based plastics?
3 Why is poly(lactic acid) becoming more financially viable in the manufacture of different goods?
4 How can poly(lactic acid) reduce the problems of plastic waste polluting the environment?
5 Draw the displayed formula of lactic acid and write its molecular formula.
6 Write an equation for the polymerisation of lactic acid to form poly(lactic acid).
7 How does poly(lactic acid) degrade when it is used to stitch an internal wound, where there are no microorganisms?
8 Films of poly(lactic acid) and poly(glycolic acid) are used to coat the tablets and capsules of certain drugs. How does this help with the delivery of the drug into the body?

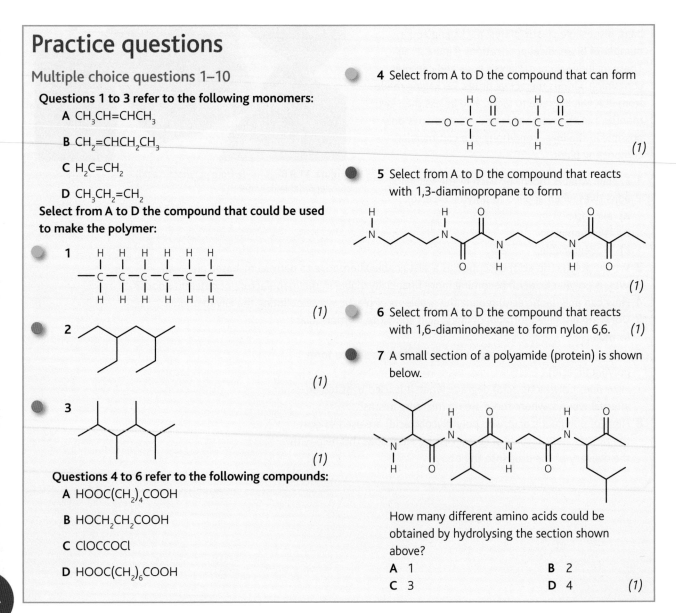

Test yourself

5 Polyhydroxyalkanoates (PHAs) are linear polyesters. There are more than
150 monomers that are produced in nature by bacterial fermentation
of sugars and lipids. The monomers are used in the production of
biodegradable plastics.
   a) Draw the following monomers:
      i) 3-hydroxybutanoic acid
      ii) 4-hydroxypentanoic acid.
   b) Draw sections of the polymer formed from each of the monomers in
      part a). Show two repeat units.
6 Name the monomers used to make the biodegradable polymers shown below.
   a)   b)

Practice questions

Multiple choice questions 1–10

Questions 1 to 3 refer to the following monomers:
   A CH₃CH=CHCH₃

   B CH₂=CHCH₂CH₃

   C H₂C=CH₂

   D CH₃CH₂=CH₂
Select from A to D the compound that could be used
to make the polymer:

1

   (1)

2

   (1)

3

   (1)

Questions 4 to 6 refer to the following compounds:
   A HOOC(CH₂)₄COOH

   B HOCH₂CH₂COOH

   C ClOCCOCl

   D HOOC(CH₂)₆COOH

4 Select from A to D the compound that can form

   (1)

5 Select from A to D the compound that reacts
   with 1,3-diaminopropane to form

   (1)

6 Select from A to D the compound that reacts
   with 1,6-diaminohexane to form nylon 6,6.   *(1)*

7 A small section of a polyamide (protein) is shown
   below.

   How many different amino acids could be
   obtained by hydrolysing the section shown
   above?
   A 1   B 2
   C 3   D 4   *(1)*

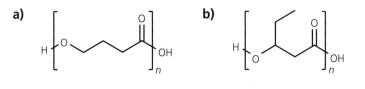

Chapter 11 Polymers

224

Use the key below to answer questions 8, 9 and 10.

A	B	C	D
1, 2 & 3 correct	1 & 2 correct	2 & 3 correct	1 only correct

8 Which of the following statements are true?

1 Nylon-6,6 is more flexible than nylon-6,10.

2 Terylene® is more flexible than Kevlar®.

3 Nylon-6,6, nylon-6,10, Terylene® and Kevlar® can all be hydrolysed. *(1)*

9 There are three classes of polymers: poly(alkenes), polyesters and polyamides. Which of the following statements are true?

1 Polyesters and polyamides have permanent dipole–dipole intermolecular forces but poly(alkenes) do not.

2 Polyesters and polyamides can be hydrolysed but poly(alkenes) cannot be hydrolysed.

3 Polyesters and polyamides can only be prepared from two different monomers whereas poly(alkenes) are prepared from one monomer. *(1)*

10 Poly(phenylethene) also known as polystyrene can be prepared from phenylethanol.

A possible reaction sequence is:

$$C_6H_5CH_2CH_2OH \longrightarrow C_6H_5CHCH_2 \longrightarrow \left(\begin{array}{cc} H & H \\ | & | \\ -C & -C- \\ | & | \\ C_6H_5 & H \end{array} \right)_n$$

Which of the following statements are true concerning this reaction sequence?

1 Phenylethanol could be refluxed with concentrated sulfuric acid to make $C_6H_5CHCH_2$.

2 The polymer is non-biodegradable.

3 The polymer is a good insulator. *(1)*

11 Draw a section of the polymer, showing two repeat units, which could be formed from each of the following pairs of monomers:

a) 1,6-diaminohexane and decanedioic acid

b) benzene-1,4-dioic acid and 1,3-diaminobenzene

c) ethane-1,2-diol and benzene-1,3-dioic acid. *(6)*

12 Draw two repeat units of the polymer that could be formed from:

a)

$$HOH_2C-CH_2-CH_2-CH_2OH + HO-\overset{\overset{\displaystyle O}{\|}}{C}-(CH_2)_3-\overset{\overset{\displaystyle O}{\|}}{C}-OH$$

b)

$$HO-\overset{\overset{\displaystyle O}{\|}}{C}-CH_2-\underset{\underset{\displaystyle CH_3}{|}}{CH}-NH_2$$

(4)

13 Suggest a reason why many polyesters and polyamides are degradable while poly(alkenes) are not. *(3)*

14 Identify the organic products formed when the polymer

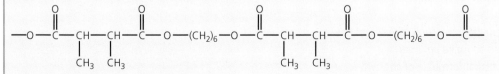

is hydrolysed by heating with:

a) an acid catalyst

b) a base catalyst. *(4)*

15 An elastic tape consists of 60% polyester and 40% neoprene. Neoprene is a polymer similar to synthetic rubber and contains the elements C, H and Cl. It is manufactured by polymerising 2-chlorobuta-1,3-diene. The equation is shown below but it is only **partially** completed.

2-chlorobuta-1,3-diene   Neoprene

a) Use the template above to write an equation for the manufacture of neoprene from 2-chlorobuta-1,3-diene by adding Cl in the appropriate places.

b) What type of polymerisation is illustrated by the manufacture of neoprene?

c) The polyester is manufactured by a reaction between benzene-1,4-dicarboxylic acid and ethane-1,2-diol.

  i) Draw the structural formulae of these two monomers.

  ii) Draw the structural formula of the molecule which forms when one molecule of each of these monomers reacts to produce an ester.

d) What reagent could you use to show the presence of neoprene in the elastic tape? State and explain what you would observe if you used the reagent with some of the tape. *(8)*

16 Short sections of the molecular structures of two polymers are shown below.

Polymer 1

Polymer 2

a) Draw the simplest repeat unit for each polymer.

b) Draw and name the structural formula of the monomer used to make polymer 1.

c) Draw the structural formulae of the two monomers that could be used to make polymer 2.

d) Name the monomers that could be used to make polymer 2.

e) During the last decade, degradable polymers have been developed to reduce the quantity of plastic waste that is dumped in landfill sites. State and explain two reasons why polymer 2 is more likely to be a degradable polymer than polymer 1. *(11)*

17 4-methylmandelic acid, $CH_3C_6H_4CH(OH)COOH$, methyl methacrylate, $CH_2C(CH_3)COOCH_3$ and cyclohexene, C_6H_{10}, can be used to form polymers.

a) **i)** Draw 4-methylmandelic acid and show two repeat units of the polymer that could be formed from it.

  ii) Draw methyl methacrylate and show two repeat units of the polymer that could be formed from it.

  iii) Draw cyclohexene and show two repeat units of the polymer that could be formed from it.

b) The polymers made from 4-methylmandelic acid and methyl methacrylate can both be hydrolysed. Identify the organic products when the polymer made from:

  i) methyl methacrylate is hydrolysed by HCl(aq)

  ii) 4-methylmandelic acid is hydrolysed by NaOH(aq).

c) Suggest why the polymer made from 4-methylmandelic acid is likely to be '*better for the environment*' than the polymer made from methyl methacrylate.

d) 4-methylmandelic acid can be used to prepare two different esters, which are shown below.

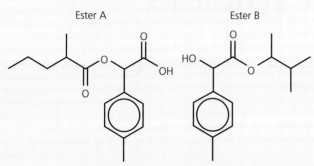

Ester A   Ester B

Identify the organic chemical used to make

i) ester A

ii) ester B. *(13)*

18 A bottle made of poly(phenylethene) can be used to store dilute potassium hydroxide, but holes gradually appear in a polyester lab coat which has soaked up splashes of the same reagent. Account for the difference in the behaviour of the two polymers. *(4)*

19 Nylon is not a single material, but a group of chemicals made from diamines and dioic acids (or dioyl chlorides).
Identify the monomers used to make

a) nylon-4,6

b) nylon-6,4

c) nylon-6,10. *(6)*

20 Nylon-6 can be made from caprolactam:

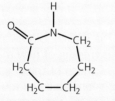

Caprolactam

Draw two repeat units of nylon-6 and explain how it differs from nylon-6,6. *(5)*

Challenge

21 Polymer science is relatively new and was developed during the twentieth century.
The father of polymer science was Hermann Staudinger who, in the 1920s, investigated a series of 'macromolecules'. Colleagues of Staudinger referred to his work as 'grease chemistry' and maintained that the waxy, rubbery substances were impure. Staudinger persisted with his work and developed a range of polymers made from methanal, HCHO. In 1953, he received the Nobel Prize. The diagram below shows a section of a poly(methanal) pioneered by Staudinger.

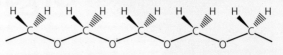

Copy the diagram below and add curly arrows to show the movement of electrons in the formation of poly(methanal).

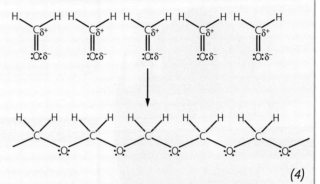

(4)

22 Polylactic acid, PLA, is a polymer of lactic acid, $CH_3CH(OH)COOH$. It is made by first forming a cyclic dimer from lactic acid (2-hydoxypropanoic acid), rather than directly from $CH_3CH(OH)COOH$.

a) Write an equation that would represent the formation of PLA from $CH_3CH(OH)COOH$. Suggest why PLA is not manufactured directly from the monomer $CH_3CH(OH)COOH$.

b) Write a balanced equation for the dimerisation and draw the displayed formula of the cyclic dimer.

c) Write an equation for the formation of PLA from the cyclic dimer. Suggest why PLA is manufactured directly from the cyclic dimer.

d) Draw a section of PLA showing two repeat units. *(11)*

Chapter 12

Organic synthesis

Test yourself on prior knowledge

1 Draw the displayed formula of each of the following and name the functional group(s) present:
   a) pentan-2-one
   b) ethane-1,2-diol
   c) methyl propanoate
   d) 1-propyl ethanamide
   e) $CH_3CHCHCHO$
   f) $HOCH_2(CH)_2CH_2OH$.

Making new materials

One of the main purposes of chemistry is the making of new materials such as pigments, perfumes, drugs and dyes. This making of new materials is called **synthesis**. Synthesis is at the heart of much of the chemical research that goes on today. We depend on synthesis for processed foods, for our fuels, for the clothes we wear and for many of the modern materials we use every day. Synthesis is also important to our understanding of reactions and molecular structure, particularly those of organic molecules.

Many features of modern life depend on the skills of chemists and their ability to synthesise new and complex materials. New colours and fabrics for the fashion industry are synthetic organic molecules. Also there are many compounds synthesised every day for testing in pharmaceutical laboratories as potential drugs. Lightweight, flat-screen computer monitors depend on liquid organic crystals. These organic compounds in the computer screen have been tailor-made by chemists, so that they respond to an electric field and affect light. Sportswear now includes specialist molecules and the chemist is every bit as important as the rest of the sports team.

Figure 12.1

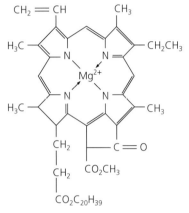

Figure 12.2 The structure of chlorophyll.

The essential job of synthetic organic chemists is to consider the proposed structure for a target molecule and then devise a way of making it from simpler, readily available starting materials. Sounds easy, but it is very complex! For instance, a group of scientists at Harvard University, led by Robert Woodward, investigated the synthesis of chlorophyll (Figure 12.2).

The team of 17 scientists started work in 1956 and 4 years later in 1960 Woodward's team published a paper describing the successful synthesis of chlorophyll.

Organic analysis

After complex molecules, such as chlorophyll, have been synthesised, chemists use a variety of methods to analyse them in order to identify their precise composition and structure.

Traditionally, chemical tests were used to identify functional groups in organic molecules, together with combustion and quantitative analysis. Nowadays, modern laboratories rely on a range of highly sensitive automated and instrumental techniques to identify products of synthesis. These include chromatography, mass spectrometry and various kinds of spectroscopy, including infrared and nuclear magnetic resonance spectroscopy.

Functional groups – the keys to organic molecules

Functional groups provide the key to organic molecules. Knowledge of the properties and reactions of a limited number of functional groups has opened up our understanding of most organic compounds.

Chemists often think of an organic molecule as a relatively unreactive hydrocarbon skeleton with one or more functional groups in place of one or more hydrogen atoms. The functional group in a molecule is responsible for most of its reactions. In contrast, the carbon–carbon bonds and carbon–hydrogen bonds are relatively unreactive, partly because they are strong and have very little polarity. If the skeletal formula is used, the functional groups in cortisone are displayed on the hydrocarbon chain (Figure 12.3).

> **Key term**
>
> A **functional group** is the atom or set of atoms which give a series of organic compounds their characteristic properties and reactions.

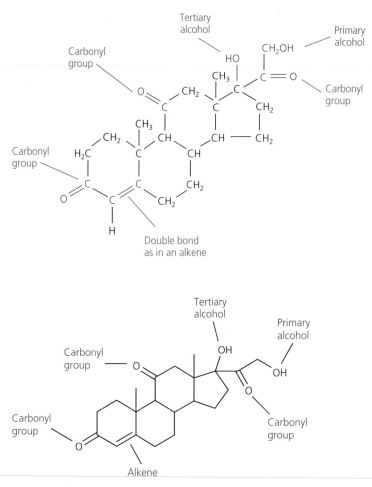

All three carbonyl groups are in fact ketones

Figure 12.3 The structure of the steroid cortisone, labelled to show the reactive functional groups and the hydrocarbon skeleton.

Your knowledge of functional groups should now cover: alkanes, alkenes, alcohols, haloalkanes, aldehydes, ketones, carboxylic acids, esters, acid chlorides, amines and amides. Table 12.1 summarises the reactions of these groups.

Table 12.1 A summary of functional groups and their typical reactions.

Functional group		Type of reactions	Reagents that react
Name	Group		
alkene	C=C (structure)	electrophilic addition	H_2, HBr, Br_2 $H_2O(g)$
alcohol	R — OH	oxidation	$H^+/Cr_2O_7^{2-}$
		esterification	RCOOH (carboxylic acids)
		elimination	H_2SO_4
haloalkane	R — Cl	nucleophilic substitution	common nucleophiles include :OH⁻, :NH₃, :CN⁻
aldehyde	(structure) —C双键O with H	nucleophilic addition	:H⁻/NaBH₄, :CN⁻
		oxidation	$H^+/Cr_2O_7^{2-}$
ketone	(structure) —C双键O	nucleophilic addition	:H⁻/NaBH₄, :CN⁻
carboxylic acid	(structure) —C双键O with OH	acidic reactions	Na, NaOH, Na_2CO_3
		esterification/condensation	alcohols, ROH
ester	(structure) —C双键O with O—	hydrolysis	H^+(aq), OH⁻(aq)
acyl chloride	(structure) —C双键O with Cl	condensation	H_2O, RCOOH, NH_3, RNH_2
amine	R—NH₂ or R—N with H and R	basic reactions	HCl
		nucleophilic reactions	haloalkanes, R–Cl
amide	(structure) —C双键O with NH₂ or —C—N— with O and H	hydrolysis	H^+(aq), OH⁻(aq)
nitrile	—C≡N	hydrolysis	H^+(aq), OH⁻(aq)
		reduction	Na in ethanol (or $LiAlH_4$)
Using the nitrile allows the carbon chain length to be increased and then converted into either a carboxylic acid or an amine			

Test yourself

1 a) Write the empirical, molecular, structural, displayed and skeletal formulae of the hydrocarbon in Figure 12.4.

b) What is the name of this compound?

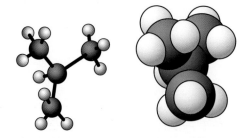

Figure 12.4 Ball-and-stick and space filling models of a hydrocarbon.

2 Anaerobic respiration in muscle cells breaks down glucose to simpler compounds including the following two molecules. Identify the functional groups in these molecules.

a) $CH_2OH-CHOH-CHO$

b) $CH_3-CO-COOH$

3 Pheromones are messenger molecules produced by insects to attract mates or to give an alarm signal. Identify the functional groups in the pheromone below produced by queen bees.

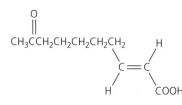

4 The molecule below, called carvone, is responsible for the taste of spearmint.

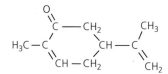

Use your knowledge of the functional groups to predict six reactions of carvone. Identify the organic product in each reaction.

Extending carbon chain length

When synthesising new compounds, it is often necessary to extend the carbon chain. This can be achieved in a number of ways but usually involves the formation of a nitrile. Nitriles can be formed by reacting cyanide ions, CN^-, with either haloalkanes or with carbonyl compounds.

The reaction can be achieved by refluxing a solution of the haloalkane and potassium cyanide in ethanol. The reaction proceeds via a nucleophilic substitution mechanism and the nucleophile is the cyanide anion, $C\equiv N^-$.

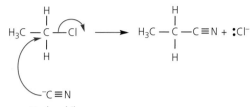

Nucleophile

The reaction between HCN and a carbonyl is covered on page 171 and is a nucleophilic addition reaction.

Reaction of nitriles

Nitriles can readily be converted into amines or carboxylic acids.

Nitriles can be reduced to form an amine:

$$CH_3C\equiv N + 4[H] \rightarrow CH_3CH_2NH_2$$

The reduction can be achieved by using hydrogen gas and a nickel catalyst at high temperature.

Nitriles can also be hydrolysed, via the amide, to form a carboxylic acid. The hydrolysis requires an acidic, H^+ condition.

The reaction between HCN and a carbonyl is covered on page 171

> **Tip**
>
> The net reaction for hydrolysis of a nitrile can be written as:
> $R-CN + 2H_2O + H^+ \rightarrow R-COOH + NH_4^+$

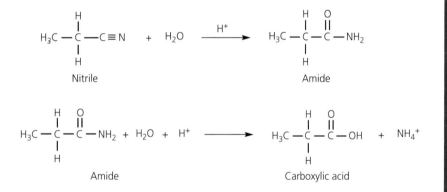

Nitrile   Amide

Amide   Carboxylic acid

Organic routes

Organic chemists synthesise new molecules using their knowledge of functional groups, reaction mechanisms and molecular shapes – as well as the factors which control the rate and extent of chemical change. They often start by examining the 'target molecule'. Then, they work backwards through a series of steps to find suitable starting chemicals that are readily available and cheap enough. In recent years, chemists have developed computer programs to help with the process of working back from the target molecule to a range of possible starting molecules. A simple example is shown below.

In this case, the 'target molecule' is butanoic acid and the starting molecule is 1-bromobutane.

Step 1 – Start with the target molecule and identify the compounds that could readily be converted directly into the target. Concentrate on the functional group.

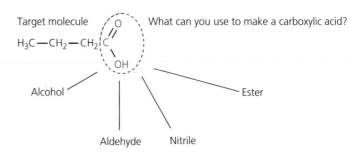

Target molecule — What can you use to make a carboxylic acid?

Alcohol — Ester — Aldehyde — Nitrile

Step 2 – Look at your starting molecule, 1-bromobutane. What relevant reactions of haloalkanes do you know?

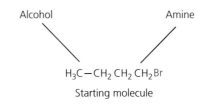

Alcohol   Amine

$H_3C-CH_2\ CH_2\ CH_2\ Br$

Starting molecule

You should now see a possible two-stage synthetic route from your starting molecule to the target molecule. In this case, the route can go via the alcohol.

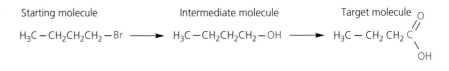

Starting molecule — Intermediate molecule — Target molecule

$H_3C-CH_2CH_2CH_2-Br \longrightarrow H_3C-CH_2CH_2CH_2-OH \longrightarrow H_3C-CH_2\ CH_2\ C$

Chemists normally seek a synthetic route that has the least number of stages which, therefore, produces an optimal yield of the product. It is rare for any one reaction to be 100% efficient – normally the percentage yield is significantly below the theoretical yield.

Changing the functional groups

All reactions in organic chemistry convert one compound to another, but there are some reactions which are particularly useful for developing synthetic routes. These useful reactions are shown in Table 12.2.

Table 12.2 Useful synthetic routes.

Starting functional group → Reagent used → Target functional group

Functional group	Reagent	Target functional group
alkene	hydrogen halides	haloalkanes
	steam	alcohol
	hydrogen	alkanes
haloalkane	NaOH(aq)	alcohol
	NH_3(ethanol)	amine
	cyanide, $^-C\equiv N$	nitrile
alcohols	carboxylic acids	esters
	$H^+/Cr_2O_7^{2-}$	aldehyde, ketone or carboxylic acid
	hot concentrated H_2SO_4	alkene
aldehyde/ketone	$NaBH_4$	alcohols
	$^-C\equiv N$	hydroxynitriles
carboxylic acids	alcohols	esters
acyl chlorides	H_2O	carboxylic acids
	alcohols	esters
	carboxylic acids	acid anhydrides
	ammonia	amides
acid nitriles	H_2O/ H^+(aq)	carboxylic acid
	Na in ethanol or $LiAlH_4$	amine

Before starting any investigation into a two-stage synthesis, always examine the starting and the target molecules to check that they contain the same number of carbon atoms. If the number of carbon atoms increases, at some point in the synthesis cyanide ions, $^-C\equiv N$, have to be used. Alternatively, another organic reagent is needed to create the desired number of carbon atoms. To prepare an ester would require two organic reagents.

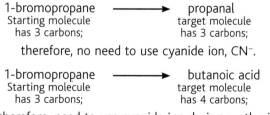

1-bromopropane ⟶ propanal
Starting molecule has 3 carbons; target molecule has 3 carbons;
therefore, no need to use cyanide ion, CN⁻.

1-bromopropane ⟶ butanoic acid
Starting molecule has 3 carbons; target molecule has 4 carbons;
therefore, need to use cyanide ion during synthesis.

Converting one functional group to another

Make a copy of the flow chart in Figure 12.5. Beside each arrow, write the reagents and conditions needed for the conversion. You may need to refer to previous chapters to do this.

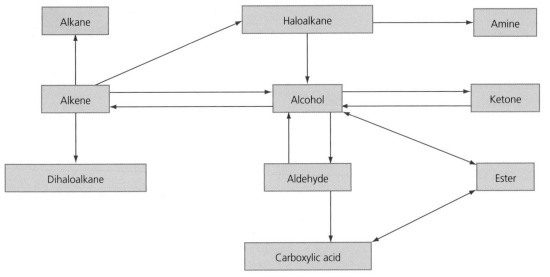

Figure 12.5 A flow diagram summarising the methods for converting one functional group to another.

Using your completed copy of Figure 12.5, suggest two-stage syntheses showing the reagents and conditions for each of the following conversions:

1 ethene to ethanoic acid
2 butan-1-ol to butan-2-ol
3 ethanol to ethyl ethanoate (using ethanol as the only carbon compound)
4 propan-2-ol to propane.

Example 1

Each of the following examples involves a two-stage conversion (or synthesis).

In each example:

● check the number of carbon atoms in the target and the starting molecules

● start with the target molecule and identify the functional groups that can be used to make the target functional group

● look at the starting molecule and identify the functional groups that can be made from the starting functional group.

benzene →   ?   → phenylamine

Number of carbon atoms

Both the target and the starting molecule have the same number of carbon atoms.

Target molecule

The only reaction you know for making $C_6H_5NH_2$ is the reduction of $C_6H_5NO_2$.

Starting molecule

C_6H_6 reacts by electrophilic substitution and can be nitrated, chlorinated, alkylated or acylated to give $C_6H_5NO_2$, C_6H_5Cl, $C_6H_5CH_3$ and $C_6H_5COCH_3$ respectively.

Clearly, the connecting intermediate molecule is $C_6H_5NO_2$, such that the two-step synthesis is:

benzene → nitrobenzene → phenylamine

Step 1

benzene → nitrobenzene

- reagents: concentrated HNO_3 and concentrated H_2SO_4
- conditions: temperature approximately 55 – 60 °C
- equation:

$C_6H_6 + HNO_3 \rightarrow C_6H_5NO_2 + H_2O$

- type of reaction: electrophilic substitution.

Step 2

nitrobenzene → phenylamine

- reagents: Sn and concentrated HCl
- conditions: heat (under reflux)
- equation:

$C_6H_5NO_2 + 6[H] \rightarrow C_6H_5NH_2 + 2H_2O$

- type of reaction: reduction

Example 2

1-chloropropane $\rightarrow$? $\rightarrow$ propanoic acid

Number of carbon atoms

Both the target and the starting molecule have the same number of carbon atoms.

Target molecule

The carboxylic acid functional group can be made from a primary alcohol, aldehyde, ester, acid chloride or a nitrile.

Starting molecule

The haloalkane functional group can be used to form a primary alcohol, amine or nitrile.

Clearly, the connecting intermediate molecule is a primary alcohol, such that the two-step synthesis is:

1-chloropropane $\rightarrow$ propan-1-ol $\rightarrow$ propanoic acid

Step 1

1-chloropropane $\rightarrow$ propan-1-ol

- reagent: NaOH(aq)
- conditions: warm
- equation:

$CH_3CH_2CH_2Cl + NaOH \rightarrow CH_3CH_2CH_2OH + NaCl$

- type of reaction: nucleophilic substitution

Step 2

propan-1-ol $\rightarrow$ propanoic acid

- reagents: acidified dichromate, $H^+/Cr_2O_7^{2-}$
- conditions: heat (under reflux), with excess $H^+/Cr_2O_7^{2-}$
- equation:

$CH_3CH_2CH_2OH + 2[O] \rightarrow CH_3CH_2COOH + H_2O$

- type of reaction: oxidation

Example 3

1-chloromethane →   ?   → ethanoic acid

Number of carbon atoms

The target molecule has two carbon atoms, but the starting molecule only has one carbon atom. The carbon chain can be increased by using $^-C\equiv N$.

Target molecule

The carboxylic acid functional group can be made from a primary alcohol, aldehyde, ester, acid chloride or a nitrile.

Starting molecule

The haloalkane functional group can be used to form a primary alcohol, amine or nitrile.

Clearly, the connecting intermediate molecule is a nitrile such that the two-step synthesis is:

1-chloromethane → ethanenitrile → ethanoic acid

Step 1

1-chloromethane → ethanenitrile

- reagent: KCN(alcohol)
- conditions: heat under reflux
- equation:

$CH_3Cl + KCN \rightarrow CH_3CN + KCl$

- type of reaction: nucleophilic substitution

Step 2

ethanenitrile → ethanoic acid

- reagents: H_2O, H^+(aq)
- conditions: heat (under reflux) with a dilute mineral acid such as H_2SO_4(aq)
- equation:

$CH_3CN + 2H_2O + H^+ \rightarrow CH_3COOH + NH_4^+$

- type of reaction: hydrolysis

The carboxylic acid formed in the second step of this synthesis is formed via an acid amide.

$H_3C-C\equiv N$ $\xrightarrow[\text{adds to form an amide}]{\text{H}_2\text{O / H}^+\text{(aq)}}$ $H_3C-\overset{\overset{\textstyle O}{\|}}{C}-NH_2$

hydrolysis occurs and water adds to form an amide

The amide is then hydrolysed to form the carboxylic acid and an ammonium salt

$H_3C-\overset{\overset{\textstyle O}{\|}}{C}-NH_2$ $\xrightarrow[\text{adds to form a carboxylic acid}]{\text{H}_2\text{O / H}^+\text{(aq)}}$ $H_3C-\overset{\overset{\textstyle O}{\|}}{C}-OH + NH_4^+$

hydrolysis occurs again and water adds to form a carboxylic acid

5 Draw the structural formula of the main organic product in each of the following reactions.
Classify each reaction as addition, substitution or elimination and classify the reagent attacking the organic reactant as a free radical, nucleophile, electrophile or base.

a) $CH_2 = CH_2(g) \xrightarrow{HBr(g)}$

b) $CH_3CH_2CH_2Br(l) \xrightarrow{KOH(aq)}$

c) $CH_3CHOHCH_3(l) \xrightarrow{H_3PO_4(l) \text{ catalyst and heat}}$

d) $C_6H_6 \xrightarrow{\text{conc. } HNO_3 \text{ and conc. } H_2SO_4}$

6 Give the reagents and conditions for converting:

a) butanone to butane in three stages

b) ethanol to ethane-1,2-diol in three stages.

7 a) Identify substances A, B and C in the flow diagram below.

$$CH_3CH_2OH \xrightarrow[\text{then heat}]{\text{Excess } A} \boxed{B} \xrightarrow{HCl(g)} CH_3CH_2Cl \xrightarrow[\text{in ethanol}]{\text{conc. } NH_3} \boxed{C}$$

b) Identify the reagents and the conditions, if any, in each stage of the reaction scheme shown below:

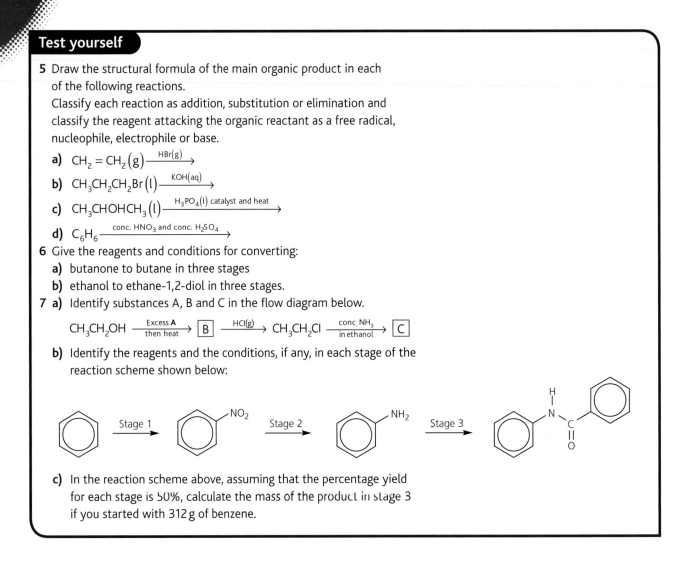

c) In the reaction scheme above, assuming that the percentage yield for each stage is 50%, calculate the mass of the product in stage 3 if you started with 312 g of benzene.

Practical skills – synthetic techniques

Chemists have developed a range of practical techniques for the synthesis of solid and liquid organic compounds. These methods allow for the fact that reactions involving molecules with covalent bonds are often slow and that it is difficult to avoid side reactions which give by-products. There are several stages in the preparation of an organic compound which are summarised in Figure 12.6.

In stage 2, when carrying out the reaction, one of the more common techniques is to heat the reaction mixture in a flask fitted with a reflux condenser (Figure 12.7).

In stage 3, when separating the product from the reaction mixture, solids are removed by filtration using a Buchner or Hirsch funnel with suction from a water pump (Figure 12.8).

Liquids can often be separated by simple distillation, fractional distillation or steam distillation (Figure 12.9).

Distillation with steam at 100 °C allows the separation of compounds which decompose if heated near their boiling points. The technique only works with compounds that do not mix with water. When used to separate the products of organic preparations, steam distillation leaves behind those reagents and products which are soluble in water.

In stage 4 the 'crude' product is usually contaminated with unreacted reagents or with by-products. The method of purifying this 'crude' product depends on whether it is a solid or a liquid.

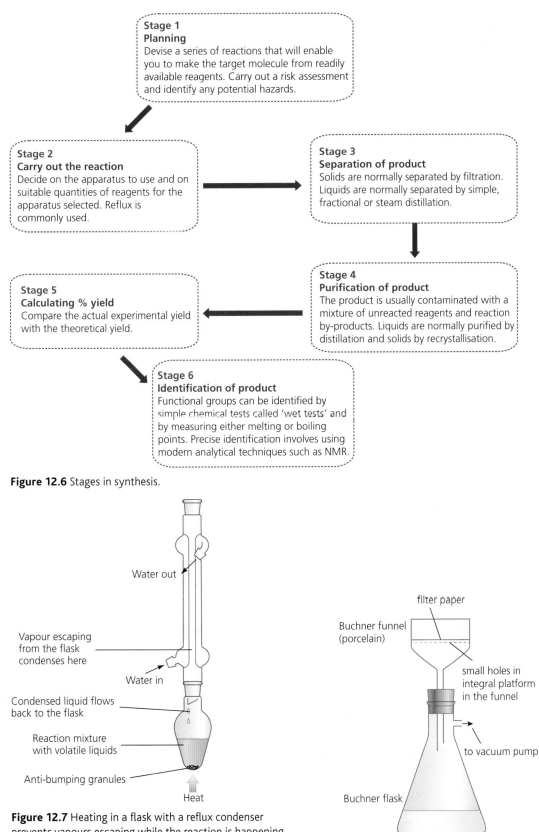

Stage 1
Planning
Devise a series of reactions that will enable you to make the target molecule from readily available reagents. Carry out a risk assessment and identify any potential hazards.

Stage 2
Carry out the reaction
Decide on the apparatus to use and on suitable quantities of reagents for the apparatus selected. Reflux is commonly used.

Stage 3
Separation of product
Solids are normally separated by filtration. Liquids are normally separated by simple, fractional or steam distillation.

Stage 4
Purification of product
The product is usually contaminated with a mixture of unreacted reagents and reaction by-products. Liquids are normally purified by distillation and solids by recrystallisation.

Stage 5
Calculating % yield
Compare the actual experimental yield with the theoretical yield.

Stage 6
Identification of product
Functional groups can be identified by simple chemical tests called 'wet tests' and by measuring either melting or boiling points. Precise identification involves using modern analytical techniques such as NMR.

Figure 12.6 Stages in synthesis.

Water out

Vapour escaping from the flask condenses here

Water in

Condensed liquid flows back to the flask

Reaction mixture with volatile liquids

Anti-bumping granules

Heat

Figure 12.7 Heating in a flask with a reflux condenser prevents vapours escaping while the reaction is happening. Vapours from the reaction mixture condense and flow back (reflux) into the flask.

filter paper

Buchner funnel (porcelain)

small holes in integral platform in the funnel

to vacuum pump

Buchner flask

Figure 12.8 Buchner filtration for product purification.

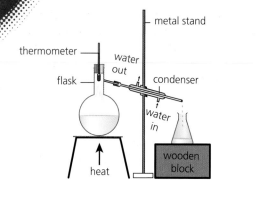

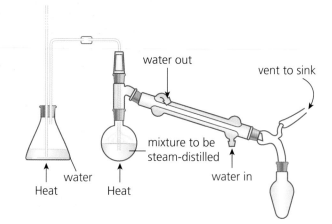

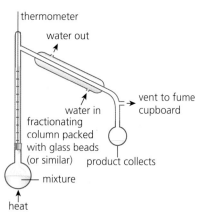

Figure 12.9 Simple distillation, fractional distillation and steam distillation.

Solids are normally purified by recrystallisation. Recrystallisation requires use of a solvent that dissolves the product when hot, but not when cold. The choice of solvent is usually made by trial and error.

Recrystallisation can be broken down into several steps:

1 Dissolve the impure solid in the minimum volume of hot solvent.

2 If the hot solution is not clear, filter the hot mixture through a heated funnel to remove insoluble impurities, leaving the product dissolved in the hot solution.

3 Cool the filtrate so that the product recrystallises, leaving the smaller amounts of soluble impurities in solution.

4 Filter the cold solution to recover the purified product.

5 Wash the purified solid with small amounts of pure cold solvent, to wash away any solution containing impurities.

6 Allow the solvent to evaporate from the purified solid in the air.

Purifying organic liquids

Chemists often begin to purify organic liquids which are insoluble in water by shaking with aqueous reagents in a separating funnel to extract impurities. This is followed by washing with pure water, drying and finally fractional distillation.

Fractional distillation separates mixtures of liquids with different boiling points. On a laboratory scale, the process takes place in a distillation apparatus which has been fitted with a fractionating column between the flask and the still-head (Figure 12.9). Separation is improved if the column is packed with inert glass beads or rings to increase the surface area, so that the rising vapour can mix with the condensed liquid that runs back to the flask. The column is hotter at the bottom and cooler at the top. The thermometer reads the boiling point of the compound passing over into the condenser.

If the flask contains a mixture of liquids, the boiling liquid in the flask produces a vapour which is richer in the most volatile of the liquids present – the one with the lowest boiling point. Most of the vapour condenses in the column and runs back. As it does so, it meets more of the rising vapour. Some of the vapour condenses and some of the liquid evaporates. In this way, the mixture evaporates and condenses repeatedly as it rises up the column. But every time it does so, the vapour becomes richer in the most volatile liquid present. At the top of the column, the vapour contains 100% of the most volatile liquid. So, during fractional distillation, the most volatile liquid with the lowest boiling point distils over first, then the liquid with the next lowest boiling point and so on.

Percentage yield

The percentage yield can be obtained by comparing the actual yield with the yield expected. This was discussed fully in Chapter 12, page 196, of the Year 1 book.

Identifying the product and checking its purity

Qualitative tests

Functional groups can be identified by simple qualitative tests. These 'wet tests' are summarised in Table 12.3.

Table 12.3 'Wet tests'.

Test	Observation	Conclusions
pH of solution (add litmus)	red	carboxylic acid or phenol
Br_2	decolourises white ppt	alkene phenol or phenylamine
Na	gas (H_2) given off, bubbles, fizzes	carboxylic acid or phenol or alcohol
Na_2CO_3	gas (CO_2) given off, bubbles, fizzes	carboxylic acid
$Ag(NH_3)_2^+$ (aq) in water bath at about 60 °C	white ppt cream ppt yellow ppt	chloroalkane bromoalkane iodoalkane
2,4 DNPH	orange ppt	aldehyde or ketone
Tollens' reagent $Ag^+(NH_3)_2$	silver mirror	aldehyde
Heat with $H^+/Cr_2O_7^{2-}$	orange to green	primary alcohol, secondary alcohol or aldehyde
water	white fumes of HCl	acyl chloride
warm with NaOH	NH_3 gas evolved which turns litmus blue	amide

There is no simple test for an ester. Esters are detected by first eliminating all other functional groups. Smell is not accepted as a chemical test.

ppt = precipitate

Measuring melting points and boiling points

Pure solids have sharp melting points. Since databases now include the melting points of all known compounds, it is possible to check the identity and purity of a product by checking its melting point. If the solid is impure, the melting point will be lower than expected. In addition, it will not have a sharp melting point, but will soften and melt over a temperature range.

Boiling points can be used to check the purity and identity of liquids. If a liquid is pure, it should all distil over a narrow range, at the expected boiling point. The boiling point can be measured as the liquid distils over during fractional distillation. Impurities increase the boiling point and the range over which the liquid boils.

Chromatography and spectroscopy

The use of chromatography and spectroscopy in identifying products and checking purity is covered in Chapter 13.

Test yourself

8 A possible two-stage synthesis of 1,2-diaminoethane first converts an alkene to a dihaloalkane, and then reacts this with ammonia.
   a) Write out a reaction scheme for the synthesis giving reagents and conditions.
   b) Calculate the mass of the alkene needed to make 2 g of the 1,2-diaminoethane assuming a 60% yield in stage 1 and a 40% yield in stage 2.
       Which chemicals should be in excess?
   c) What hazards does the synthesis pose and what safety precautions should be taken?

9 A sample of benzoic acid was contaminated with a mixture of potassium dichromate and carbon. Use the information below to explain how you could obtain a pure sample of benzoic acid.

Chemical	Appearance	Solubility in cold water	Solubility in hot water
benzoic acid	white crystalline solid	insoluble	soluble
potassium dichromate	orange crystalline solid	soluble	soluble
carbon	black solid	insoluble	insoluble

10 13.14 g of ethanol is heated, under distillation, with excess acidified potassium dichromate and 9.43 g of ethanal was obtained.
   a) What is the percentage yield of ethanal?
   b) Suggest why the percentage yield was significantly below the theoretical yield.
   c) Suggest the identity of any likely impurities.
   d) State a simple chemical test that would confirm the presence of the impurities.

11 Benzene can be converted into phenylamine via the two-stage synthesis shown below.

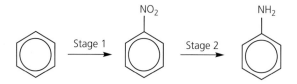

18 g of benzene was first converted into 22 g of nitrobenzene and then to 12 g of phenylamine.
   a) State the reagents and conditions for each stage.
   b) Calculate the percentage yield for each stage.
   c) What is the overall percentage yield?

Stereochemical synthesis

When organic compounds are prepared synthetically in the laboratory, the product is sometimes a mixture of two optical isomers. This happens fairly frequently in the reaction of carbonyl compounds with hydride ions, :H$^-$ (sodium tetrahydridoborate(III), NaBH$_4$) and with cyanide ions, $^-$CN, in potassium cyanide. Both of these reactions involve nucleophilic addition. The initial product in each case is an intermediate anion, which is converted to the final product containing an $-$OH group by adding dilute sulfuric acid. The final product contains a central chiral carbon atom.

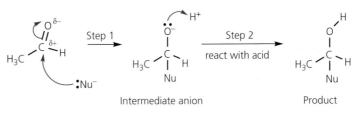

Intermediate anion   Product

The bond angle around the carbon atom in the carbonyl group is approximately 120° and the shape around the carbon is trigonal planar. In a reaction with ethanal, the nucleophile can attack from either side of the plane.

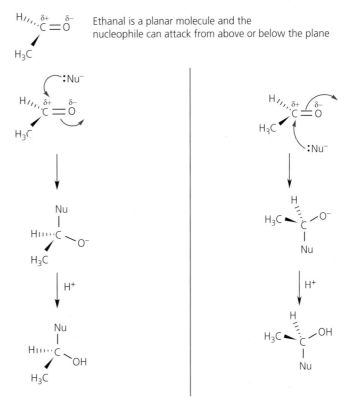

Ethanal is a planar molecule and the nucleophile can attack from above or below the plane

The central C is now joined to four different atoms or groups and the two organic products are mirror images of each other

There are now four different groups attached to the carbon atom. This means that the final product will be a 1:1 mixture of two **optical isomers**. Mixtures of this kind are called **racemic mixtures**.

The reaction which we have just considered is used as the first stage in the two-stage laboratory synthesis of lactic acid (2-hydroxypropanoic acid) from ethanal. In the second stage, the product formed from the nucleophilic addition, CH$_3$CH(OH)CN, is refluxed with the dilute acid and this converts the $-$CN group to a carboxylic acid group, $-$COOH, forming CH$_3$CH(OH)COOH.

Tip

You should appreciate the importance of nitriles: nucleophilic reactions using $^-$CN increase the carbon chain length. The nitrile is readily hydrolysed by refluxing with aqueous sulfuric acid.

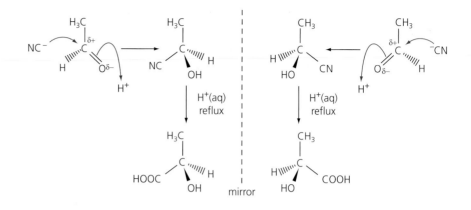

Formation of a carboxylic acid by hydrolysis of the nitrile via an amide

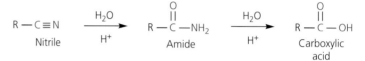

$$R-C\equiv N \xrightarrow[H^+]{H_2O} R-\overset{\overset{\displaystyle O}{\|}}{C}-NH_2 \xrightarrow[H^+]{H_2O} R-\overset{\overset{\displaystyle O}{\|}}{C}-OH$$

Nitrile   Amide   Carboxylic acid

Nitriles are also readily reduced to form the amine.

$$R-C\equiv N \ + \ 4[H] \xrightarrow[\text{agent}]{\text{reducing}} R-CH_2-NH_2$$

Nitrile   Amine

LiAlH$_4$ or Na dissolved in ethanol can be used as the reducing agent

The laboratory preparation of lactic acid always produces a 50:50 mixture of both optical isomers of lactic acid. By contrast, the lactic acid that is formed in our muscles during excessive activity is the single (+) isomer. The reason for this is that the lactic acid formed in our muscles is produced naturally through processes which are catalysed by enzymes.

In fact, most optically active compounds in living systems are present as just one optical isomer. Almost all the chemical reactions in living organisms are controlled by enzymes, which are natural catalysts consisting of proteins. When enzymes act as catalysts, the reactions take place at an active site somewhere on the enzyme molecule. The shape of the active site and the functional groups around it are very stereospecific and give the enzyme its catalytic properties. The stereospecific nature of the active site means that enzymes can usually accommodate only one particular molecule at their active site and this molecule is often only one of the optical isomers.

Test yourself

12 The compounds shown below are chemicals that have pharmaceutical applications.
Copy the molecules and identify the chiral centres by marking with an asterisk, *.

Salbutamol
(used to treat asthma)

Levodopa (one of the optical isomers is
used to treat Parkinson's disease)

Chloramphenicol (an optical isomer is the active
antibiotic used to treat typhoid)

Practice questions

Multiple choice questions 1–10

1 An organic compound X is found to react as follows:
- when refluxed with $H^+/Cr_2O_7^{2-}$ it produces a compound that forms a precipitate when 2,4-dinitrophenylhydrazine is added
- with hot concentrated H_2SO_4 it produces a compound that decolourises Br_2.

Compound X is:

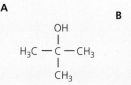

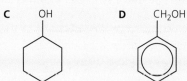

(1)

2 The functional groups present in this molecule are:

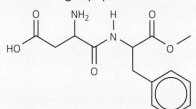

- **A** carboxylic acid, arene, ketone, amine and ester
- **B** carboxylic acid, arene, ketone, amide and ester
- **C** carboxylic acid, arene, amide, amine and ester
- **D** alcohol, arene, amine, amide, ester. *(1)*

3 An organic compound reacts vigorously with water and immediately produces a white precipitate when $AgNO_3(aq)$ is added to the mixture. The compound could be

A CH_3CH_2COCl

B $ClCH_2CH_2COOH$

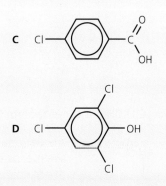

(1)

4 0.01 mol of an organic compound reacts with exactly 200 cm³ of 0.1 mol dm⁻³ NaOH(aq). It also decolourises Br_2. The organic compound could be:
- **A** $CH_2CHCOOH$
- **B** $HOOC-C_6H_4-CHCH_2$
- **C** $HOOCCH_2COOH$
- **D** HOOCCHCHCOOH *(1)*

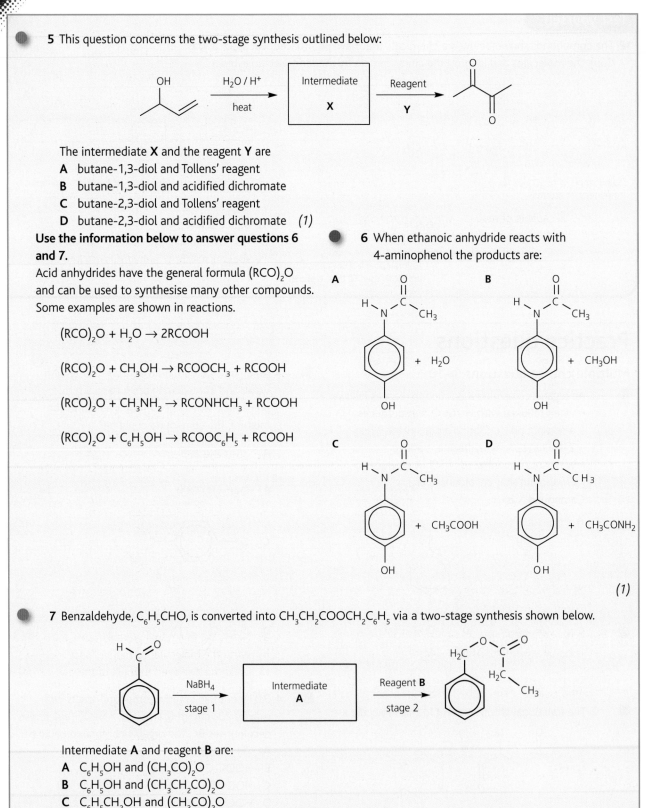

5 This question concerns the two-stage synthesis outlined below:

OH / H₂O / H⁺ heat → Intermediate **X** → Reagent **Y** →

The intermediate **X** and the reagent **Y** are
A butane-1,3-diol and Tollens' reagent
B butane-1,3-diol and acidified dichromate
C butane-2,3-diol and Tollens' reagent
D butane-2,3-diol and acidified dichromate *(1)*

Use the information below to answer questions 6 and 7.

Acid anhydrides have the general formula $(RCO)_2O$ and can be used to synthesise many other compounds. Some examples are shown in reactions.

$$(RCO)_2O + H_2O \rightarrow 2RCOOH$$

$$(RCO)_2O + CH_3OH \rightarrow RCOOCH_3 + RCOOH$$

$$(RCO)_2O + CH_3NH_2 \rightarrow RCONHCH_3 + RCOOH$$

$$(RCO)_2O + C_6H_5OH \rightarrow RCOOC_6H_5 + RCOOH$$

6 When ethanoic anhydride reacts with 4-aminophenol the products are:

A + H_2O

B + CH_3OH

C + CH_3COOH

D + CH_3CONH_2

(1)

7 Benzaldehyde, C_6H_5CHO, is converted into $CH_3CH_2COOCH_2C_6H_5$ via a two-stage synthesis shown below.

NaBH₄ / stage 1 → Intermediate **A** → Reagent **B** / stage 2 →

Intermediate **A** and reagent **B** are:
A C_6H_5OH and $(CH_3CO)_2O$
B C_6H_5OH and $(CH_3CH_2CO)_2O$
C $C_6H_5CH_2OH$ and $(CH_3CO)_2O$
D $C_6H_5CH_2OH$ and $(CH_3CH_2CO)_2O$ *(1)*

Use the key below to answer questions 8–10.

A	B	C	D
1, 2 & 3 correct	1 & 2 correct	2 & 3 correct	1 only correct

8 This molecule

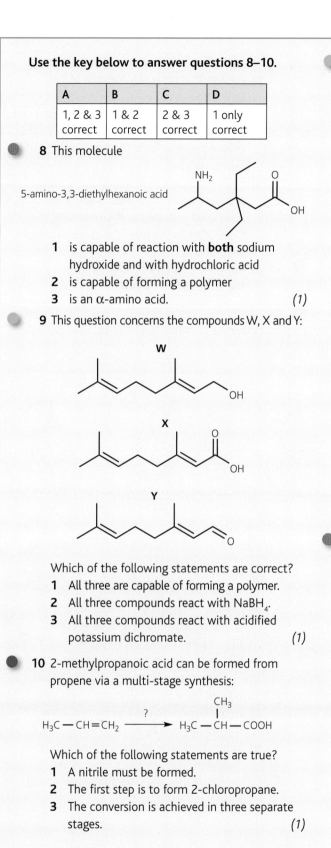

5-amino-3,3-diethylhexanoic acid

1 is capable of reaction with **both** sodium hydroxide and with hydrochloric acid
2 is capable of forming a polymer
3 is an α-amino acid. *(1)*

9 This question concerns the compounds W, X and Y:

W

X

Y

Which of the following statements are correct?
1 All three are capable of forming a polymer.
2 All three compounds react with NaBH₄.
3 All three compounds react with acidified potassium dichromate. *(1)*

10 2-methylpropanoic acid can be formed from propene via a multi-stage synthesis:

$$H_3C - CH = CH_2 \xrightarrow{?} H_3C - \underset{\underset{CH_3}{|}}{CH} - COOH$$

Which of the following statements are true?
1 A nitrile must be formed.
2 The first step is to form 2-chloropropane.
3 The conversion is achieved in three separate stages. *(1)*

11 The following steps are taken from one method of synthesising methyl ethanoate:

$$CH_3COOH + CH_3OH \rightleftharpoons CH_3COOCH_3 + H_2O$$

A Heat methanol and ethanoic acid under reflux for about 45 minutes with a little concentrated sulfuric acid.
B Then distil the reaction mixture, collecting all the liquid which distils below 65 °C.
C Shake the distillate with aqueous sodium carbonate solution and then discard the aqueous layer.
D Add two spatula measures of anhydrous sodium sulfate or anhydrous calcium chloride to the organic product.
E Finally, redistill the organic product collecting the liquid which boils between 57 °C and 64 °C.

Chemical	CH₃COOH	CH₃OH	CH₃COOCH₃	H₂O
Boiling point/°C	118	65	57	100

a) Draw a diagram of the apparatus for heating under reflux in step A.
b) Suggest why the sulfuric acid used in step A was concentrated and not dilute.
c) State the reasons for each of the procedures in steps B to E. *(12)*

12 Salicylic acid has been used as a painkiller. Its displayed formula is shown below.

a) Name the functional groups in salicylic acid.
b) Write the molecular formula of salicylic acid.
c) Draw the displayed formula of the organic product that forms when salicylic acid:
   i) is warmed with aqueous sodium hydroxide
   ii) reacts with bromine water
   iii) is heated under reflux with ethanol and concentrated sulfuric acid
   iv) is heated under reflux with ethane-1,2-diol and concentrated sulfuric acid. *(9)*

13 Each of the following conversions involves a multi-stage synthesis. In each case, explain how the synthesis could be achieved. Write equations and state the conditions (if any) for each stage in the synthesis.

   a) propene → propanone
   b) 3-chloropropan-1-ol → 3-hydroxypropene
   c) phenylethanone → poly(phenylethene)   *(16)*

14 During the late 1950s and early 1960s, thalidomide, shown below, was given to pregnant women to help combat morning sickness.

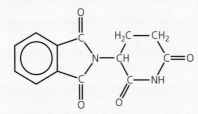

   a) What is the molecular formula of thalidomide?
   b) Draw the organic product when thalidomide is treated with:
   i) an excess of hot NaBH$_4$(aq)
   ii) HCl.
   c) Copy the structure of thalidomide and include an asterisk, *, beside the chiral carbon.   *(6)*

15 Compound **X**, C$_6$H$_5$(CH)$_3$(CH$_3$)CHO exhibits stereoisomerism.
   a) Define stereoisomerism.
   b) State the essential structural/stereochemical features for:
   i) *E/Z* isomerism
   ii) optical isomerism.
   c) Draw the displayed formula of compound **X**.
   d) Circle and label structural/stereochemical features responsible for *E/Z* isomerism.
   e) Mark with an asterisk, *, the structural/ stereochemical features responsible for optical isomerism.

16 Salbutamol is used in inhalers to relieve the symptoms of asthma. Its displayed formula is shown below.

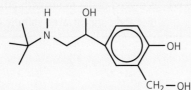

   a) Name the functional groups in salbutamol.
   b) Deduce the molecular formula of salbutamol.
   c) Draw a displayed formula of the organic products form when salbutamol is refluxed with acidified potassium dichromate(VI).
   d) Draw a displayed formula of the organic products which form when salbutamol is refluxed with excess ethanoic acid, CH$_3$COOH, in the presence of concentrated sulfuric acid.
   e) Copy the structure of salbutamol and identify the chiral centre by using an asterisk, *.

17 Aspirin and paracetamol and ibuprofen, shown below, are common 'over-the-counter' painkillers.

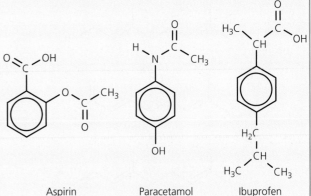

Aspirin   Paracetamol   Ibuprofen

   a) Name the functional groups, other than arenes, in aspirin, in paracetamol and in ibuprofen.
   b) Identify a reagent that would react with:
   i) both aspirin and with paracetamol
   ii) aspirin but not with paracetamol
   iii) both aspirin and ibuprofen but not with paracetamol
   iv) all three.
   c) Draw the organic product formed in each of the four parts in **b)**.
   d) Suggest why the laboratory synthesis of ibuprofen might be more difficult and expensive than the laboratory synthesis of aspirin and paracetamol.

18 Devise a multi-stage synthesis for each of the following conversions. For each synthesis write an equation, state the reagents and conditions.
   a) 1-bromopropane → propanoic acid
   b) propene → 2-propylethanoate
   c) ethanal → ethylethanoate
   (ethanal is the only organic chemical you are allowed to use in this two-stage synthesis – you can use other common laboratory chemicals, but no other organic chemicals). *(19)*

Challenge

19 The skeletal formula of compound X is shown below. X is a constituent of jasmine oil and it is partly responsible for the taste and smell of black tea.

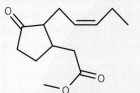

   a) What is the molecular formula of X?
   b) Name the functional groups in X.
   c) Compound X is a stereoisomer.
      i) Draw the structure of X. Label each stereochemical component using a circle to show any E/Z isomerism and an asterisk to identify the chiral centre.
      ii) How many stereoisomers are there with the skeletal formula shown? Explain your answer.
   d) Design a possible two-stage synthesis of compound Y, shown below, starting with compound X.

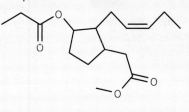

(14)

20 Carbonyl compounds that contain at least one hydrogen on the carbon atom adjacent to the carbonyl group (CHC=O) can, in the presence of a base such as NaOH, undergo a condensation reaction, for example:

propanone → 4-hydroxy-4-methylpentan-2-one

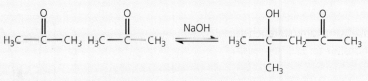

4-hydroxy-4-methylpentan-2-one

Use this information and your knowledge of the reactions covered in the specification to suggest how but-2-enal, $CH_3CHCHCHO$, could be prepared from ethanal. *(5)*

Chapter 13

Analysis

Prior knowledge

In this chapter it is assumed that you are familiar with simple analysis using:
- empirical and molecular formula calculations
- mass spectrometry
- infrared spectroscopy.

You should be familiar with definitions of both empirical formula and molecular formula, and be able to calculate both from given data and using a mass spectrum. You should be able to use the characteristic infrared absorptions given on the data sheet to identify key functional groups, including alcohols, aldehydes and ketones, as well as carboxylic acids.

Test yourself on prior knowledge

1 Deduce the empirical formula of each of the following:
   a) Compound A is $C_4H_8O_2$
   b) Compound B is C_4H_9OH
   c) Compound C is $C_6H_5COOCH_2CH_3$.
2 Find the empirical formula and the molecular formula of a compound that has:
   C = 59.23%, H = 11.18%, O = 29.59% , M_r = 162
3 Cumene is a hydrocarbon, a compound containing only carbon and hydrogen. It has 89.94% carbon, and has a molar mass of 120 g mol^{-1}. Find its empirical and molecular formula.
4 Compound E has C = 62.1%, H = 10.3%, O = 27.6%. The mass spectrum and the infrared spectrum of compound E are shown below. Identify compound E. Show your working.

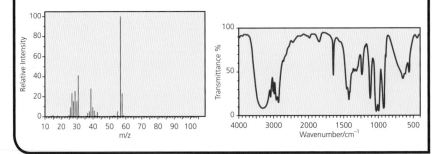

Chromatography

You will be familiar with chromatography from studying science at GCSE and you may have used paper chromatography to separate out the components of black ink (Figure 13.1).

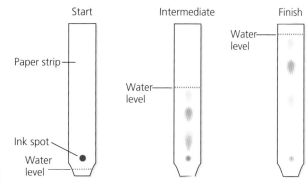

Figure 13.1 Chromatographic separation of black ink.

Black is a mixture of several coloured components, as can be seen from the colour chart (Figure 13.2).

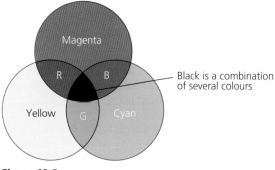

Figure 13.2

Chromatography involves the small-scale separation of components within a mixture. All types of chromatography contain a stationary phase and a mobile phase.

Chemicals in the mixture separate for one or more of the following reasons.

- they differ in the extent to which they are soluble in the mobile phase
- if the stationary phase is a liquid (on a solid support), separation depends on the relative solubility in the stationary phase and in the mobile phase
- in the case of a solid support, they adsorb (stick) to the stationary phase by a different amount.

Chemicals in the mixture separate because they differ in the extent to which they mix with the mobile phase (due to their relative solubilities or tendency to partition) or in the extent to which they stick to the stationary phase (adsorption).

Key terms

The stationary phase in chromatography may be a solid or a liquid held by a solid support. The mobile phase moves through the stationary phase and may be a liquid or a gas.

Tip

A sponge **ab**sorbs into its pores and soaks up the liquid. A solid **ad**sorbs very thin films of liquid or gas onto its surface. Be careful not to confuse the two. In chromatography, **ad**sorption onto the surface of the stationary phase occurs.

Thin layer chromatography, TLC

Thin layer chromatography uses a thin layer of either aluminium oxide, Al_2O_3, or silicon oxide, SiO_2, which is supported on a glass or plastic plate. The thin layer is the stationary phase. The sample is spotted onto the plate and is placed in a solvent (Figure 13.3). The container is covered with a lid to ensure that the air inside the beaker is saturated with solvent vapour. This stops the solvent evaporating as it rises up the plate.

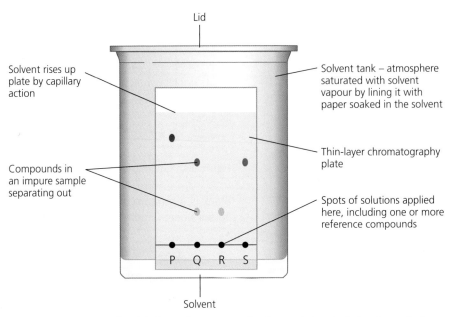

Lid

Solvent rises up plate by capillary action

Solvent tank – atmosphere saturated with solvent vapour by lining it with paper soaked in the solvent

Thin-layer chromatography plate

Compounds in an impure sample separating out

Spots of solutions applied here, including one or more reference compounds

P Q R S

Solvent

Figure 13.3 Apparatus for thin-layer chromatography. Spots of chemicals for the analysis are placed on the start line at P, Q, R and S.

Coloured compounds are easy to see. Colourless compounds can be made visible by using either iodine crystals or ninhydrin. It is also possible to impregnate the thin layer with a fluorescent chemical so that, when exposed to UV light, the whole plate glows except where the spots are.

As the solvent travels up the plate, the components in the mixture travel at different rates. The solvent is allowed to rise until it almost reaches the top of the plate, which is then removed, and the height of the solvent front is marked in pencil (Figure 13.4).

The separated components may be identified by using R_f values.

R_f stands for **retardation factor**. It is measured by using the equation:

$$R_f = \frac{\text{distance moved by spot/solute}}{\text{distance moved by solvent}} = \frac{x}{y}$$

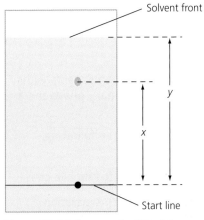

Solvent front

y

x

Start line

Figure 13.4 A diagram of a TLC plate showing the distances used to calculate R_f values. R_f values can help to identify components of mixtures, so long as the conditions are carefully controlled. The values vary with the type of TLC plate (or paper) and the nature of the solvent.

Tip

Similar compounds have similar R_f values and a 'spot' might contain more than one compound.

Test yourself

1 Why is the start line for spotting TLC samples on a plate drawn in pencil and not with ink?
2 Why is the air inside the container of the TLC saturated with the vapour of the solvent?
3 Which of the four samples, P, Q, R and S, in Figure 13.3 above
   a) is the mixture being analysed
   b) are reference compounds also present in the mixture
   c) is a reference compound not present in the mixture?

Activity

Food colours and children

The Food Standards Agency recommends that parents should avoid letting their children eat foods with some colours in them, especially if their children show signs of hyperactivity. The colours to avoid include these dyes (with their E-numbers):

- sunset yellow (E110)
- quinoline yellow (E104)
- carmoisine (E122)
- tartrazine (E102).

These colours are used in a foods such as soft drinks, sweets, cakes and ice cream (Figure 13.5).

Manufacturers must state the colours they use in the list of ingredients by giving the 'colour' and either naming the chemical or showing the E-number. Analysts can use thin-layer chromatography to check whether or not the claims made by manufacturers are correct.

The chromatogram in Figure 13.6 shows the results of analysing the colours extracted from two soft drinks. The colours from the soft drinks are compared with reference compounds.

1 Which of the reference dyes is not a pure compound?
2 What is the R_f value for carmoisine under the conditions used to make this TLC plate?
3 What can you conclude from the chromatogram about:
   a) soft drink A
   b) soft drink B?
4 What advice could the analyst give to parents of a hyperactive child about the two soft drinks, based on this chromatogram?
5 What further investigations should the analyst carry out to confirm the advice given?

Figure 13.5 Soft drinks at a party for children.

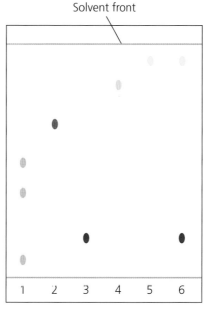

Figure 13.6 TLC plate showing the results of the analysis for two soft drinks and four dyes for reference: 1 quinoline yellow; 2 colour from soft drink A; 3 carmoisine; 4 sunset yellow; 5 tartrazine; 6 colour from soft drink B.

Gas chromatography, GC

Gas chromatography is a sensitive technique for analysing complex mixtures (Figure 13.7). The technique is used for compounds which vaporise on heating without decomposing. This type of chromatography not only separates the chemicals in a sample, but also gives a measure of how much of each is present.

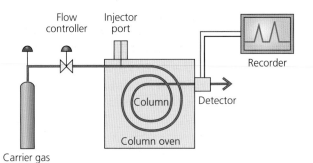

Figure 13.7 Diagram representing the apparatus for gas chromatography.

In GC a sample is vaporised and injected onto the head of the chromatographic column; the sample travels through the column in the gaseous state. The mobile phase is an unreactive gas, such as nitrogen or one of the noble gases. This is known as the **carrier gas** and it flows under pressure through the column. The column contains either a solid (gas–solid chromatography) or a liquid adsorbed onto an inert solid (gas–liquid chromatography). The way in which separation is achieved depends on whether the stationary phase is a liquid or a solid.

If the stationary phase is a liquid, separation depends on the relative solubility of the components in the stationary phase and in the mobile phase. If the stationary phase is a solid, separation depends on the adsorption of the components onto the stationary phase.

The components in the mixture separate as they pass through the column. After a time the chemicals emerge one by one. They pass into a detector, which sends a signal to a recorder as each compound appears. A series of peaks, one for each compound in the mixture, make up the chromatogram. A typical print out from a GC is shown in Figure 13.8.

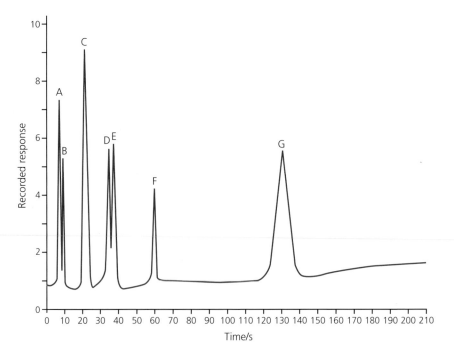

Figure 13.8

According to Figure 13.8, the first component to separate out and emerge from the column, A, took about 5 minutes, while the last component, G, did not emerge from the column until about 130 minutes. So, the position of the peak on the GC print-out is a record of how long it took a component to pass through the column. This is known as the retention time for that component.

The area underneath each peak gives an indication of the relative amount of each component present. The relative amounts of the components can be estimated by assuming that the peak is a triangle and by using the equation:

$$\text{area} = \text{base} \times \tfrac{1}{2}\text{height}$$

If the peak is very narrow, it is sufficient to estimate the amount by measuring the height of the peak.

Test yourself

4 Which of the following gases is most likely to be used as the carrier gas in GC?

hydrogen   nitrogen   oxygen.

Explain your choice.

5 Why must the liquid for the stationary phase in gas–liquid chromatography have a high boiling point?

6 What are the implications of the fact that similar compounds may have very similar retention times in gas chromatography?

7 Refer to the GC print out in Figure 13.8.

a) How many components were in the mixture?

b) Why are components D and E likely to be similar chemicals?

c) Which is the least abundant component?

d) Which is the most abundant component?

e) If the relative amount of component B is 4.5, estimate the relative amount of component G.

Analysis by GC has its limitations, in that similar compounds often have similar retention times; it is likely that in a mixture of methanal, ethanal, propanal and butanal some of the peaks would overlap. Identification of unknown compounds is difficult because reference times vary depending on flow rate of the carrier gas and on the temperature of the column. These limitations have been largely overcome by coupling GC with mass spectrometry.

The chromatogram in Figure 13.9 shows the breath alcohol levels in a sample taken from a suspected drink-driver. This is a complex mixture containing many different chemicals, but ethanol is clearly the most abundant.

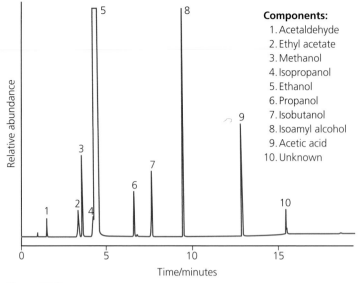

Figure 13.9

Comparing TLC and GC

A summary of the two types of chromatography is shown in Table 13.1.

Table 13.1 Summary of chromatography.

Type	Stationary phase	Mobile phase	Method of separation
TLC	thin layer	solvent	adsorption
GC	liquid on a solid support	carrier gas	relative solubility (partition)
	solid on a solid support	carrier gas	adsorption

Forensic investigations of arson

Arsonists sometimes use flammable liquids, such as petrol or paraffin, to accelerate fires. Firefighters collect samples from burned-out homes which forensic scientists can analyse in the search for clues as to how the fire started. Suitable samples for analysis come from areas where furniture and fittings have not been completely destroyed. Useful samples include carpet underlays, soil from pot plants, bedding, clothing and material collected from underneath floor boards (Figure 13.10).

1 Suggest a reason why firefighters collect only partially burned materials for analysis.
2 Suggest a reason why soil from pot plants can provide good evidence that flammable liquids were present.

Figure 13.10 Firefighters searching through the wreckage of a burned-out house. They are looking for evidence of how the fire started to determine whether it was an accident or arson.

Analysts use solvents to extract chemicals from the samples, and then investigate the solutions by gas chromatography. They compare the chromatograms with those from standard samples of common flammable substances.

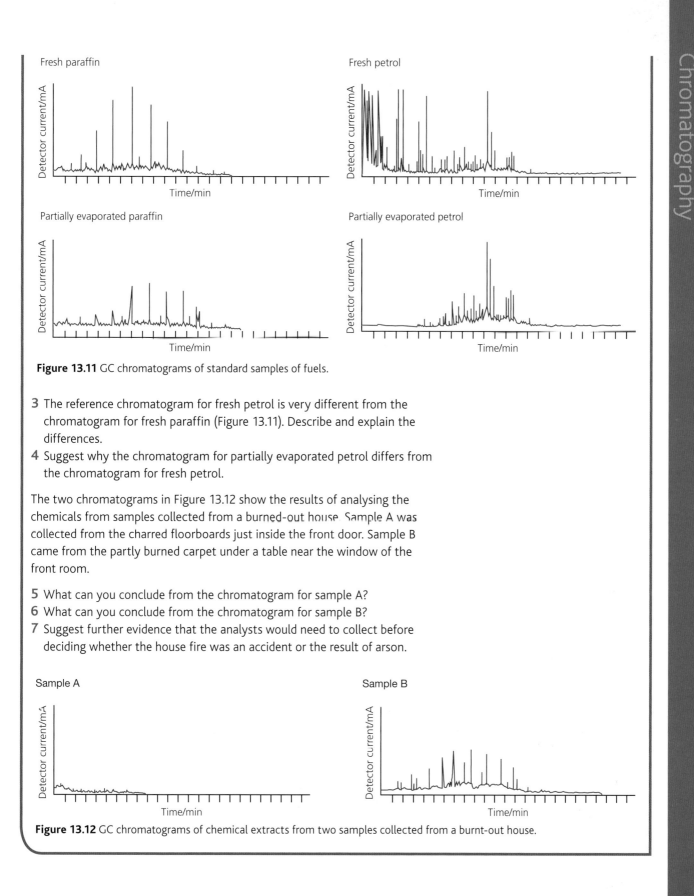

Figure 13.11 GC chromatograms of standard samples of fuels.

3 The reference chromatogram for fresh petrol is very different from the chromatogram for fresh paraffin (Figure 13.11). Describe and explain the differences.

4 Suggest why the chromatogram for partially evaporated petrol differs from the chromatogram for fresh petrol.

The two chromatograms in Figure 13.12 show the results of analysing the chemicals from samples collected from a burned-out house. Sample A was collected from the charred floorboards just inside the front door. Sample B came from the partly burned carpet under a table near the window of the front room.

5 What can you conclude from the chromatogram for sample A?

6 What can you conclude from the chromatogram for sample B?

7 Suggest further evidence that the analysts would need to collect before deciding whether the house fire was an accident or the result of arson.

Figure 13.12 GC chromatograms of chemical extracts from two samples collected from a burnt-out house.

Combining gas chromatography with mass spectrometry (GC–MS)

The combination of gas chromatography and mass spectrometry provides a powerful analytical tool that is used widely in areas such as forensics, environmental analysis and airport security (Figure 13.13).

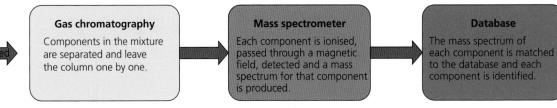

Gas chromatography	Mass spectrometer	Database
Components in the mixture are separated and leave the column one by one.	Each component is ionised, passed through a magnetic field, detected and a mass spectrum for that component is produced.	The mass spectrum of each component is matched to the database and each component is identified.

mixture injected

Figure 13.13 How GC and MS are combined.

Activity

Solving a pollution problem with GC–MS

With complex mixtures, a computer can help an analyst look for a 'needle in a haystack'. Figure 13.14 shows the chromatogram from an investigation of the air in a home where the family was feeling very sick. During the analysis by GC–MS, the computer stored 700 mass spectra as the mixture of chemicals emerged from the chromatography column.

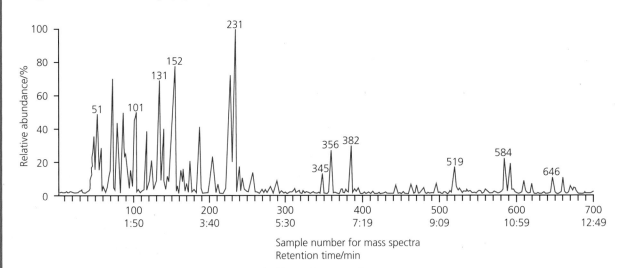

Figure 13.14 Gas chromatogram for chemicals sampled from the air in a house.
The numbers 1–700 on the time axis indicate the points at which mass spectra were recorded and stored. Underneath these numbers are the retention times.

The analysts suspected that the chemicals might have come from petrol, so they asked the computer to plot a chromatogram showing only those chemicals producing a peak with mass/charge ratio of 91 in their spectra. The result is shown in Figure 13.15, which indicates that methylbenzene, ethylbenzene and three dimethylbenzenes were present in the mixture. These are all chemicals that are distinctive for the mixture of hydrocarbons found in petrol. With this evidence, the investigators carried out further searches and tracked down the source of the petrol vapour.

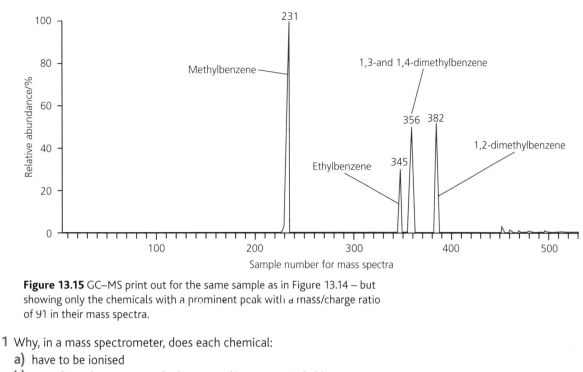

Figure 13.15 GC–MS print out for the same sample as in Figure 13.14 – but showing only the chemicals with a prominent peak with a mass/charge ratio of 91 in their mass spectra.

1 Why, in a mass spectrometer, does each chemical:
   a) have to be ionised
   b) pass through a region with electric and/or magnetic fields
   c) produce a spectrum with several peaks?
2 Suggest the identity of the ion fragment with a mass/charge ratio of 91 in the mass spectra of methylbenzene and related compounds.
3 How does the computer identify a chemical with a mass spectrum recorded at a particular retention time?
4 Suggest two reasons why forensic scientists find GC–MS particularly valuable.

Nuclear magnetic resonance spectroscopy (NMR)

NMR is a powerful tool that is used to determine the structure of a compound. You are expected to be able to predict and recognise NMR spectra of simple organic compounds.

If the nucleus of an atom has a net nuclear spin, it can be detected using particular radio frequencies, for example, ^1H and ^{13}C can both be detected. However other common atoms such as ^{16}O, ^{12}C and ^{14}N cannot be detected.

The net spin on a nucleus that has an odd number of protons or neutrons means that the nucleus behaves like a tiny bar magnet and, as it spins, it generates a magnetic moment. Adjacent nuclei also have magnetic moments. Therefore, each nucleus is affected by neighbouring nuclei. The frequency at which each nucleus absorbs radio waves depends on its environment.

The radio waves are at the low-energy end of the electromagnetic spectrum. ^1H (H-1) and ^{13}C (C-13) both absorb energy in the radio wave part of the spectrum. However, the frequency of the radio waves absorbed depends on the surrounding atoms, i.e. the exact frequency absorbed depends on the chemical environment. This variation in the frequency absorbed is the key to the determination of structure. It is known as the chemical shift, δ. All absorptions are measured relative to tetramethylsilane, $Si(CH_3)_4$ (TMS).

Key term

The horizontal scale of an NMR spectrum shows the chemical shift, δ, of the peaks measured in parts per million (ppm).

The symbol δ stands for the chemical shift relative to the zero on the scale, which is given to the signal obtained from standard tetramethylsilane, TMS.

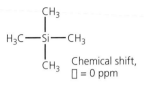

Chemical shift, $\delta = 0$ ppm

The chemical shift, δ, of TMS is set at zero. TMS is used as a standard because:

- All 12 hydrogens are equivalent and the chemical shift is standardised at $\delta = 0$ ppm.

- All four carbons are equivalent and the chemical shift is standardised at $\delta = 0$ ppm.

- It is chemically inert and does not react with the sample.

- It is volatile and easy to remove at the end of the procedure.

- It absorbs at a higher frequency than other organic compounds. Therefore, its mass spectrum does not overlap with that of the sample.

It is unlikely that you will have access to an NMR spectrometer but you will be expected to interpret the spectra that they produce. A cross section of a NMR spectrometer is shown in Figure 13.16:

Tip

All other peaks are measured relative to TMS. You do not have to learn these. All relevant absorptions for ^{1}H and ^{13}C-NMR are listed in the data sheet, which you will be given in the examination.

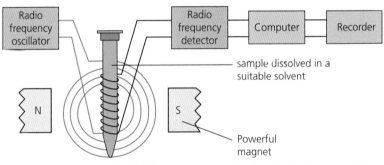

Figure 13.16 Diagram of an NMR spectrometer showing the key features of the technique.

In order to obtain a spectrum, the sample is:

- dissolved in a suitable solvent and TMS is added as a reference

- placed in a strong magnetic field and the sample is submitted to a range of radio frequencies.

Different radio frequencies are absorbed depending on the environment, the chemical shift, of the ^{13}C or the ^{1}H. When the radio frequency is absorbed, a signal is detected, a resonance frequency is recorded and a spectrum produced.

Test yourself

8 State four properties of TMS that enable it to be added to any sample and to be used as a reference point in NMR.

Carbon-13, ^{13}C, NMR spectroscopy

Carbon-12, ^{12}C, is the most abundant isotope of carbon. It does not have spin because it has an even number of protons and an even number of neutrons. The second isotope of carbon, carbon-13, C-13, can be detected using low-energy radio waves and it is possible to generate ^{13}C-NMR spectra, but C-13 accounts for only about 1.1% of all naturally occurring carbon. Interactions between adjacent ^{13}C atoms and adjacent protons do occur but because of the low abundance of ^{13}C atoms, all peaks in C-13 NMR spectra appear as single peaks. Each peak represents a different carbon environment and each environment will have a different chemical shift.

It is essential that you are able to recognise different environments within carbon compounds.

Example 1

There are three isomers of C_5H_{12} – pentane, 2-methylbutane and 2,2-dimethylpropane.

In pentane:

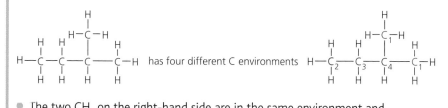

- The carbons on each end of the chain are in the same environment.

- The central carbon is in a different environment and is sandwiched between two CH_2CH_3 groups.

- The two remaining carbons are in the same environment, sandwiched between a CH_3 group and a CH_2.

Therefore, a ^{13}C spectrum of pentane will produce three peaks.

In 2-methylbutane:

- The two CH_3 on the right-hand side are in the same environment and will, therefore, produce a single peak.

- The CH_3 on the left-hand side is in a different environment.

- The CH_2 has its own environment.

- The CH has its own environment.

2-methylbutane, therefore, produces four peaks on a ^{13}C spectrum.

▶▶▶

263

In 2,2-dimethylpropane:

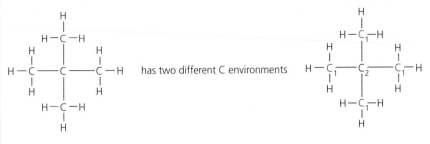

has two different C environments

- There are four CH_3 groups which are all in the same environment.

- The central carbon atom is in a different environment.

2,2-dimethylpropane, therefore, produces two peaks on a ^{13}C spectrum.

^{13}C-NMR easily distinguishes between the isomers of C_5H_{12} because each isomer has a different number of carbon environments and, hence, a different number of peaks in the spectra.

Test yourself

9 Why is it so important that both ^{13}C and ^1H can be detected by NMR?

10 Identify the number of carbon environments in each of the following compounds.

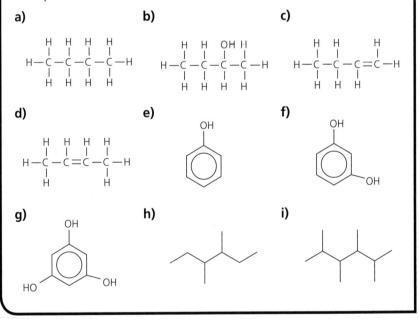

Different carbon environments absorb different radio frequencies and have different chemical shifts. Although chemical shifts are provided on the data sheet supplied in the examination, it is important that you practice using the data sheet by identifying different carbon environments and their associated chemical shifts. Table 13.2 summarises the essential carbon environments and their chemical shift range (which are given a δ value relative to TMS).

Table 13.2

Carbon environment	Possible functional group(s)	Symbol used by OCR	Range, δ /ppm
C bonded to another carbon	alkanes and any C that is part of a carbon chain	C — C	5–55
C bonded to a halogen	haloalkanes	C — Cl, C — Br	20–50
C bonded to a nitrogen	amines	C — N	35–75
C bonded by a single bond to an oxygen	alcohol, ester, carboxylic acid	C — O	50–90
C bonded by a double bond to a carbon	alkenes	C = C	110–165
C as part of an arene	aromatic compounds	(aromatic ring)	110–165
C with double bond to oxygen and single bond to another oxygen	carboxylic acids, esters	(C=O single bond O)	160–220
C with double bond to oxygen and single bond to a nitrogen	amides	(C=O single bond N)	160–220
C with double bond to oxygen	aldehydes or ketones	(C=O aldehyde/ketone)	160–220

Ethanol has two carbon atoms, C_1 and C_2:

- C_1 is joined to a CH_2 and is part of an alkyl group. The chemical shift should be between 5 ppm and 55 ppm.

- C_2 is joined to an alcohol OH. The chemical shift should be between 50 ppm and 90 ppm.

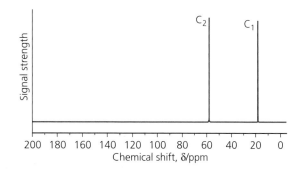

The key to interpreting ^{13}C-NMR spectra is to identify the number of different carbon environments and then to match them with the groups in the data sheet.

Example 2

Determine the number of carbon environments in propanoic acid. For each environment, predict the chemical shift (the δ value).

Answer

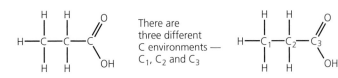

The ^{13}C-NMR spectrum for propanoic acid should contain three peaks:

- C_1 is part of an alkyl group and should, therefore, have a δ value of between 5 ppm and 55 ppm.

- C_2 is also part of an alkyl group, but because it is next to a carbonyl group the δ value will be towards the higher end of the range.

- C_3 is part of a carboxylic acid group and should, therefore, have a δ value between 160 ppm and 220 ppm.

The ^{13}C spectrum of propanoic acid is shown below.

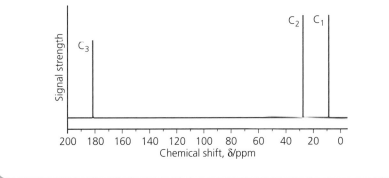

Example 3

Determine the number of carbon environments in propan-2-ol. For each environment, predict the chemical shift (the δ value).

Answer

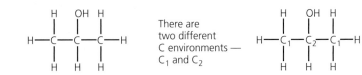

The ^{13}C-NMR spectrum for propan-2-ol should contain two peaks:

- C_2 is next to an OH group and should, therefore, have a δ value of between 50 ppm and 90 ppm.

- C_1 is part of an alkyl group and should, therefore, have a δ value of between 5 ppm and 55 ppm. However, because the adjacent carbon is bonded to an OH the δ value will be towards the high end of the range.

The ^{13}C spectrum of propan-2-ol is shown below.

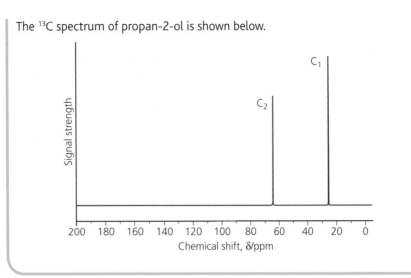

Test yourself

11 State the range of the chemical shift, δ, in ppm, for each C highlighted in red in the following structures.

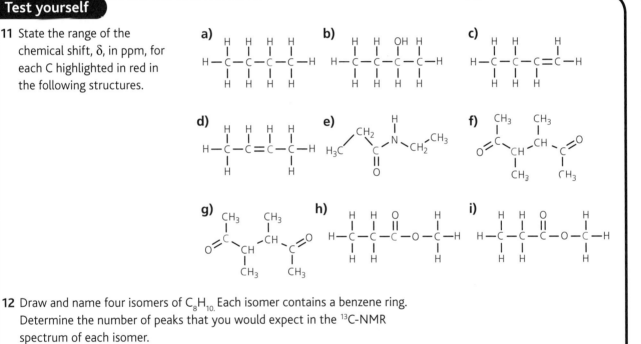

12 Draw and name four isomers of C_8H_{10}. Each isomer contains a benzene ring. Determine the number of peaks that you would expect in the ^{13}C-NMR spectrum of each isomer.

H-1 (proton) NMR spectroscopy

Proton NMR relies on the magnetic properties of ^1H, (H-1), and, like ^{13}C-NMR, it is essential that you can recognise different hydrogen environments and assign different chemical shifts to the different environments.

Consider a molecule of ethanol, C_2H_5OH:

The six hydrogen atoms are not identical:

● The three hydrogen atoms in the CH_3 group are in the same environment and can be labelled H_a.

- The two hydrogen atoms in the CH$_2$ group are in the same environment (labelled H$_b$).

- The hydrogen in the OH group is different from all of the rest (labelled H$_c$).

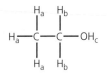

The six hydrogens in ethanol are, therefore, in three different environments. In ^{1}H (proton) NMR spectroscopy, this leads to three different absorptions and, hence, three different peaks (H$_a$, H$_b$ and H$_c$) in the spectrum at three different chemical shifts.

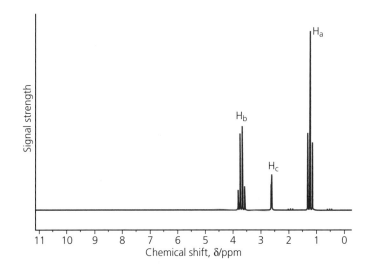

As expected, in the ^{1}H-NMR spectrum of ethanol there are three peaks, but you will notice that the peaks are of different sizes and that the peaks may be split.

Chemical shift, δ/ppm

As with ^{13}C-NMR, you do not need to learn the chemical shifts as you will be provided with a data sheet in the examination.

In the ^{1}H (proton) NMR of ethanol:

- The CH$_3$ (H$_a$) has a chemical shift between 0.7 ppm and 2.0 ppm.

- The CH$_2$ (H$_b$) has a chemical shift between 3.0 ppm and 4.3 ppm.

- The OH (H$_c$) has a chemical shift between 0.5 ppm and 12.0 ppm.

Table 13.3

Hydrogen environment	Possible functional group(s)	Symbol used by OCR	Range, δ/ppm
H bonded to a C that is part of a chain	alkanes and any H that is attached to C in a carbon chain	R — CH	0.6–2.0
H bonded to a C that is next to a carbonyl	aldehydes, ketones, carboxylic acid, esters, amides	HC — C (=O)	2.0–3.0
H bonded to a C that is next to a nitrogen	amines	HC — N	2.0–3.0
H bonded to a C that is next to a benzene ring	substituted arenes	(benzene) — CH	2.0–3.0
H bonded to a C that is next to a Cl or a Br	haloalkanes	HC — Cl / HC — Br	3.0–4.3
H bonded to a C that is next to an oxygen	alcohols, carboxylic acids, esters	HC — O	3.3–4.3
H bonded to a C that is next to a C with a double bond	alkenes	C=C with H	4.5 6.0
H bonded to a C that is part of a benzene ring	arenes	(benzene) H	6.2–8.0
H bonded to a C that has a double bond to an oxygen	aldehydes	—C(=O)H	9.0–10
H that is part of a COOH group	carboxylic acids	—C(=O)OH	11–12
H that is bonded to either an oxygen or a nitrogen	alcohols, carboxylic acids, esters, amines	H — O / H — N	0.5–12.0*
H that is part of an OH group that is attached to a benzene	phenols	(benzene) — OH	0.5–12.0*
H that is attached to a N next to a C=O	primary and secondary amides	—C(=O)NH$_2$ / —C(=O)HN—	0.5–12.0*

* OH and NH chemical shifts are very variable, cover a wide range and may be sometimes outside these limits.

Relative peak area

When interpreting ^{13}C-NMR spectra, you cannot draw any simple conclusions from the heights of the various peaks, but in proton NMR the areas under the various peaks do give useful information about the number of hydrogen atoms in a particular environment. An NMR instrument is set up to integrate the curve and to work out the ratio of the areas under each peak. There are two common ways of showing the peak ratios:

- as an integration curve – see NMR 1 in Figure 13.17

- placing numbers at either the top or the bottom of each peak – see NMR 2 in Figure 13.17.

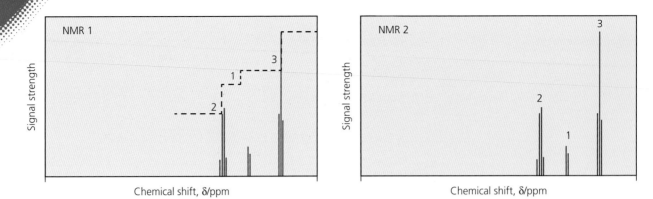

Figure 13.17

The relative size of each peak reflects the number of hydrogens in each environment:

H_a – there are three hydrogens in this environment

H_b – there are two hydrogens in this environment

H_c – there is one hydrogen in this environment.

It follows that the relative intensity of the peaks H_a, H_b and H_c is 3:2:1.

In the examination, the relative numbers of each type of proton present will either be given as a ratio or presented in the form of an integration trace.

Example 4

Propane, C_3H_8, has two different proton environments, H_a and H_b:

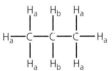

How many peaks are there in its NMR spectrum and what is the relative intensity of the peaks?

Answer

There are two different proton environments, so there are two peaks.

There are six protons in the environment H_a and two protons in the environment H_b. The simplest ratio is, therefore, 3:1.

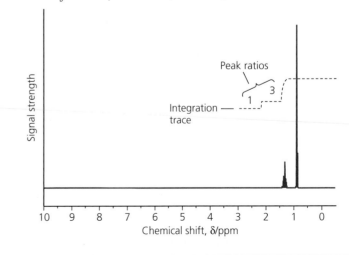

Spin–spin coupling or splitting

The hydrogens attached to one carbon atom influence the hydrogens on adjacent carbon atoms. This is called **spin–spin coupling** and it causes peaks to split into a close bunch of smaller peaks. The easiest way to predict this splitting (called the **splitting pattern**) is to count the number of hydrogens on the adjacent carbon atoms and then use what is known as the **'n + 1'** rule, where n is the number of hydrogens on the adjacent carbon atoms. A split peak is known as a doublet if the peak is split into two, triplet if into three, and so on (Table 13.4).

Tip

The n + 1 rule can only be applied if the protons are equivalent (in the same environment).

Table 13.4 Naming of split peaks in ^{1}H-NMR spectroscopy.

Number of hydrogens on adjacent carbons	Splitting	Type of peak
0	1 (the peak is not split)	singlet
1	2	doublet
2	3	triplet
3	4	quartet
4	5	pentet
5	6	sextet

In the NMR spectrum of ethanol, each of the peaks is split differently:

- H_a is next to two hydrogens in CH_2 and, hence, the peak is split into (2 + 1) – a **triplet**.

- H_b is next to three hydrogens in CH_3 and, hence, the peak is split into (3 + 1) – a **quartet**.

- H_c is not attached to a carbon atom and, hence, does not undergo spin–spin coupling. It is, therefore, a **singlet**.

Tip

Signals for OH and NH are usually singlets and are not split.

We would, therefore, expect the high-resolution NMR spectrum of ethanol to have three peaks of relative intensity 3:2:1 and split into a triplet, a quartet and a singlet.

Test yourself

13 For each of the compounds below determine:
- the number of H environments
- the relative ratios of the peaks
- the splitting of each peak
- the chemical shift range of each peak.

a)

b)

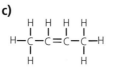

c)

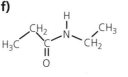

d)

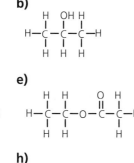

e)

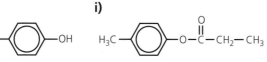

f)

g)

h)

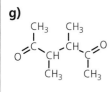

i)

In **h)** and **i)** the protons on the benzene ring will be split into a multiplet. You are not expected to determine the exact splitting for aromatic protons but you will be expected to identify them by their chemical shift.

Proton exchange using D_2O

The O–**H** and the N–**H** peaks have chemical shifts that differ between compounds and, unlike other **H** peaks, are found over a wide range of chemical shifts. Therefore, they are difficult to assign. When alcohols, carboxylic acids, amines or amides are dissolved in water, there is a rapid exchange between the protons in the functional groups (**labile protons**) and the protons in the water:

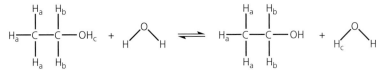

If water is replaced by deuterated water, ^2H_2O, the peak at H_c disappears. The H_c proton is replaced by deuterium, ^2H, which does not absorb in this region of the spectrum.

Deuterated water, ^2H_2O can be written as D_2O. Ethanol dissolved in deuterated water can be represented as:

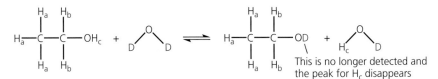

The use of ^2H_2O to identify labile protons is a valuable technique in proton (^1H) NMR.

The use of deuterated solvents

When samples are prepared for ^1H-NMR, it may be necessary to dissolve the sample in a suitable solvent. Solvents containing protons are unsuitable because the protons would be detected by the technique and would interfere with the spectrum. This is overcome by using a deuterated solvent, such as $CDCl_3$.

Example 5

Draw the structure of propan-1-ol and determine:

- the number of different H environments
- the relative ratio of the peaks
- the splitting of each peak
- the chemical shift of each peak.

Tip

At first glance it might appear that H_b would be split into a sextet but this is not the case as the five adjacent protons are not equivalent. Three are labelled H_a and the other two are H_c. The splitting is therefore described as a 'multiplet'.

Answer

	Number of protons	Ratio	Splitting	Chemical shift, δ ppm
H_a H_b H_c H_a—C—C—C—OH$_d$ H_a H_b H_c	Four peaks H_a, H_b, H_c, H_d	H_a, H_b, H_c, H_d 3 : 2 : 2 : 1	H_a — triplet H_b — multiplet H_c — triplet H_d — singlet	H_a 0.7–1.6 H_b 1.2–1.4 H_c 3.3–4.3 H_d 0.5–12.0* *OH peaks are variable and should be confirmed by using D_2O as the solvent

Hint: H_a and H_b are both part of an alkyl chain and fall in the region (0.5–2.0 ppm) – the one closest to the O in the OH will have the higher chemical shift.

Example 6

Compound **A** has the empirical formula C_2H_4O and a molar mass of $88\,g\,mol^{-1}$. The ^1H-NMR spectrum of compound **A** is shown below.

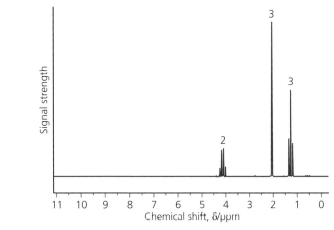

Deduce the identity of compound **A**.

Answer

Molar mass of $C_2H_4O = 24 + 4 + 16 = 44\,g\,mol^{-1}$

Therefore, the molecular formula is $C_4H_8O_2$.

The NMR spectrum shows three peaks, so there are three different proton environments.

The relative peak areas are $2:3:3$. This adds up to 8, which indicates that the molecule is likely to contain two CH_3 groups and a CH_2.

The peak at 1.2 is a triplet, indicating that the proton is next to a CH_2.
The peak at 4.2 is a quartet, indicating that the proton is next to a CH_3.
This indicates the presence of a CH_3–CH_2– grouping. The peak at 1.2 ppm is in the range 0.5–2.0 ppm whilst the peak at 4.2 ppm is within the range 3.0–4.3 ppm indicating that the CH_2 is bonded to O.

The peak at 2.1 is a singlet, suggesting that it is a CH_3– and that the adjacent carbon has no hydrogen. The data sheet confirms that CH_3– bonded to a carbonyl (C=O) absorbs in the region 2.0–3.0 ppm.

Compound **A** is known to contain CH_3–CH_2– and

The molecular formula is $C_4H_8O_2$, so the extra oxygen is likely to be part of an ester group:

Compound A is ethyl ethanoate:

14 The ^{1}H-NMR spectrum of an ester is shown below. The molar mass of the ester is 102. Identify the ester. Show all of your working and assign each peak in the NMR to a particular ^{1}H environment.

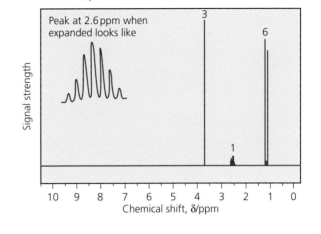

There are many different ways of working through questions like this. It is essential to use all of the information provided and that you make sufficient points to match the number of marks allocated to the question.

Combined techniques

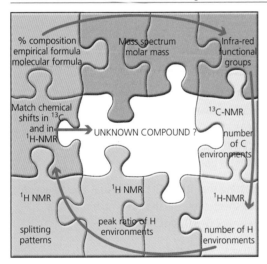

Analytical chemistry is rather like detective work – pieces of evidence are gathered from different places. Usually no one piece of evidence is conclusive but when the pieces of evidence are slotted together, the combination is definitive. Gathering together the evidence is a bit like doing a jigsaw.

There are numerous ways of piecing together the evidence. Often questions in examinations guide you step by step, but some questions are open and leave it to you to decide how best to answer them.

You must develop a method for solving open-ended questions. One such method is outlined below.

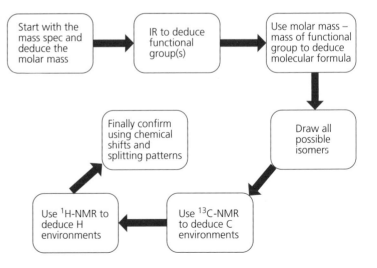

Example 7

The spectra below are for compound **X**. Use the spectra to identify compound **X**. Show all of your working. (12)

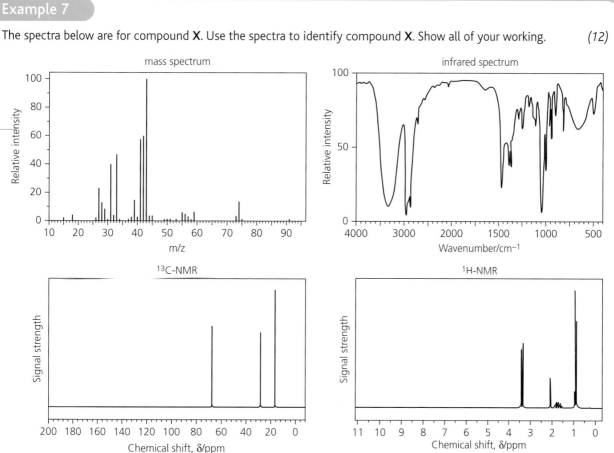

This is a very open ended question and it is difficult to know where to start. However, following the method outlined above helps.

Answer

Marks have been assigned to parts of the answer to show you how the question might be marked in an exam.

The mass spectrum shows the molecular ion at m/z = 74, hence molar mass is 74. (1)

The infrared spectrum has a broad absorption at about 3300 cm⁻¹ and does not have an absorption at about 1700 cm⁻¹. Compound X is an alcohol. (1)

Molar mass = 74, alcohol group has mass = 17, hence the rest of the molecule has a mass (74 − 17) = 57. The rest of the molecule is likely to be made of C and H. Four C has a mass of 48. (57 − 48) = 9, hence there are also 9Hs. The formula of compound X is C_4H_9OH. (1)

Next, draw all possible isomers of C_4H_9OH and identify the number of different C environments.

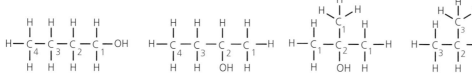

The ¹³C-NMR has three peaks. Therefore, compound X is 2-methylpropan-1-ol. (1)

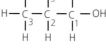

Having identified compound X, it might appear that you have finished, BUT you have only scored six marks and you have not used chemical shifts and the ^{1}H-NMR spectrum.

> **Tip**
>
> Check the mark allocation to make sure that you have made enough points to match the total number of marks.

In the ^{13}C-NMR, the peak at chemical shift $\delta = 70$ ppm is due to the C that is bonded to the OH. *(1)*

The ^{1}H-NMR shows four H environments, which matches with proposed identity of compound X:

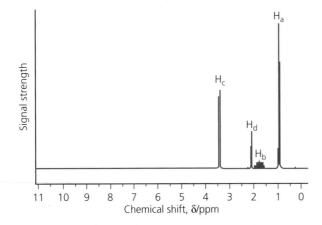

4 H environments

(1)

H_a will be doublet because it is next to a C–H, according to the $(n + 1)$ rule. *(1)*

H_b will be a multiplet* because it is next to two CH_3 and a CH_2. *(1)*

H_c will be doublet because it is next to a C–H, according to the $(n + 1)$ rule. *(1)*

H_d will be single as Hs attached to an O in a O–H are not split. *(1)*

TIP

*The adjacent protons are not equivalent so you are not expected to determine the exact splitting.

Chemical shifts can distinguish the two doublets:

- H_c is next to an O–H and will fall in the range 3.5–4.5 ppm. *(1)*
- H_a is next to a C and will fall in the range 0.8–2.0 ppm. *(1)*

You could even label the ^{1}H NMR spectrum:

In an exam it is not always easy to work out the exact mark allocation. The answer above has made 13 distinct points. It does not matter that the mark allocation is 12 marks. It is good exam technique to make sure that you have covered all aspects.

Example 8

Compound **Y** has the following percentage composition by mass: carbon 54.5%; hydrogen 9.1%; oxygen 36.4%.

Use this information and the spectra below to identify compound **Y**.

Mass spectrum

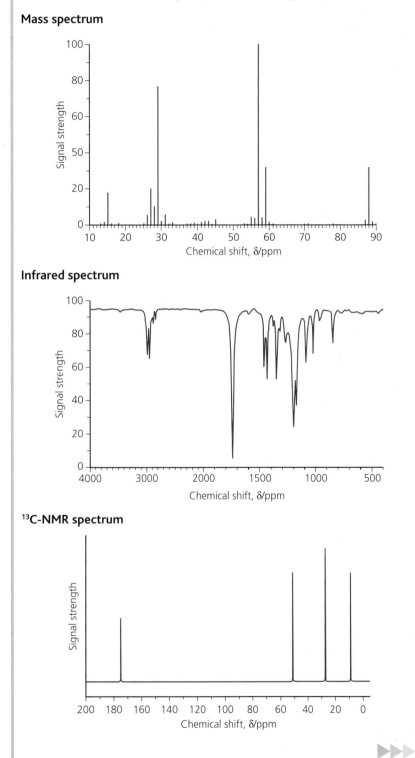

Infrared spectrum

¹³C-NMR spectrum

▶▶▶

¹H-NMR spectrum

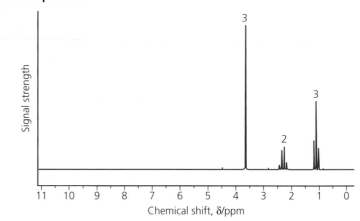

Answer

Empirical formula

Method	Carbon	Hydrogen	Oxygen
Percentage	54.5	9.1	36.4
Divide by relative molecular mass	54.5/12.0 = 4.54	9.1/1.0 = 9.1	36.4/16.0 = 2.28
Divide by smallest	4.54/2.28 = 1.99 ≈ 2	9.1/2.28 = 3.99 ≈ 4	2.28/2.28 = 1

The simplest ratio of C:H:O is 2:4:1. The empirical formula is C_2H_4O.

Empirical formula mass = 24.0 + 4.0 + 16.0 = 44.0

Molecular formula

Mass peak in the mass spectrum = 88

Empirical formula = C_2H_4O, empirical formula mass = 44.0

Hence, the molecular formula is $C_4H_8O_2$

Information from the spectra

1 The peak at 175 in the ¹³C-NMR spectrum is due to

This suggests that compound **Y** is either an ester or a carboxylic acid.

2 The infrared spectrum has a peak at about $1700\,cm^{-1}$ confirming a C=O group. However, there is no broad absorption in the range 2500–$3300\,cm^{-1}$. This indicates that compound **Y** is not a carboxylic acid. Hence, it is probably an ester.

3 The molecular formula is $C_4H_8O_2$ and the ¹³C-NMR shows four peaks. This confirms that all four carbons are in different environments.

4 The ^{1}H-NMR spectrum has only three peaks. This means that one of the carbon atoms has no hydrogens attached. There is a familiar pattern of a triplet and a quartet, indicating a CH_3 next to a CH_2. This accounts for five of the eight hydrogens. The third peak in the ^{1}H-NMR spectrum is a singlet, which suggests that there is a CH_3 next to a carbon with no hydrogens. There is now enough information to conclude that compound **Y** is methyl propanoate:

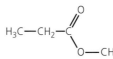

- In the mass spectrum, the fragment ions at 29 and 57 confirm the $CH_3CH_2^+$ and the $CH_3CH_2CO^+$ ions respectively.

- The absorption at about $3000\,cm^{-1}$ in the infrared spectrum confirms C–H.

- The chemical shifts in the ^{13}C-NMR spectrum can be used to confirm the identity of the carbons.

Tip

This is not the only way of deducing the identity of a compound – there are many others.

Test yourself

15 Compound **Z** contains 31.37% oxygen by mass. Use the four spectra below to identify compound **Z**. Show all of your working .

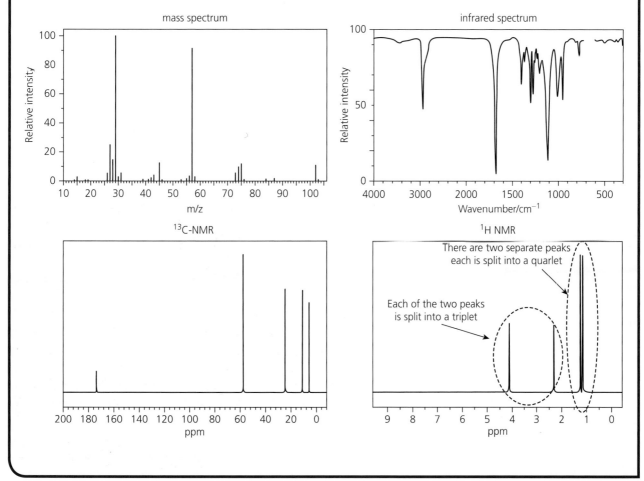

Magnetic resonance imaging (MRI)

The body is made up of about 70% water. Since NMR can detect the protons in water, it should be possible to detect the water in the body. Water in the body is in a very different environment from free water and these differences can be detected by NMR.

NMR is a common diagnostic tool used in medicine, where it is known as magnetic resonance imaging (Figure 13.18). Use is made of a magnetic field that is varied across the object being examined. It is able to detect differences in water content and other soft tissue deep inside the object. It is non-invasive and, unlike X-rays, does not damage any tissue. It is used extensively in the study of tissues, muscles and blood flow.

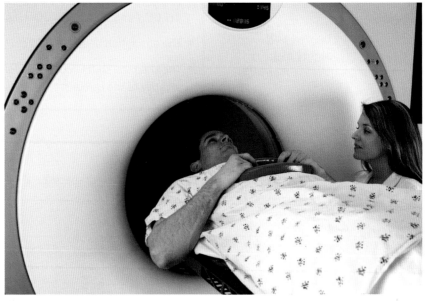

Figure 13.18 An MRI scan in progress.

Hydrocephalus means water (hydro) in the head (cephalus). It is sometimes called 'water on the brain'. In Figure 13.19, the MRI scan on the left shows the build up of extra fluid in the brain compared with the normal MRI scan on the right. MRI scans are now often coloured (Figure 13.20).

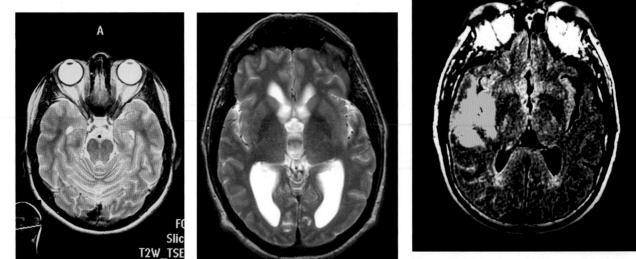

Figure 13.19 MRI scans of a brain with hydrocephalus (right) and a normal brain (left).

Figure 13.20 A coloured MRI scan of the brain of a patient after a stroke. A large area of dead tissue can be seen on the left of the image (orange area).

Practice questions

Multiple choice questions 1–10

1 A mixture of compounds was separated using thin layer chromatography. The TLC obtained is shown here.

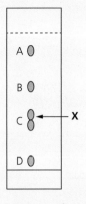

X is known to be glycine (H_2NCH_2COOH). Which spot is most likely to be alanine ($H_2NCH(CH_3)COOH$)? *(1)*

2 The ^{13}C-NMR of ethyl benzoate has seven different peaks. Which carbon atom, numbered here,

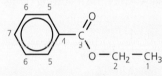

is responsible for a peak at 60.9 ppm? (You will need a data sheet for this question.)
A C_1 **C** C_3
B C_2 **D** C_4 *(1)*

3 The ^1H-NMR spectrum of this molecule

contains a number of peaks, some of which are split and some that are not. How many singlets would the ^1H-NMR spectrum contain?
A 1 singlet
B 2 singlets
C 3 singlets
D 4 singlets *(1)*

4 Compound W has a molar mass of 73 g mol^{-1}. The ^1H-NMR spectrum of compound W is shown below. The peak with chemical shift 9.5 disappears in the presence of D_2O.

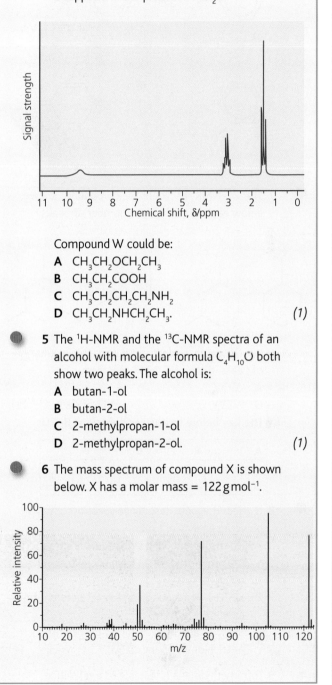

Compound W could be:
A $CH_3CH_2OCH_2CH_3$
B CH_3CH_2COOH
C $CH_3CH_2CH_2CH_2NH_2$
D $CH_3CH_2NHCH_2CH_3$. *(1)*

5 The ^1H-NMR and the ^{13}C-NMR spectra of an alcohol with molecular formula $C_4H_{10}O$ both show two peaks. The alcohol is:
A butan-1-ol
B butan-2-ol
C 2-methylpropan-1-ol
D 2-methylpropan-2-ol. *(1)*

6 The mass spectrum of compound X is shown below. X has a molar mass = 122 g mol^{-1}.

Use the fragmentation ions in the spectrum on page 283 to identify compound X from the list of isomers given below.

a)

b)

c)

d)

(1)

7 The ^{13}C-NMR spectra of the compounds shown below each has a different number of peaks.

W   X   Y   Z

Put the compounds in order of increasing number of peaks in their ^{13}C-NMR spectrum.

A WXYZ   **B** XYZW

C WZYX   **D** ZYWX   *(1)*

Use the key below to answer questions 8–10.

A	B	C	D
1, 2 & 3 correct	1 & 2 correct	2 & 3 correct	1 only correct

8 Nuclear magnetic resonance spectroscopy is a powerful analytical tool.

Which of the following statements are true?

1 ^{1}H, ^{13}C, ^{19}F and ^{31}P all have nuclei that have a net spin.

2 $Si(CH_3)_4$ is added to all samples before running an NMR spectrum.

3 Samples are often dissolved in solvents such as trichloromethane, $CHCl_3$.   *(1)*

9 Tetramethylsilane is suitable as a reference standard and is added to all samples before running an NMR spectrum because:

1 it only produces one peak in both ^{1}H-NMR and in ^{13}C-NMR

2 it has a low boiling point

3 it is stable and does not react with the samples.   *(1)*

10 Glycolic acid, propanoic acid, ethyl methanoate and 1,3-diaminopropane are shown below.

Glycolic acid   Propanoic acid

Ethyl methanoate   1,3-diaminopropane

All four can easily be distinguished by:

1 using ^{1}H NMR only.

2 using ^{13}C-NMR only

3 using the mass spectra without reference to any fragmentation ions   *(1)*

11 Explain each of the following terms which are used in chromatography:

a) adsorption

b) relative solubility

c) R_f value in TLC

d) retention time in GC.   *(4)*

12 Describe, with the aid of a diagram, how TLC could be used to identify the individual amino acids present in a mixture of amino acids.   *(7)*

13 The structure of cyclopentanone, C_5H_8O, is shown below.

a) Determine the number of:

   i) hydrogen environments detected in the ^{1}H-NMR spectrum

   ii) carbon environments detected in the ^{13}C-NMR spectrum.

b) Predict the chemical shift (δ value) for each peak in the ^{13}C-NMR spectrum.

c) Predict the chemical shift (δ values), the splitting pattern and the relative peak area for each peak in the ¹H-NMR spectrum. *(6)*

14 Lactic acid (2-hydroxypropanoic acid, $CH_3CH(OH)COOH$) was analysed using a combination of analytical techniques.

a) Calculate the percentage composition by mass of each of the following in lactic acid:
i) carbon
ii) hydrogen
iii) oxygen.

b) What would you expect the following to reveal about lactic acid?
i) the infrared spectrum
ii) the mass spectrum. *(9)*

15 Two unbranched isomeric ketones with the molecular formula $C_5H_{10}O$ have the mass spectra shown below.

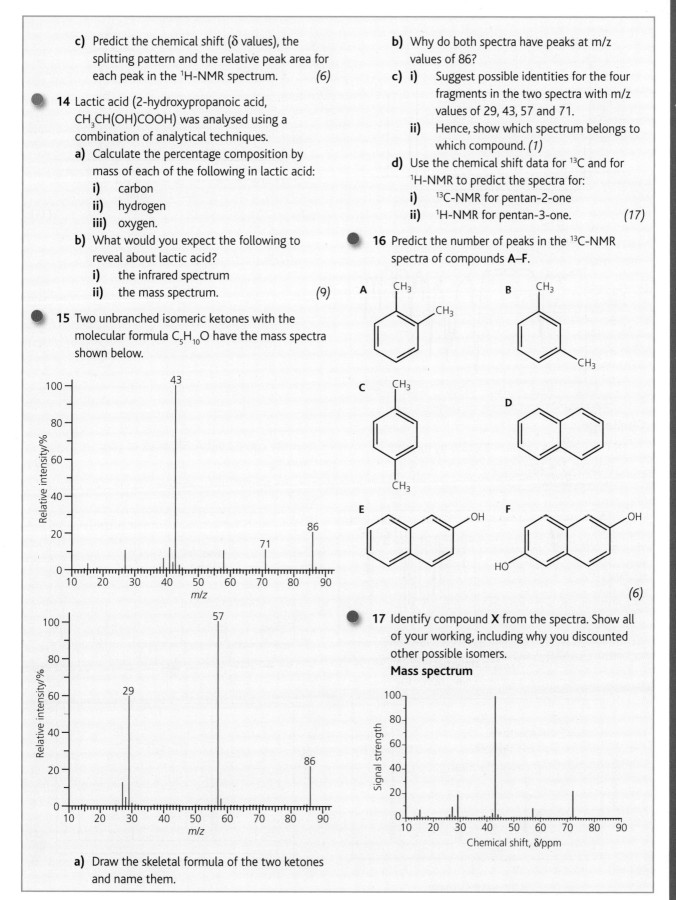

a) Draw the skeletal formula of the two ketones and name them.

b) Why do both spectra have peaks at m/z values of 86?

c) i) Suggest possible identities for the four fragments in the two spectra with m/z values of 29, 43, 57 and 71.

ii) Hence, show which spectrum belongs to which compound. *(1)*

d) Use the chemical shift data for ¹³C and for ¹H-NMR to predict the spectra for:
i) ¹³C-NMR for pentan-2-one
ii) ¹H-NMR for pentan-3-one. *(17)*

16 Predict the number of peaks in the ¹³C-NMR spectra of compounds **A–F**.

(6)

17 Identify compound **X** from the spectra. Show all of your working, including why you discounted other possible isomers.

Mass spectrum

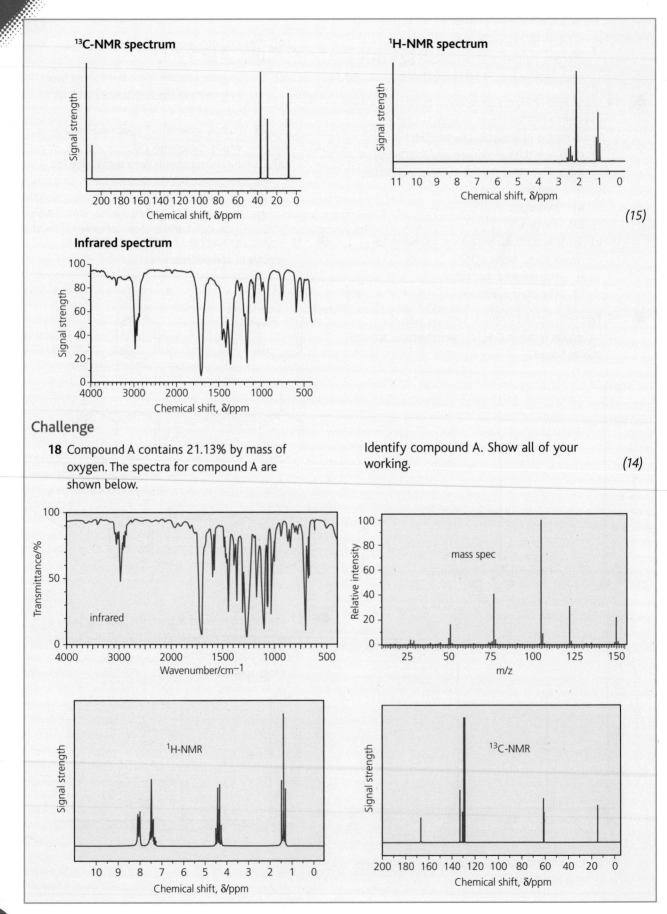

¹³C-NMR spectrum

Signal strength

200 180 160 140 120 100 80 60 40 20 0
Chemical shift, δ/ppm

¹H-NMR spectrum

Signal strength

11 10 9 8 7 6 5 4 3 2 1 0
Chemical shift, δ/ppm

(15)

Infrared spectrum

Signal strength

100
80
60
40
20
0
4000   3000   2000   1500   1000   500
Chemical shift, δ/ppm

Challenge

18 Compound A contains 21.13% by mass of oxygen. The spectra for compound A are shown below.

Identify compound A. Show all of your working.

(14)

Transmittance/%

100

50

0
4000   3000   2000   1500   1000   500
Wavenumber/cm⁻¹

infrared

Relative intensity

100
80
60
40
20
0

mass spec

25   50   75   100   125   150
m/z

Signal strength

¹H-NMR

10   9   8   7   6   5   4   3   2   1   0
Chemical shift, δ/ppm

Signal strength

¹³C-NMR

200 180 160 140 120 100 80 60 40 20 0
Chemical shift, δ/ppm

Chapter 14

Maths in chemistry

Using logarithms

Very little extra maths is required to complete the A Level Chemistry course, but you will have to be broadly familiar with the use of logarithms. These are required both for the section of the specification concerned with the effect of temperature on rates of reaction and for that concerning the use of pH as a measure of acidity. You will not be expected to manipulate logarithmic expressions in any detailed way, but you should be confident that – with the aid of a calculator – you can handle their use as explained in this chapter.

Logarithms to base 10 (log)

Many numbers are easier to express in standard (index) form.

100 is 1×10^2 and 0.01 is 1×10^{-2}. The number 20 000 can be written as 2×10^4 and 0.000 32 as 3.2×10^{-4}. In each case, the power to which the 10 has been raised represents the number of zeros that are present in the number. A minus sign in the power of 10 indicates that the number is less than zero.

The advantage of this system is that scientists do not have to write out all the zeros, with the possible risk of making a mistake.

Any number can be written down in the form 10^n where the '10' is called the **base number** and the 'n' is the **power** to which the base number, 10, has been raised. The number 'n' can also be referred to as the logarithm (log) of the number to base 10. For example 100 is 10^2 and you can write $\log(100) = 2$.

The logarithm to base 10 of 0.01 is -2 as this is the power of 10 when the number is written as 10^{-2}. It is written down as $\log(0.01) = -2$.

Table 14.1 gives more examples.

Table 14.1 The logarithms to base 10 of some powers of 10.

Number	Standard (index) form	Log (of number)
100 000	1×10^5	$\log(100\,000) = 5$
10 000	1×10^4	$\log(10\,000) = 4$
1000	1×10^3	$\log(1000) = 3$
100	1×10^2	$\log(100) = 2$
10	1×10^1	$\log(10) = 1$
1	1×10^0	$\log(1) = 0$
0.1	1×10^{-1}	$\log(0.1) = -1$
0.01	1×10^{-2}	$\log(0.01) = -2$
As the number changes by a factor of 10	⟶	The log changes by 1

It is possible to write numbers such as 20 000 as 10^y and 'y' will then be the logarithm of 20 000. However, in this case, it is not obvious what the value of 'y' will be. In fact, to decide what 'y' is you need to use a calculator; all scientific calculators have a button labelled 'log' (or log$_{10}$). In this case, entering 'log' followed by 20000 gives a value of 4.301 (or, on some calculators, you first enter 20000 and then press 'log'). So, log(20000) = 4.301.

If numbers are written in standard index form, the first part of the log is always the same as the power of 10 (i.e. for 20000 it will be 4 because 20000 is 2×10^4) and the decimal part includes the logarithm of 2. Entering log and then 2 into a calculator gives 0.301. So the log of the whole number is 4 + 0.301 = 4.301.

The procedure is:

- write down the first part of the log (i.e. the power to which 10 has been raised)

- use a calculator to find the log of the rest of the number

- add the two together to give the overall log.

Example 1

What is the log of 2.324×10^5?

Answer
The first part of the log is 5 as this is the power of 10.

Using a calculator, log(2.324) = 0.366

Therefore, log(2.324×10^5) = 5 + 0.366 = 5.366

If the number has a power of 10 that is negative, the procedure is the same. Care has to be taken when adding the two parts to obtain the overall log as one part will be negative and the other will be positive.

Example 2

What is the log of 7.89×10^{-9}?

Answer
The first part of the log is −9.

Using a calculator, log(7.89) = 0.897

Therefore, log(7.89×10^{-9}) = −9 + 0.897 = −8.103

Example 3

What is the log of 3.2×10^{-12}?

Answer
The first part of the log is −12.

Using a calculator, log(3.2) = 0.505

Therefore, log(3.2×10^{-12}) = −12 + 0.505 = −11.495

Example 4

What is the log of 0.0000463?

Answer

In standard index form $0.0000463 = 4.63 \times 10^{-5}$

Using a calculator, $\log(4.63) = 0.666$

Therefore, $\log(4.63 \times 10^{-5}) = -5 + 0.666 = -4.334$

Some more examples of logs are given in Table 14.2.

Table 14.2 The logs of some numbers.

Number	Standard (index) form	Log (of number)
23460	2.346×10^4	4.37
654321	6.54321×10^5	5.82
1234.5	1.2345×10^3	3.09
0.00654	6.54×10^{-3}	−2.18
0.000678	6.78×10^{-4}	−3.17
0.00006565	6.565×10^{-5}	−4.18

Converting 'log x' to 'x'

To convert a logarithm back into the number from which it came, you must remember that the log was obtained by determining the number as a power of 10. The log of 100 is 2, because $100 = 10^2$. Therefore, given a logarithm, n, a calculator must be used to determine the value of 'n' in 10^n. A calculator will have a button to allow you to do this and it is usually labelled '10^x'.

Example 5

What is x if $\log x = 2.610$?

Answer

Using the 10^x button on the calculator, $x = 407$ (the calculator gives 407.38)

Example 6

What is x if $\log x = -4.710$?

Answer

Using the 10^x button on the calculator, $x = 0.0000195$ which is 1.95×10^{-5}.

The second example requires you to enter −4.71 into the calculator. Your calculator will probably have a button labelled '+/−'. First, enter the number as 4.71 and then, when the '+/−' button is pressed, it will convert to −4.71.

The procedure is:

● enter the log into the calculator. If necessary using the '+/−' button to convert a positive number into a negative number

● press the '10^x' button on the calculator to find the number to base 10.

pH

When you study the chemistry of acids and bases (Chapter 3), you will encounter the use of logarithms because the acidity of a substance is defined as $-\log[H^+]$ and this quantity is known as the pH. In fact the 'p' of pH is a standard way of writing '$-\log$' of a quantity. So $pH = -\log[H^+]$. You can change the variable from $[H^+]$ to K_a and in which case pK_a is equal to $-\log K_a$.

So, the pH of a solution with a hydrogen ion concentration of 0.01 (10^{-2}) will be

$$-\log(10^{-2}) = -(-2) = 2$$

If the pH of a substance is 3.15, this means $-\log[H^+] = 3.15$ or $\log[H^+] = -3.15$.

You can convert this log into the concentration of hydrogen ions using the procedure described earlier (page 287). You enter 3.15, then use button labelled '+/−' to convert it to −3.15, before pressing the '10^x' button. If you do this, pH = 3.15 gives $[H^+] = 0.000\,71$ (rounded to 2 significant figures).

Example 7

What is $[H^+]$ if the pH of a solution = 2.8?

Answer

Enter 2.8 into the calculator and use the '+/−' to change to −2.8.

Using the 10^x button on the calculator, $[H^+] = 0.0016$ (the calculator gives 0.00158), which in standard form is 1.6×10^{-3}.

Example 8

What is $[H^+]$ if the pH of a solution = 8.10?

Answer

Enter 8.10 into the calculator and use the '+/−' to change to −8.1.

Using the 10^x button on the calculator, $[H^+] = 7.9 \times 10^{-9}$.

Test yourself

2 What is x if:
  a) $\log x = 2.3$
  b) $\log x = 7.1$
  c) $\log x = 4.0$
  d) $\log x = -1.6$
  e) $\log x = -3.8$
  f) $\log x = -8.8$?

3 What is $[H^+]$ if:
  a) pH = 3.0
  b) pH = 5.2
  c) pH = 1.5
  d) pH = 9.6?

Other features of logarithms

100 multiplied by 1000 equals 100 000. If you did the calculation with the numbers written in standard index form, this would be $10^2 \times 10^3$ and the answer would be 10^5. You can see that the powers of 10 are simply added together to give the answer.

> **Example 9**
>
> What is $10^5 \times 10^7$?
>
> **Answer**
> Since $5 + 7 = 12$, the answer is 10^{12}.

> **Example 10**
>
> What is $10^5 \times 10^{-3}$?
>
> **Answer**
> Since $5 - 3 = 2$, the answer is 10^2.

Logarithms relate to the powers of 10 and a similar rule applies when you multiply them together.

So

$$\log(m \times n) = \log(m) + \log(n)$$

In chemistry, you meet this when taking the logarithms of some equations.

For example, in water

$$[H^+][OH^-] = 10^{-14}$$

If logs are taken for both sides of the equation, it becomes

$$\log\{[H^+][OH^-]\} = \log\{10^{-14}\} = -14$$

which is the same as

$$\log[H^+] + \log[OH^-] = -14$$

or, using the pH notation (which means that both sides of the equation are multiplied by -1), this becomes:

$$pH + pOH = 14$$

You may find it useful to also note that

$$\log(m/n) = \log(m) - \log(n)$$

> **Tip**
>
> You should be able to calculate the pH from $[OH^-]$ using K_w and should not simply use the equation $pH + pOH = 14$ as a shortcut.

Logarithms to base 'e'

It is possible to write numbers to bases other than 10 and you will meet this when studying the Arrhenius equation in Chapter 1.

Instead of numbers being expressed as powers of 10, the Arrhenius equation expresses numbers as powers of 'e' where e (to 4 significant figures) is 2.718. So, in this case, a number is written as $(2.718)^n$.

This may seem very strange but, for various mathematical reasons, some problems lead to an answer expressed in this form. Logarithms based on 'e' are called 'natural logarithms'. There is no need to be concerned about why at this stage. Logarithms based on 'e' are conventionally given the symbol 'ln'.

So, $\ln(2.718) = 1$ because 2.718 to the power 1 is 2.718.

$(2.718)^2 = 7.3875$, so $\ln(7.3875) = 2$.

A scientific calculator will have a button labelled 'ln' so other values can be determined.

Of more immediate use is to note that

$$\log(n) = \ln(n)/\ln(10)$$

Since $\ln(10) = 2.303$, this gives a formula for converting natural logs into logs to the base 10, as $\ln(n) = 2.303\log(n)$ or $\log(n) = 0.434\ln(n)$. There is no need to learn this formula.

To convert '$\log_{10}x$' to give the value of x you use the 10^x button on your calculator. To convert $\ln(x)$ to x, you use a button labelled 'e^x' on your calculator.

Example 11

What is $\ln(15.142)$?

Answer

Using the $\ln$ button on the calculator, $\ln(15.142) = 2.717$.

Example 12

What is x if $\ln(x) = 7.80$?

Answer

Using the e^x button on the calculator, $x = 2441$ or 2.4×10^3 (the calculator gives 2440.6).

Example 13

What is x if $\ln(x) = -3.500$?

Answer

Enter 3.5 into the calculator and use the '+/−' to change to −3.5.

Using the e^x button on the calculator, $x = 0.0302$.

The Arrhenius equation

In the 'Other features of logarithms' section (page 288), it was noted that

$$\log(m \times n) = \log(m) + \log(n)$$

The same thing applies to logarithms to base 'e'. So,

$$\ln(m \times n) = \ln(m) + \ln(n)$$

If you have studied the section on the effect of temperature on the rate constant, k, you will have met the Arrhenius equation which is:

$$k = Ae^{-E_a/RT}$$

If you take 'ln' of both sides of this equation, the equation becomes:

$$\ln(k) = \ln(Ae^{-E_a/RT})$$

Remembering that $\ln(m \times n) = \ln(m) + \ln(n)$, if you think of m as being 'A' and n as being '$e^{-E_a/RT}$', the right-hand side of this equation can be written as:

$$\ln A + \ln(e^{-E_a/RT})$$

which is the same as $\ln A + (-E_a/RT)$ or $\ln A - E_a/RT$.

Therefore, the Arrhenius equation can be stated as:

$$\ln(k) = \ln A - E_a/RT$$

and this is often a more useful form of the equation.

Test yourself

4 What is the logarithm to base 'e' (ln) of the following numbers:
   a) 54.2
   b) 706.1
   c) 0.000129?
5 What is x to 2 significant figures if:
   a) $\ln(x) = 5.40$
   b) $\ln(x) = 2.40$
   c) $\ln(x) = -2.80$
   d) $\ln(x) = -1.90$?
6 a) If $\ln(x) = 1.6$, what is $\log(x)$?
   b) If $\log(x) = 3.6$, what is $\ln(x)$?

Index

Free online resources

Answers for the following features found in this book are available online:

- Test yourself questions
- Activities

You'll also find Practical skills sheets and Data sheets. Additionally there is an Extended glossary to help you learn the key terms and formulae you'll need in your exam.

Scan the QR codes below for each chapter.

Alternatively, you can browse through all chapters at:
www.hoddereducation.co.uk/OCRChemistry2

How to use the QR codes

To use the QR codes you will need a QR code reader for your smartphone/tablet. There are many free readers available, depending on the smartphone/tablet you are using. We have supplied some suggestions below, but this is not an exhaustive list and you should only download software compatible with your device and operating system. We do not endorse any of the third-party products listed below and downloading them is at your own risk.

- for iPhone/iPad, search the App store for Qrafter
- for Android, search the Play store for QR Droid
- for Blackberry, search Blackberry World for QR Scanner Pro
- for Windows/Symbian, search the store for Upcode

Once you have downloaded a QR code reader, simply open the reader app and use it to take a photo of the code. You will then see a menu of the free resources available for that topic.

1 How fast? Rates of reaction

3 Acids, bases and buffers

2 How far? Equilibrium

4 Enthalpy, entropy and free energy

5 Redox and electrode potentials

10 Nitrogen compounds

6 Transition elements and qualitative analysis

11 Polymers

7 Benzene and aromatic compounds

12 Organic synthesis

8 Carbonyl compounds

13 Analysis

9 Carboxylic acids and derivatives

14 Maths in chemistry

The Periodic Table

Key
atomic number
symbol
relative atomic mass

1	2	3	4	5	6	7	8	9	10	11	12	13	14	15	16	17	18
1 H 1.0																	2 He 4.0
3 Li 6.9	4 Be 9.0											5 B 10.8	6 C 12.0	7 N 14.0	8 O 16.0	9 F 19.0	10 Ne 20.2
11 Na 23.0	12 Mg 24.3											13 Al 27.0	14 Si 28.1	15 P 31.0	16 S 32.1	17 Cl 35.5	18 Ar 39.9
19 K 39.1	20 Ca 40.1	21 Sc 45.0	22 Ti 47.9	23 V 50.9	24 Cr 52.0	25 Mn 54.9	26 Fe 55.8	27 Co 58.9	28 Ni 58.7	29 Cu 63.5	30 Zn 65.4	31 Ga 69.7	32 Ge 72.6	33 As 74.9	34 Se 79.0	35 Br 79.9	36 Kr 83.8
37 Rb 85.5	38 Sr 87.6	39 Y 88.9	40 Zr 91.2	41 Nb 92.9	42 Mo 95.9	43 Tc	44 Ru 101.1	45 Rh 102.9	46 Pd 106.4	47 Ag 107.9	48 Cd 112.4	49 In 114.8	50 Sn 118.7	51 Sb 121.8	52 Te 127.6	53 I 126.9	54 Xe 131.3
55 Cs 132.9	56 Ba 137.3	57–71	72 Hf 178.5	73 Ta 180.9	74 W 183.8	75 Re 186.2	76 Os 190.2	77 Ir 192.2	78 Pt 195.1	79 Au 197.0	80 Hg 200.6	81 Tl 204.4	82 Pb 207.2	83 Bi 209.0	84 Po	85 At	86 Rn
87 Fr	88 Ra	89–103	104 Rf	105 Db	106 Sg	107 Bh	108 Hs	109 Mt	110 Ds	111 Rg	112 Cn		114 Fl		116 Lv		

57 La 138.9	58 Ce 140.1	59 Pr 140.9	60 Nd 144.2	61 Pm 144.9	62 Sm 150.4	63 Eu 152.0	64 Gd 157.2	65 Tb 158.9	66 Dy 162.5	67 Ho 164.9	68 Er 167.3	69 Tm 168.9	70 Yb 173.0	71 Lu 175.0
89 Ac	90 Th 232.0	91 Pa	92 U 238.1	93 Np	94 Pu	95 Am	96 Cm	97 Bk	98 Cf	99 Es	100 Fm	101 Md	102 No	103 Lr